9/97

Volume 31

Ceramic Transactions

POROUS MATERIALS

Edited by Kozo Ishizaki, Nagaoka University
of Technology; Laurel Sheppard,
The American Ceramic Society;
Shojiro Okada, Japan Grain
Institute, Inc.; and Toyohiro
Hamasaki and Ben Huybrechts,
Nagaoka University of Technology

The American Ceramic Society
Westerville, Ohio

Proceedings of the symposium on Porous Materials presented at the International Forum of Materials Engineers at Sanjo Tsubame (IFMEST) '92, held in Niigata, Japan, September 28-29, 1992.

Library of Congress Cataloging-in-Publication Data

Porous materials / Edited by Kozo Ishizaki . . . [et al.].
    p.   cm. -- (Ceramic transactions, ISSN 1042-1122 ; v. 31)
   Selected papers presented at the International Forum for Materials Engineering at Sanjo Tusbame (IFMEST) '92, held Sept. 28-29, 1992 in Niigata, Japan.
   Includes index.
   ISBN 0-944904-59-9
   1. Porous materials--Congresses. 2. Ceramic materials--Congresses. I. Ishizaki, Kozo, 1946- . II. Series.
TA418.9.P6P67 1993
620.1'16--dc20   93-118
                                                                                    CIP

Copyright © 1993, The American Ceramic Society. All rights reserved.

No part of this book may be reproduced, stored in a retrieval system, or transmitted in any form or by any means, electronic, mechanical, photocopying, microfilming, recording, or otherwise, without written permission from the publisher.

Permission to photocopy for personal or internal use beyond the limits of Sections 107 and 108 of the U.S. Copyright Law is granted by the American Ceramic Society, provided that the base fee of US$2.50 per copy, plus US$.50 per page, is paid directly to the Copyright Clearance Center, 27 Congress Street, Salem MA 01970, USA. The fee code for users of the Transactional Reporting Service for *Ceramic Transactions* is 0-944904-59-9/93 $2.50+$.50. This consent does not extend to other kinds of copying, such as copying for general distribution, for advertising or promotional purposes, or for creating new collective works. Requests for special photocopying permission and reprint requests should be directed to the Director of Publications, The American Ceramic Society, 735 Ceramic Place, Westerville OH 43081, USA.

Printed in the United States of America.

1 2 3 4–97 96 95 94 93

ISSN 1042-1122
ISBN 0-944904-59-9

Preface

The science and technology of porous materials are progressing steadily and expanding in many new directions of application and processing methods. These expanding usages and capabilities, as well as reconsideration of old processes, demanded a forum that could provide an opportunity for interaction between scientists and engineers interested in porous materials.

This volume contains selected and refereed papers on porous materials presented at the International Forum for Materials Engineering at Sanjo Tsubame (IFMEST) '92, held September 28-29, 1992 in Niigata, Japan. This forum brought scientists and engineers from all over the world and from a variety of disciplines together to focus on the application and processing of porous materials, with an emphasis on ceramics. The attendance of nearly 1000 participants from eight countries is indicative of the active interest in this topic. The Central Niigata Prefecture Regional Industries Promotion Center provided an excellent location for the forum, which was the first one in Japan to focus on porous materials. This collection of high-quality papers presents the current developments in the area of porous materials, with an emphasis on ceramics, and will benefit those who could not attend the forum.

We wish to express our gratitude to Sumio Sakka and Sridhar Komarneni for the invited lectures, and to Douglas M. Smith, Shoji Yamanaka, Hideo Hosono, Joaquin Lira-Olivares, Hiroaki Katsuki, Tetsuo Yazawa, Takao Kokugan, Lieh-Jiun Shyu, Kan-Sen Chou, Chikao Kanaoka, Masayuki Yamane, and A. Atkinson for chairing sessions and refereeing papers. We are also grateful for the contributions of Tokimune Shioura (organizing committee chairperson), Anatolijs Kuzjukevics (program committee chairperson), and Kazuo Sato (secretariat).

IFMEST '92 was organized by the Central Niigata Prefecture Regional Industries Promotion Center, and endorsed and/or subsidized by the American Ceramic Society, the Ceramic Society of Japan, the Society of Gypsum and Lime, Niigata Prefecture, Sanjo City, Tsubame City, and Shinanogawa Techno-Polis Development Organization. The forum could not have been successful without the support provided by these organizations.

Kozo Ishizaki
Laurel Sheppard
Shojiro Okada
Toyohiro Hamasaki
Ben Huybrechts

Contents

Section I. Introduction

Porous Ceramics: Processing and Applications 3
L.M. Sheppard

Section II. Production Methods

Preparation of Porous Materials by the Sol-Gel Method 27
S. Sakka, H. Kozuka, and T. Adachi

Gel Processing Routes To Porous Ceramics 41
A. Atkinson

Macropore Structure Design of Sol-Gel-Derived Silica by Spinodal Decomposition 51
K. Nakanishi, H. Kaji, and N. Soga

Porous Materials Prepared from Silica Gel by Hydrothermal Hot Pressing 61
K. Yanagisawa, M. Nishioka, K. Ioku, and N. Yamasaki

Preparation of Low-Density Aerogels at Ambient Pressure for Thermal Insulation 71
D.M. Smith, R. Deshpande, and C.J. Brinker

Ceramic Membranes and Application to the Recovery of Soy Sauce 81
N. Maebashi

Developmental Studies on Porous Alumina Ceramics 89
P.K. Tantry, S.N. Misra, and A.L. Shashi Mohan

Microstructure Evolution During Fabrication of a Porous Ceramic Filter 101
K.-S. Chou, H.C. Liu, and K.L. Chang

Production of Porous Alumina by Hot Isostatic Pressing 111
A. Kuzjukēvičs and K. Ishizaki

Sintering of Porous Materials by a Capsule-Free HIP Process 117
M. Nanko, K. Ishizaki, and A. Takata

Sintering Behavior of Tile Materials Under Normal and High-Gas-Pressure Atmospheres 127
T. Kurushima

Fabrication and Properties of Mullite Ceramics with Needlelike Crystals 137
H. Katsuki, H. Takagi, and O. Matsuda

Sintering Behavior and Thermal Shock Resistance of Porous ZrO_2 Ceramics with ZrO_2 Fiber, Al_2O_3 Fiber, or SiC Whisker 147
K. Hattori and T. Kurushima

Microporous Metal Intercalated Clay Nanocomposites 155
S. Komarneni

Micro- and Mesoporous Materials Derived from Two-Dimensional Silicate Layers of Clay 169
S. Yamanaka and K. Takahama

Properties and Applications of Functional Porous Glass-Ceramics Composed of a Titanium Phosphate Crystal Skeleton 181
H. Hosono and Y. Abe

Preparation of Porous Glasses for Photonic Materials 191
M. Umino, T. Yano, A. Yasumori, S. Shibata, and M. Yamane

Section III. Properties

Characterization of Porous Silica Gel by Means of Adsorption-Desorption Isotherms of Water Vapor and Nitrogen Gas 203
H. Naono, M. Hakuman, H. Joh, M. Sakurai, and K. Nakai

Permeation Characteristics of CO_2 Through Surface-Modified Porous Glass Membrane 213
T. Yazawa and H. Tanaka

Fractal Analysis on Pore Structure in Relation To Frost Durability of Fiber-Reinforced Building Cement Materials 223
M. Nakamura, S. Urano, and T. Fukushima

Mechanical Properties of HIPed Porous Copper 233
A. Takata and K. Ishizaki

Bending Strength and Electric Resistivity of Porous Si-SiC Ceramics 243
H. Osada, A. Kani, S. Katayama, and T. Koga

Electrical and Electrochemical Properties of TiO_2-SiO_2 Porous Glass-Ceramics 253
T. Kokubu and M. Yamane

Corrosion Resistance of Permeable Refractories 263
A. Tsuchinari, T. Hokii, and C. Kanaoka

Hexaaluminate-Related Compounds as Thermally Stable Catalyst Materials 273
M. Machida, T. Shiomitsu, H. Inoue, K. Eguchi, and H. Arai

Section IV. Applications

Application of Ceramic Foam 285
T. Masuda, K. Tomita, and T. Iwata

Microporous Gels as Desiccants 295
S. Komarneni and P.B. Malla

Electrostatic Formation of Ceramic Membrane 305
H. Yamamoto and S. Masuda

Application of HIPed Porous Metals for an Electrochemical Detector 315
Y. Okumoto, K. Ishizaki, and A. Yamada

Application of Porous Alumina Ceramics for Casting Molds 325
Y. Kondo, Y. Hashizuka, S. Okada, and M. Shibayama

Prospects for Obtaining a Superconducting Filter To Purify Oxygen from Argon 335
Y. Sawai, K. Ishizaki, S. Hayashi, and R. Jain

Effect of Porous Materials on the Generation and the Growth of Bubbles in Aeration 343
N. Ueno

Development of Porous Ceramics for a Negative Pressure Difference Irrigation System 353
M. Kubota and T. Kojima

Mercuric Ion Sensor with FIA System Using Immobilized Mercuric Reductase on Porous Glass 361
M. Uo, M. Numata, I. Karube, and A. Makishima

Bimodal Porous Cordierite Ceramics for Yeast Cell Immobilization 371
H. Abe, H. Seki, A. Fukunaga, and M. Egashira

A New Approach That Uses Bioreactors With Inorganic Carriers (Ceramic) in the Production of Fermented Foods and Beverages 381
H. Horitsu

Porous Ceramic Carrier for Bioreactor 391
M. Kawase, Y. Kamiya, and M. Kaneno

Amylose Separation from Starch and Glucose Separation from Starch Saccharified Solution by Filtration Through a Ceramic Membrane 401
F. Takahashi and Y. Sakai

Preparation of Thin Porous Silica Membranes for Separation of Nonaqueous Organic Solvent Mixtures by Pervaporation 411
M. Asaeda, K. Okazaki, and A. Nakatani

Applications of Microporous Materials to Membrane Reactors 421
T. Kokugan and G. Keitoh

Index 433

Section I. Introduction

POROUS CERAMICS: PROCESSING AND APPLICATIONS

Laurel M. Sheppard
Editor, Ceramic Bulletin and Ceramic Source
The American Ceramic Society
735 Ceramic Place
Westerville, OH 43081
U.S.A.

Porous ceramics based on Al_2O_3, SiC, aluminosilicate, mullite, cordierite, and other compositions have been used as filters and honeycomb elements for applications in aeration, nuclear generation plants, dust collectors, acoustic absorbers, catalyst carriers for chemical plants and automobiles, engine components, and heat exchangers. Even electronic materials are being used: a method has been patented for selective filtering of a fluid using porous piezoelectrics (lead zirconate titanate or PZT)[1]. Porous ceramics with a foamlike structure are used in some of these same applications, as well as for thermal insulation, electrodes, lightweight structural laminates, and as burner materials.

The biotechnology and biomedical industries are also using porous ceramics. MgO-SiO_2 and diatomite ceramics are being used as ceramic carriers, with pores ranging from 0.02 - 100 μm, for bioreactors used in fermentation processes. Reaction times can be significantly reduced. For bone replacement, porous hydroxyapatite combined with partially stabilized zirconia has been successfully used. Though there are many other applications, this article will focus on membranes, filters, and sensor applications.

PROCESSING METHODS

Porous ceramics can be made of either a reticulate or foam structure[2]. A reticulate material consists of interconnected voids surrounded by a web of ceramic and is usually made by burning out a polymeric sponge impregnated with a ceramic slurry, Figs. 1 and 2. A foam structure has closed voids within a continuous ceramic matrix and is usually made by producing a foam from evolved gas. The polymer carrier is burned out, leaving a porous ceramic. This foaming method can also produce reticulate materials. Pore sizes for foamed ceramics are typically around 3 mm.

A number of ways have been developed to improve strength of foam or reticulate ceramics[3]. Fibers, usually aluminosilicate, or other reinforcements such as SiC whiskers can be added. An agent can be used to partially remove the sponge prior to burnout or the sponge can be pretreated with an adhesive. To improve filtering characteristics of reticulate ceramics, microcracking can be produced (which improves the gas diffusion rate) or surface roughening can be produced by an etching treatment. Containerless hot isostatic pressing has also been used as discussed elsewhere in this proceedings.

Several other methods are used for fabricating porous ceramics, including chemical leaching, solid-state sintering, and sol-gel processing. For chemical leaching, pore sizes are

---

To the extent authorized under the laws of the United States of America, all copyright interests in this publication are the property of The American Ceramic Society. Any duplication, reproduction, or republication of this publication or any part thereof, without the express written consent of The American Ceramic Society or fee paid to the Copyright Clearance Center, is prohibited.

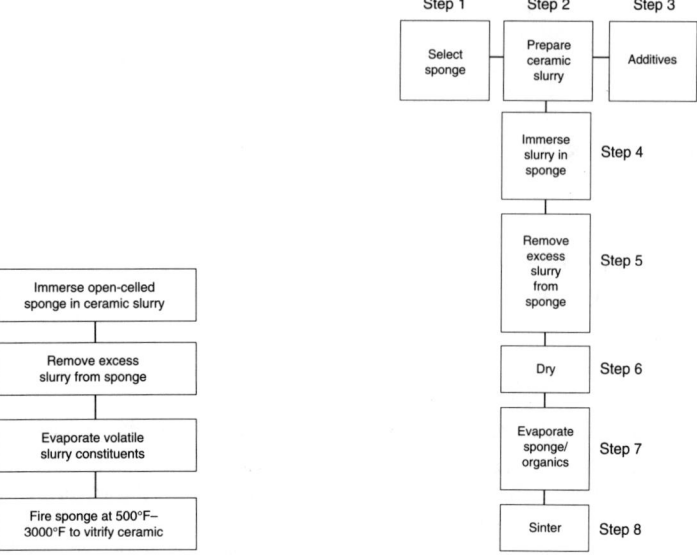

Figure 1 Schematic of polymeric sponge method.   Figure 2 Details of method in Fig. 1.

typically 10 to 100 nm. Sintering produces larger pores, in the size range 100-1000 nm. Pore size can be controlled by adjusting both particle size of the starting materials, as well as the firing temperature[4]. For increasing open pore volume of cordierites, the addition of coal powder has been found effective [5].

Schott Glaswerke has patented a method of producing open-pore sintered borosilicate glass filters by mixing two glass powders together with additives and then sintering[6]. These filters have pore volumes between 60% and 75%, high flexure tensile strength (11 MPa), and good flow velocities with precisely adjustable pore diameters between 10-15 μm, 30-40 μm, 80-100 μm, and 110-150 μm. The different pore sizes are obtained by varying the composition of the parent material and the grain size of the second glass powder added.

On the other hand, sol-gel processing can produce much finer pore sizes, less than 50 nm, as well as narrower particle size distributions[7], compared to conventional methods. Sol-gel technology can also provide simpler processing, as well as providing the capability of tailoring microstructure, including pore volume, pore size, and thickness[8]. For instance, Shinko Pantec Co. has used sol-gel technology to develop a SiO_2-ZrO_2 glass with surface pores of 1-2 nm uniformly distributed, and the porosity can be controlled from 10% to 80%. Non-supported microporous silica and titania thin films have been made by the polymeric gel route as well, with pores <1 nm in diameter.

Rutgers University has investigated processing parameters in order to synthesize free-standing unsupported membranes with a geometry of thin foils. Silica gels have been made with both symmetric and asymmetric microstructures that mimic those in polymer membranes, using a solution of a high-density, nonpolar organic liquid[9]. Thermal analysis has been used to study the difference between alumina, silica, and zirconia. Both alumina and zirconia underwent dehydroxylation and phase transformation below 500°C[10]. Silica, on the other hand, remained amorphous and therefore had better thermal stability.

A number of other novel techniques are being developed to prepare porous materials. Rapid expansion of supercritical fluid solutions followed by drying allows formation of highly porous aerogel products. Self-propagating high-temperature synthesis, otherwise known as combustion synthesis, can also be used to make porous filters as large as 400 mm in diameter and 30 mm in thickness, with a porosity of 50%, and a pore diameter ranging from 0.02- 5 mm. Rigid Al_2O_3 insulation with 90% porosity has been made using a combination of freeze drying slurries and firing the resultant powders.

MOLTEN METAL FILTRATION

For production of castings, porous ceramic filters are helping to improve quality and productivity by removing nonmetallic inclusions, Table 1[11]. The filters must be able to resist attack at high temperatures by a variety of molten metals, which can contain such reactive elements as aluminum, titanium, hafnium, and carbon. In one case study, ceramic filters reduced scrap rates by 40%, salvage rates by 30%, and increased yield by 10%.

Table 1. Ceramic Filter Performance During Vacuum Melting*

PART	Damper seal	Jet engine swirl plat	Engine shroud	Jet engine bearing hub
PROBLEM	Unrepairable	Oxide inclusions at surface	Oxide inclusions at surface	Oxide inclusions at surface
RESULTS	13% increase in yield	Decrease in rework 4.5 welds/part vs 1.5 welds/part	10% increase in yield	Decrease in rework: 6 welds/part vs 2.5 welds/part

*Partially stabilized zirconia, 10 pores per inch

The material is usually a metal oxide of various compositions. These include alumina, silicates, mullite, steatite, cordierite, and zirconia, to name a few. Ceramic fibers--alumina, aluminum silicates, zirconia, boron, SiC, or carbon--have also been incorporated with these materials to form a two- or three-layer system. Both inorganic (such as alumina hydrate) and/or organic binders are usually required. A coating is often applied to the filter to improve resistance to slag attack.

These filters generally come in two forms: a porous foamlike structure or an extruded porous cellular structure with cells of various shapes (square or triangular). The properties of a solid foam depend on both the cellular geometry (density, cell size) and the properties of the material. Advantages include high-temperature stability and low weight.

Solid foams can have either open cells or closed cells. Open cell foams consist of a reticulated network of interconnected beams. Closed cell foams consist of a similar network, but the beams are bridged by thin faces which isolate the individual. The open porosity in an open cell structure is critical in applications in which contact with the environment is needed or for filtration[12].

Ceramiques et Composites has also recently patented alveolar ceramic filters based on alumina, zirconia, and silica[13]. Carborundum Co. has developed a filter based on partially stabilized zironia and alumina[14]. Manufacturers of molten metal filters include Corning Inc.,

Hi-Tech Ceramics, Selee Corporation, and Drackekeramikfilter Productions Gmbh (distributed by Christy Refractories Co., St. Louis, MO), Table 2.

Table 2. Selected Manufacturers of Molten Metal Filters

Company	Tradename	Composition	Applications	Benefits
Corning Inc. Foseco Inc.	Celtrex	55% Al_2O_3, 38% SiO_2, 7% MgO	Iron alloys	Reduction in scrap rate
Corning Inc.	None	77% Al_2O_3 23% SiO_2	Carbon, low alloy, stainless steel	Pouring temperatures up to 1675°C
Drackekeramick Productions GmbH	Cerapor	Alumina, SiC, cordierite, ZrO_2	Aluminum, iron, copper, bronze, steel, zinc	Laminated duplex and triplex construction
Hi-Tech Ceramics	Udicell	Alumina, mullite, ZTA, PSZ	Superalloys, low-carbon stainless steel	Large volumes up to 120 tons
	Alucel	92% alumina with mullite phase	Nonferrous alloys	Improved thermal shock resistance, smaller filters required
Selee Corp.	Selee	Alumina, PSZ, others	Aluminum, iron, steel, others	High flow rates

Thermal shock behavior for foam filters is obviously important and has been investigated at Pennsylvania State University's Center for Advanced Materials. The thermal shock resistance of the open cell materials (Fig. 3 shows pore structure) was found to be strongly dependent on cell size (increasing with increasing cell size) and weakly dependent on density (increasing with increasing density)[15]. Strength was retained after thermal shock, then showed a gradual decrease with increasing quench temperature, Fig. 4, reflecting an increase in mechanical damage throughout the materials. It was also determined that a bulk temperature gradient existed after quenching, which was attributed to a heating of the quenching media as it infiltrated the cellular structure.

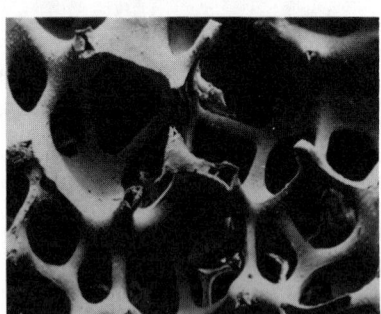

Figure 3 Typical pore structure.

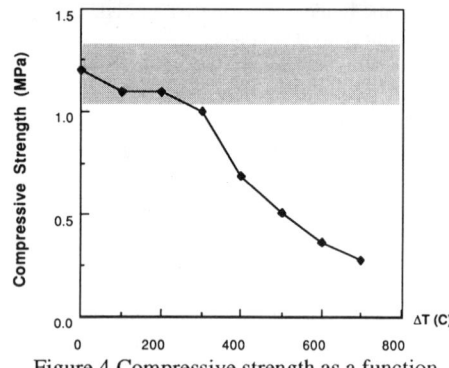

Figure 4 Compressive strength as a function of quench temperature.

MEMBRANES FOR SEPARATION PROCESSES

Ceramic membranes are receiving more attention lately because of their advantages over polymeric and metallic membranes--they are resistant to high temperatures, abrasion, and chemical attack. Generally stable to 1000°C, they not only have high-temperature applications but they can also be steam cleaned. Ceramic membranes tolerate cleaning with caustics, acids, and any detergent (such as bleaching with agents, e.g., sodium hypochlorite), readily accommodate the abrasion encountered in slurries, and resist the high pressure (up to 30 atm (2.9 MPa)) often used in backflushing techniques.

Even though organic membranes have a much larger surface area--1000-10000X greater than inorganic--the permeance of inorganic membranes can be expected to be 1000-10000X larger than the permeance of organic membranes. Therefore, even if the cost of the inorganic membranes were 1000-10000X the cost of organic membranes per unit surface area, the costs of the separation units could be comparable[16]. Because ceramic membranes can also operate under high pressures and provide high throughput, they have the potential to provide great economy-of-scale as well.

However, improvements are still needed in the areas of mechanical properties (better fracture toughness), simpler processing, and better controlled and finer pore size[17]. There has been some progress in the area of strength. Al_2O_3-based gas-permeable ceramics have been produced with SiC active additives as bonding agents, increasing mechanical strength 2-3X over that of conventional porous materials.

Membranes are typically identified according to the pore size or the size of the materials they are used to separate. Reverse osmosis (also called hyperfiltration) or gas separation membranes separate pore sizes <1nm; ultrafiltration membranes have pore sizes in the range of 2-100 nm; microfiltration membranes have pore sizes in the range of 100-5000 nm; and membranes with pores >5000 nm are particulate filters. Inorganic membranes have been used primarily for microfiltration and ultrafiltration.

Applications are numerous and include food processing (especially for dairy products), beverage clarification, process gas production (for producing nitrogen, oxygen, argon), biotechnology purifications (for protein, yeast, and other microorganism separations), and other chemical and water-purification processes, Table 3[18]. In the biotechnology area, Japanese researchers have developed a high-performance separation membrane from porous glass, with pores ranging from 10-100 nm in diameter, used for selectively filtering viruses, proteins, and amino acids[19].

In 1989, the triregional market for advanced membranes, which primarily consist of polymeric materials, amounted to about $1.4 billion according to Chem Systems Inc. Key markets include microfiltration, reverse osmosis, ultrafiltration, and gas separation, with the latter two probably being the most promising markets for ceramics. In fact, it is predicted that the gas separation market will be $1.5 billion by the year 1995.

Another study by Kline & Company (Fairfield, NJ) predicts the Western world market for advanced membrane systems will total $4.6 billion by 1995. Companies that are trying to penetrate this market include Norton Co., Corning Inc., Ceraver, SPEC, NGK Insulators, Du Pont, Alcoa, Akzo[20], Societe des Ceramiques Techniques[21], and Osmonics. Materials

Table 3. Applications for Ceramic Membranes

Industry	Use
Food and beverages	Clarifies and sterilizes fruit juices and vinegar; homogenizes milk and eggs; removes phenols and proteins from wine
Biotechnology/pharmaceutical	Concentrates vaccines and enzymes; purifies amino acids; removes viruses from cultures
Gas separation	Removes hydrogen from refinery steam; CO_2 and HSF from natural gas; nitrogen enrichment; methane recovery in mining
Environmental control	Removes precipitated radionuclides and metal oxides; wastewater processing; recycling of machining coolants
Petrochemical	Waste-oil hydrogenation process; catalytical dehydrogenation of large molecules; coal gasification
Electronic	Purification for water, acids, solvents, organic compounds

range from alumina to carbon-zirconia, Table 4. Du Pont has already introduced a ceramic filter of porous hollow tubes made of alumina, mullite, cordierite, or combinations of all three for microfiltration applications. Pore diameter can be controlled between 0.06-1 µm, and pore volume varies from 25% to 50%.

Tech-Sep, a wholly owned subsidiary of France's Rhone-Poulenc, has acquired the membranes business of Imeca, which markets monolithic ceramic membranes to the wine and food industry, including membranes used for drinking and process water treatment. Ceramesh Ltd., a British venture spun off by Alcan International, is also marketing ceramic membranes.

Ceramic membranes from Alcoa Separations Technology Division (Warrendale, PA) have already been successfully tested for treatment of aqueous streams containing heavy metals and oil-water emulsions during manufacturing of vinyl chloride monomer and linear alkylbenzene[22]. Heavy-metal solids were reduced to nondetectable levels and oil and grease (after pretreatment with HCl and ferric chloride) were reduced to less than 5 ppm. Hollow-fiber microporous (0.8 nm) silica membranes have also been shown to have comparable separation factors (permeability ratios) for O_2/N_2 during gas separation of He, H_2, CO_2, O_2, Ar, N_2, CO, and CH_4[23].

The Department of Energy and Oak Ridge National Laboratory are commercializing their membrane technology originally developed to separate isotopes of uranium for other chemical-separation processes. The ORNL technology offers a longer operating life (up to 30X) and a tolerance of harsh environments in comparison to conventional inorganic membranes, and can reduce production costs by a factor of 4 or more. Flux, a measure of the permeability of the membrane which determines the volume throughput of a system, is improved 5X while maintaining an excellent separation factor. The separation factor is a measure of how completely the membrane separates the product and waste process streams.

Unsupported ceramic membranes have been fabricated via the sol-gel process with mean pore diameters as small as 3 nm. These membranes are being investigated for separation of solutes with molecular weights below 1000 and/or to perform reverse osmosis separations. A

Table 4. Who Is Doing What with Ceramic Membranes

Company	Product	Application
Norton	Ceraflow (alumina)	Food and beverages, environmental, industrial
Ceraver	Hollow alumina honeycomb	Food and beverages, biotechnology
SPEC	Hollow carbon-zirconia	Food and beverages, biotechnology, petrochemical
NGK Insulators	Alumina	Bacteria removal in breweries; gas and fluid filtration
Du Pont	Hollow system	Recovery of hydrogen from gases
Alcan/Anotec	Anopore: alumina	Ultrafiltration, biotechnology
Alcoa	Multichannel system (alumina) Membalox	Petrochemical, pharmaceutical, biotechnology
Osmonics	Controlled porosity ceramic microfilters (up to 1200°C)	Food processing, biotechnology, waste processing
Dutch Energy Center	No details available	High-temperature gas separation of hydrogen from methane, CO_2 from natural gas, and pollutants from hot flue gases
Air Products & Chemicals	No details available	Removal of hydrogen sulfide from coal gas in coal-fired turbines
Ceram Filtre	SiC support/ceramic membrane	Purification for water, solvents, etc.; wastewater treatment, food processing
SRI International	Hollow silica system	Removal of sulfur and nitrogen pollutants from coal gas in direct coal-fired turbines
Georgia Institute of Technology	$MgO+CaO_2+LiAlO_2$	Removal of hydrogen sulfide from fuel gases
Oak Ridge Nat. Lab.	Alumina with pore radii of 5 Å	Separate hydrogen from coal gasification gases
Rutgers University	Silica (thin foils), borosilicate	Gas separation

major problem to be solved is how to deposit the membrane onto a support without cracking.

Composite membranes also have been prepared in a number of ways. One method coats microporous ceramic filters with a thin layer of rigid polymer. Composites with selective polymer layers of less than 0.5 µm in thickness have been made and are being evaluated to selectively permeate gases such as oxygen, nitrogen, and helium. Another way to make composite membranes is to impregnate the pores of microporous Vycor glass with polysilastyrene, followed by crosslinking under UV light and pyrolysis in N_2.

Vycor tubes can also be coated with SiO_2, Al_2O_3, and B_2O_3 to make hydrogen-permselective asymmetric membranes using a chemical vapor deposition process at temperatures from 100-800°C. Japan has developed composite porous glass membranes that consist of a porous glass layer, 100 µm thick, formed on porous ceramics of about 1.5-µm pore diameter[24].

Alcan International Ltd. (Montreal, Canada) has patented a method for making composite membranes that consist of a porous metal support (1-10-μm pore size) and at least one porous inorganic film (0.05-10-μm pore size) of sintered, nonmetallic particles carried by the support[25]. A boehmite sol of desired viscosity is spray-coated as thin films onto the porous metal filter to produce coatings of desired thickness, followed by calcination. These membranes are mechanically robust and chemically inert, and therefore are suitable for separation and filtration applications.

A crack and pinhole-free composite membrane consisting of an α-Al_2O_3 support and a modified γ-Al_2O_3 top layer has also been prepared by the sol-gel method[26]. The top layer was made by dipcoating the support with a boehmite sol doped with lanthanum nitrate. After sintering at 1100°C for 30 h, average pore diameter of the top layer was 17 nm, compared to 109 μm for an undoped layer. Addition of PVA to colloid boehmite precursor solution prevented formation of defects in the γ layer. The La-doped top layer (with PVA) retained a monopore distribution after sintering at 1200°C.

Other work at Research Triangle Institute (RTI) has used a modified sol-gel process to prepare cation-doped SiO_2 porous materials for advanced desiccant cooling systems for space air conditioning[27]. These systems use heat from natural gas combustion to remove adsorbed water vapor. RTI has also developed a composite membrane, Figure 5.

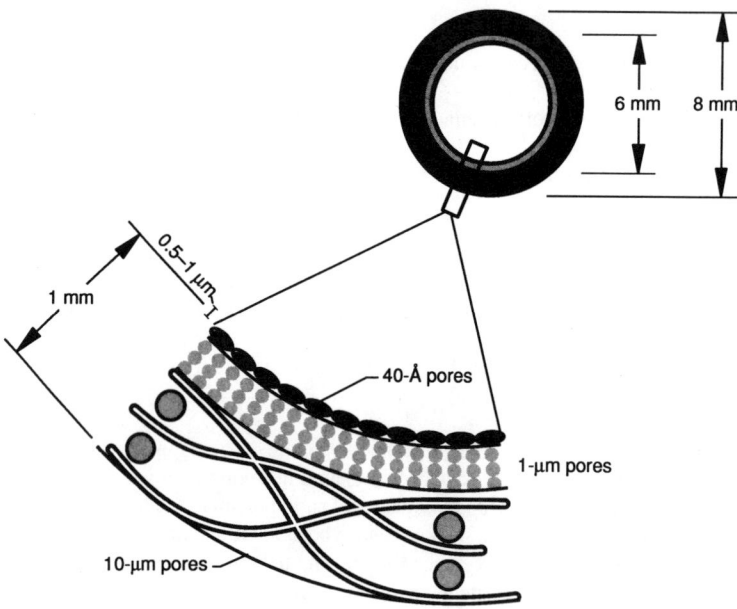

Figure 5 Schematic of composite membrane structure (1 Å=0.1 nm).

High-temperature ZnO-based desulfurization sorbents for selective removal of H_2S from coal gas are also being developed in research funded by the Department of Energy. Other researchers have used the sol-gel process. Seikei University (Tokyo, Japan) uses a metal alkoxide method, based on two benzene solutions, to produce a porous zirconia disk of 3 cm in diameter and 100 µm thickness (with the voids perpendicular to the membrane's surface), as well as alumina membranes[28].

Separation and selectivity properties are two important parameters for ceramic membranes. Separation can be improved by surface modification using graft polymerization of poly(vinylpyrrolidone)[29]. Graft polymerization is a two-step process which consists of silylation of surface hydroxyl groups followed by the growth of polymer chain from the active surface sites. Selectivity can also be improved by chemical modification of γ-alumina microporous membranes using a silanation technique[30]. This method decreases the pore size and alters the surface chemistry. By changing the latter, solute adsorption can be minimized and fouling reduced. Condensation of propylene on supported alumina films can also improve the permeability (6X) and selectivity. In addition, modification of alumina membranes with magnesia has been found to increase the separation factor by a magnitude of 5[31].

MOLECULAR SIEVES FOR CHEMICAL PROCESSES

Synthetic molecular sieves can also be considered membranes and have a market valued at a billion dollars/year. Applications range from the production of biodegradable detergents, to the removal of moisture from natural gas pipelines, to the catalytic cracking of heavy petroleum distillates into gasoline. The latter is the largest commercial application of molecular sieves--essentially 100% of the world's gasoline is now made with molecular sieve catalysts[32]. Molecular sieves have also been used to remove radioactive cesium from coolant water in the Three-Mile Island nuclear power plant accident.

Many molecular sieves are based on zeolites, a naturally occurring mineral consisting of aluminum and silicon, that has been synthesized in over 150 different structures. According to Roskill Information Services, world production capacity of synthetic zeolites has increased by 150,000 tonnes since 1990, bringing total capacity to 1.65 million tonnes/year. U.S. consumption could increase by over 100% by 1995.

Zeolites can also remove water from gases or liquids much more efficiently than most other materials. Applications include regulating humidity in dual-pane windows and removing moisture from natural gas to prevent the formation of hydrates that could plug up pipelines[33]. Other applications are in the area of pollution control, Table 5.

Table 5. Pollution Control Applications for Zeolites

Company	Application
Southwest Research Institute	Reduction of cold-start hydrocarbon automotive emission, reduction of NO_x emissions
U.S. Bureau of Mines	Removal of metal cations from mining industry wastewater
Department of Energy	Removal of hydrogen sulfide from coal-based power plants
Oak Ridge National Laboratory	Decontamination of wastewater containing ^{90}Sr and ^{137}Cs
Georgia Institute of Technology	Remove CO_2 and convert it to O_2 for submarines and orbiting space stations

Other compositions are being developed as well. Aluminophosphates are an important new class comprising aluminum and phosphorous as tetrahedral atoms. The aluminophosphate VPI-5 is of particular interest because its pores have a ring of 18 tetrahedral atoms, which is much larger than for any aluminosilicate-based zeolites. UOP (Des Plaines, IL) has recently patented a new class of molecular sieves, called ZnAPSO, that have a three-dimensional microporous crystal framework structure of ZnO_2^{-2}, AlO^{2-}, PO^{2+}, and SiO_2 tetrahedral oxide units, with pore diameters ranging from 2.65-6.2 Å[34].

Current commercial separation of synthesis gases takes place at lower temperatures, drastically reducing efficiency. The efficiency could be increased substantially if the gases can be separated at a higher downstream temperature[35]. However, inorganic membranes such as alumina that can withstand these temperatures are currently available at too large a pore size-- 30-40 Å. Fortunately, Oak Ridge has developed molecular sieves based on alumina with mean pore diameters of 10 Å for separating hydrogen from coal gas at temperatures above 500°C and at a feed pressure of 600 psi.

More recently, membranes with mean pore radii as small as 5 Å have been produced, although theoretical predictions indicate that mean pore radii smaller than 3 Å are needed. Such pore sizes are expected to produce high separation factors at elevated temperatures[36]. Because current tests cannot measure pores below 5 Å, Oak Ridge is developing a test that involves measurement of pure gas flows using test gases with ratios of flow that have sensitivity to pore size below 5 Å.

FILTERS FOR HIGH-TEMPERATURE GAS CLEANUP

For high-volume gas separation, membranes will be needed that can withstand high pressures, corrosive atmospheres, and high temperatures. Applications include nitrogen enrichment for inert blanketing, removal of carbon dioxide and hydrogen sulfide from natural gas, hydrogen recovery in refinery streams, and methane recovery in mines. Ceramics are being developed for some of these applications, especially for the harsher environments. Some of the development work is described in Table 6.

Table 6. Research in Gas Separation Membranes

Company	Type of membrane	Application
DOE/Alcoa	Alumina honeycomb	Remove >90% sulfur and nitrogen emissions from coal gasifiers
DOE/SRI International	Hollow silica fibers	Same as above
MITI	Composition unknown	Recovering CO_2 from flue gas and recycling it
Georgia Institute of Technology	MgO, ZrO_2, $LiAlO_2$ plus carbonate salt	Removal of hydrogen sulfide from fuel gases (up to 99.9%)
Research Triangle Institute	Alumina tubes with porous (3-10 nm) alumina layer	Separation of hydrogen for coal-gas (up to 450°C)

Highly efficient particulate removal from hot gas streams is also important in a number of process points, including preparation of combustion "off" gases for reuse across secondary turbines, removal of particulates from hydrocarbon-processing gas streams above the dew point, and removal and/or recovery of catalyst from "off" gases[37]. In addition, high-quality cleaning of emissions from radioactive waste and other incinerator applications requires such particulate removal. For pressurizing fluidized-bed combustion combined-cycle power plants, filtering particulates from gases is required to prevent damage and premature failure of the turbine blades.

Direct application of high-performance, high-temperature particulate control ceramic filters is expected to benefit the applications described above. Such filters must withstand variation in the effluent gas stream chemistry, variation in the nature and loadings of the entrained fines, and oscillations in the effluent gas stream temperature and possible pressure, while still maintaining high particulate removal efficiencies, high flow capacity, and relatively low pressure drop flow characteristics. During operation, the filter must also withstand a variety of mechanical, vibration, and thermal stresses. Table 7 shows parameters for coal applications[38].

Table 7. Process Parameters for Hot-Gas Filters

Parameter	Application		
	Fluidized-bed combustion	Coal gasification	Direct coal-fired turbine concept
Gas temperature, °C	500-1000	400-1000	1000-1350
Pressure, atm	3-20	10-20	10-20
Composition	Oxidizing	Reducing	Oxidizing
Particulate loading, ppm	1000-10000	1000-10000	1000
Life (h)	>10000	>10000	>10000

Ceramic filters come in a variety of configurations, including cross-flow (thin, porous ceramic plates that contain channels formed by ribbed sections), candle filters (rigid tubular filters formed by bonding ceramic fibers and/or grains with an aluminosilicate binder), and fabric filters. Cross-flow filters are usually made out of a combination of alumina and mullite, whereas most candle filters have been made of SiC. Table 8 compares fillter experience[39].

Table 8. Comparison of Filter Experience

Application	Type	Material	Hours of operation	Comments
Gasification	Cross flow	Alumina/mullite	250	Delamination
Combustion	Cross flow	Alumina/mullite	168	6/8 filters had no cracks
Combustion	Cross flow	Alumina/mullite	83	10/15 filters had no cracks
Combustion	Cross Flow	Alumina/mullite	1300	Failure along flange section
Combustion	Candle	SiC	Up to 800	1 candle broke after 300 h
Gasification	Candle	SiC	Up to 50	System failure caused filter failure
Gasification	Candle	SiC	150	---------
Combustion	Candle	SiC	5800	6 candles

Porous Materials

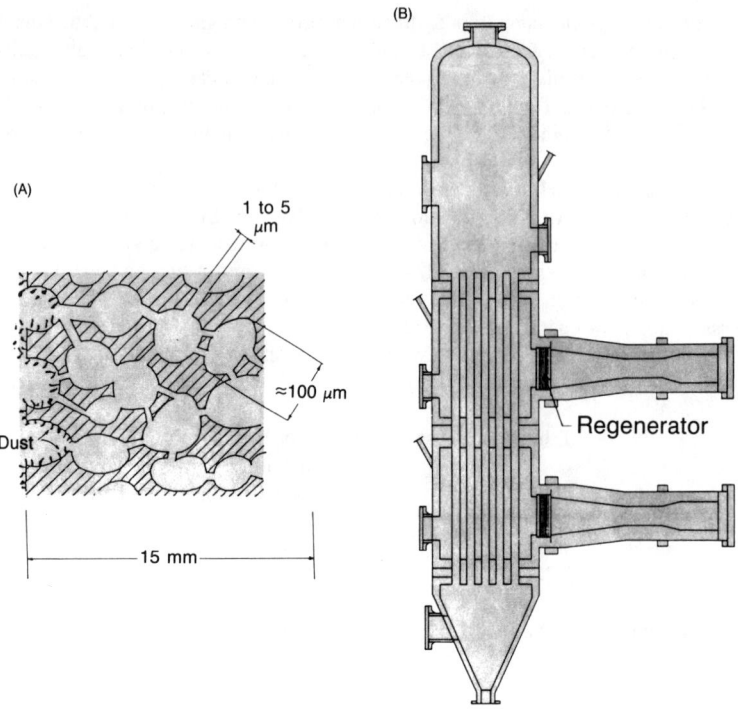

Figure 6 Schematic of pore structure, A, used in the Asahi's advanced ceramic tube filter, B.

Both alumina/mullite and cordierite have been demonstrated to have certain advantages over nonoxide materials[40]. The oxides already contain stable oxide phases which do not undergo further phase transition. They also retain their physical integrity during exposure to gas-phase alkali in comparison to the sodium silicate-glass phase, which forms along the surface of the non-oxide materials. However, potential long-term degradation mechanisms may result from chemical reactions, particularly with alkali species and/or steam, which would affect the long-term durability of the system. Westinghouse is evaluating alumina/mullite cross-flow filters after preliminary field tests and has begun subpilot- and pilot-scale field testing of integrated systems. Figure 6 shows a cordierite filter developed by Asahi Glass.

High-temperature ceramic fibers are being used in candle filters, fabric filters, and composite filters for removal of particulates from hot-gas streams in power generation systems, chemical processing, and other applications. Industrial Filter & Pump Manufacturing Co. has patented a ceramic filter manufactured by first forming a skeletal member made up of randomly disposed ceramic fibers by pulling an aqueous solution of the fibers and a liquid binder onto a mandrel and then heat treating the member to dry the binder. The dried member is then coated with alumina gel, colloidal aluminum, or colloidal silica and heat treated at a temperature within the range of 1000-1200°C to dry the coating and increase the corrosion resistance of the coating.

Industrial Filter has also tested, in conjunction with Universal Porosics, layered porous ceramic filter elements with different pore sizes and skin thicknesses[41]. Typically, the inner 85% of the wall thickness is made of a porous ceramic with a mean pore of about 125 μm, with the outer 15% having a wall thickness with 25-30-µm pores. Skin with a mean pore of 8-10 µm can be applied to support materials with mean pores of 125-150 µm without a significant amount of diffusion of the finer material into the more porous material. The degree of dust penetration decreases as mean pore size of the skin decreases. In addition, skin thickness can be decreased as the mean pore size of the skin decreases, which brings about benefits in terms of initial pressure drop at any given face velocity.

A number of companies have patented or licensed particulate filters, including Norton Co.[42], Electric Power Research Institute[43], Virginia Polytechnic Institute[44], and Babcock and Wilcox Co. The latter has licensed Asahi Glass Company's technology for application to pressurized fluidized-bed combustion power plants. This technology has been tested on several types of processes, some in excess of 4000 h, in sizes from 1-21 tubes in both two- and three-compartment configurations. Porous cordierite tubes are used in each compartment of this system. The porous structure consists of larger pores interconnected with smaller channels (see Figure 6), which achieves low-, long-term-, and stable-pressure drop, in addition to preventing dust penetration. Dust removal efficiency is on the order of 99.99% or better in most processes, and the filters are expected to withstand temperatures up to 1000°C.

CATALYSIS APPLICATIONS

Currently, the nearly $6 billion worldwide catalyst market is divided into three major markets: petroleum refining, chemical processing, and emissions control. In North America, these three areas share 35%, 31%, and 34%, respectively[45]. This market is expected to grow only about 4%/year to $2.5 billion by 1995 according to Catalyst Group. Products prepared using heterogeneous catalysis totalled more than $890 billion in 1989. Porous ceramics, including molecular sieves based on zeolites and other ceramic membranes, play an important role by acting as catalyst supports or even as the catalyst itself. Their transport and reaction properties are strongly dependent upon the connectivity of the pore network. Table 9 lists applications and materials[46-56].

Perhaps because the European market for catalysts is expected to have a higher annual growth rate (10%), AEA Technology's Harwell Laboratory has established a £2 million research program of European partners to develop processes for the reproducible preparation of pillared and multilayered materials of regular interlayer spacings with appropriate properties for catalysis and adsorbent applications. Materials with sheetlike structures such as clay-based materials, which will be modified by inserting pillars of inorganic oxides, such as alumina or zirconia, between the sheets in order to create pores will be tested.

Blasch Precision Ceramics (Schenectady, NY) has recently begun marketing a ceramic catalyst support system based on high-purity alumina honeycomb tiles that fit precisely together for a high degree of mechanical stability. The new catalyst support system can be used at temperatures as high as 1650°C in a variety of reaction systems that involve nitric acid, titanium dioxide, sulfuric acid, ammonia, phosphoric acid, and other corrosive chemicals. Field trials with these supports have shown that the number of annual replacement cycles can be reduced by as much as 90%. Cormetech Inc., a joint venture between Corning Inc. and Mitsubishi Heavy Industries, has opened a plant to make honeycomb ceramic catalysts for conversion of NO_x emissions from fossil fuel plants. The market for these catalysts is expected to grow from $20 million in 1992 to $150 million by 2000.

Table 9. Catalysis Applications and Material Systems

Application	Catalyst system	Manufacturers
Petroleum refining	Zeolites/aluminosilicate or titania; niobates/alumina	Davison Chemical, Akzo, Engelhard, Mobil, Exxon, Texaco, Katalistiks
Dehydrogenation of hydrocarbons	Platinum, iridium, or rhenium/alumina	UOP, Criterion
Production of polyethylene	Chrome-silica	Davison Chemical
Production of sulfuric acid	Vanadium pentaoxide	Monsanto
Synthesis of ethylene oxide	Silver oxide	Shell
Conversion of methane to C_2 hydrocarbons	Calcium, nickel, potassium/mixed oxide	Lawrence Berkeley Laboratory
Alkene isomerization	Sodium oxide/alumina	Texas Petrochemical
Removal of hydrogen sulfide from fluids	Zinc titanate	Philips Petroleum
Reduction of hydrogen sulfide emission	Platinum/alumina, ceria, germania	N.E. Chemcat
Synthesis of methane and methanol from CO_2, water vapor	Porous glass beads coated with TiO_2 (10-µm pores)	University of Osaka
Filteration of fluids	Phosphorous/silica-alumina or alumina-titania	Union Oil, Corning
Selective catalytic reduction of NO_x	Vanadium/titania	Davison Chemical
Removal of NO_x, SO_x from flue gases	Sodium carbonate/alumina	Davison Chemical, Engelhard
Removal of hydrocarbons from diesel engines	Based on sol of hydrolyzed aluminate compounds	Kyushu University
Reducing NO_x from lean-burn engines	Metal catalyst(copper)/zeolites	Allied Signal, Rhone-Poulenc, Toyota Motor
Direct conversion of coal to clean liquid fuel	Ni-Mo/hydrous titania ion exchanging thin film on porous oxide	Sandia National Laboratory

Automotive catalysts used for controlling engine emissions generally consist of precious metal (platinum or palladium or both) supported on ceramic or metal monoliths. Both Corning Inc. and NGK Insulators manufacture cordierite honeycomb monoliths with porosity ranging from 20% to 35%[57]. A washcoat is often applied to the monolith to enhance its, strength and design of the converter system must be optimized to maintain performance. Corning has shown that the ceramic monolith must be designed to have good compatibility with the washcoat and use a packaging system with positive mounting pressure under all driving conditions[58].

Because these ceramic substrates offer stronger catalyst adhesion, less sensitivity to corrosion, lower cost, and have proven reliability and performance, they are being considered for diesel oxidation catalysts for truck applications. The catalysts are dispersed in the tiny pores of the oxide washcoat. Exhaust gases and liquids diffuse into the pores and react with the catalyst.

Although diesel catalysts experience lower exhaust temperature, they must withstand a wide range of solids, liquids, and gases, as well as noncombustible additives from lubricating oil. These corrosive conditions could seriously reduce operating life. Modification of carrier properties such as surface chemistry, pore structure, and surface area is being investigated to create contamination-resistant catalysts[59].

PARTICULATE FILTERS FOR DIESEL EMISSIONS

As federal truck emissions standards become stricter, ceramic filters may be one way to achieve conformance with these restrictions. Between 1988 and 1998, NO_x limits will have dropped by about 60% and particulates limits by 80%. In 1994, particulate levels must drop by more than half to 0.1 g/(bhp·h). In 1998, NO_x must drop by 20% to 4.0 g/(bhp·h)[60]. Particulate filter or soot traps are already under investigation and look promising for certain applications, specifically for urban buses. Diesel particles are typically less than 1 μm in diameter and have a bimodal distribution (0.0075-0.056 μm and 0.056-1.0 μm)[61]. Ceramics are preferred because of their high-temperature resistance and higher trapping efficiency. However, regeneration by heating is necessary which adds cost and complexity to the system.

More than 300 million cellular ceramic substrates have been produced for automotive catalytic converters since 1974--these same substrates are now being used to manufacture diesel converter substrates and particulate filters[62]. The material of choice is usually cordierite because of its low thermal expansion, high-temperature resistance (1400°C), and high-temperature mechanical strength. A cordierite composition with improved thermal expansion has recently been developed by Corning for long-term durability over a wide range of operating conditions and temperatures. Table 10 compares several commercially available filters[63].

Table 10. Characteristics of Diesel Particulate Filters

Filter type	NGK C-286	NGK C-415	Panasonic	Corning	Ceramem
Cells/in^2	200	100	80	100	100
Length, cm	5.08	5.08	4.5	15.0	15.0
Diameter, cm	14.38	14.38	14.4	14.15	14.15
Wall thickness, mm	0.44	0.45	0.45	0.63	0.68
Filtration efficiency,%	55	80	80	70	99
Mean pore size, μm	19.5	---	28.1	35	0.2
Porosity	50	---	80	35	----

Corning is one of the world's leading producers and manufactures diesel filters of the wall flow types. Alternate channels are plugged at opposite ends so that the exhaust gases are forced to flow through the porous ceramic channel walls. These walls thus act as filters, trapping the solid particulates at efficiencies ranging from 60% to 95%. Porosities total 50% and three pore sizes are available: 12 μm, 22 μm, and 35 μm. Other companies with recent patents in diesel filters include Swiss Aluminum Ltd.[64], Daimler-Benz AG[65], and L'Institut de L'Amiante[66].

Cordierite filters, Fig. 7, have survived over 1000 regenerations. This has been attributed to filter properties, controlled regeneration conditions, and optimum packaging design. In a thermal durability study, elastic properties of the filter were reduced after successive regenerations, which was attributed to uniform and homogeneous microcracking, Fig. 8. This led to a reduction of thermal stresses by 50% to 55%, thereby improving filter durability[67].

Figure 7 Cordierite filter.

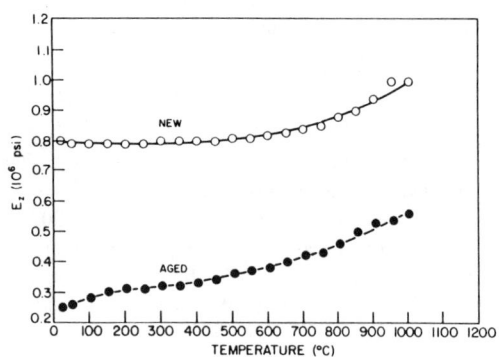

Figure 8 Variation of axial modulus of new and aged filters with temperature.

The wall flow honeycomb filter traps attain over 95% smoke-filtering efficiency and a relatively low pressure drop when not too loaded with soot. To eliminate this soot, catalysts are usually necessary, which are either impregnated into the porous ceramic wall or are added to the fuel as an additive. Catalytic coatings still require a regeneration system and are more effective with filters having both high porosity and mean pore size[68].

Catalytic fuel additives based on cerium, combined with exhaust throttling as a backup regeneration system, have been successful in bus applications in Athens, Greece, accumulating more than 100,000 km of operation[69]. The filter material remained intact, with negligible change in pore size distribution and total pore volume, and the ash deposits showed limited interaction with cordierite. More than 94% of these deposits were filtered by the monoliths and could be removed with conventional methods.

Pore size distribution of the filter material determines trapping efficiency and flow restriction. A filter with large pore material shows low flow restriction but low trapping efficiency, and a filter with small pore material shows the opposite behavior[70]. To obtain high trapping efficiency and low flow restriction, the mean pore diameter should be increased while maintaining a large pore volume within certain limits. NGK Insulators has used this approach, increasing the mean pore diameter from 13-20 µm for their filter material, while maintaining the cumulative volume of pores above 100 µm within 8% of the total pore volume. Trapping efficiencies were similar and pressure drops were lower compared to the original material.

Northeastern University has developed a rotating self-cleaning particulate trap based on ceramic monolith filter elements that can simultaneously filter the exhaust stream and remove collected particulates. Substantial removal of the gaseous hydrocarbons from the exhaust stream was attributed to condensation induced by the low temperature of the filter. Of the different filters tested, a Corning filter treated with a 50-μm-thick membrane coating with submicron-sized pores (developed by Ceramem) showed the highest collection efficiency (>90%) and excellent cleaning characteristics. Reductions of unburned gaseous hydrocarbons ranged from 50% to 90%.

Japanese researchers have developed another type of self-heating filter based on SiC[71]. Because SiC has a higher temperature resistance than cordierite, problems with filter failure due to melting or cracking caused during regeneration are significantly reduced. SiC also has a suitable electrical resistance behavior appropriate for self heating and precisely controlled temperatures. Another advantage is that a wide regeneration window is possible. However, oxidation catalysts are required to remove the CO produced during regeneration. Ibiden Co. (Ohgaki, Japan) has also developed a SiC honeycomb filter, which is planned for commercialization by 1993.

Several other novel concepts for particulate filters have been developed. One approach, for light truck applications, is based on multiple filter tubes consisting of Nextel yarn[72]. The wound yarn forms a pattern of diamond-shaped spaces between adjacent strands of yarn, which are filled with individual ceramic filaments. Another approach based on fibers developed in Germany uses ceramic microfibers (mullite or silica) knitted into a heterogeneous open pore structure. This type of filter has efficiencies in the range of 95% and operating temperatures up to 1200°C[73]. The special knitting process is economical and offers flexibility in filter design and operation. High reliability regarding mechanical, thermal, and thermal shock loading also has been demonstrated.

In order to achieve both lower NO_x emissions and particulates expected in future legislation, the particulate trap may have to be combined with the diesel catalyst[74]. Such a catalyzed particulate trap would remove solid and liquid particulates to 90% while abating CO and HC. A catalyzed ceramic particulate filter can reduce HC by up to 50%. Self regeneration may also be possible.

SENSORS

A wide range of porous materials are also being developed for different types of sensors. A carbon dioxide sensor consisting of K_2CO_3-poly(ethylene glycol) solution impregnated into porous alumina has been developed, where the resistance decrease on exposure to CO_2 is attributed to electrochemical reduction of dissolved CO_2 at the Au electrode. Resistance change is also sensitive to applied voltage and to electrode chemistry.

Humidity sensors come in a wide range of materials, including compositions based on TiO_2 combined with other oxides and α-Fe_2O_3[75,76]. A $BaTiO_3$-$SrTiO_3$ temperature-humidity sensor on a double Al_2O_3 substrate has been developed that can measure temperatures from -20 to 100°C, humidities from 10% to 100% at a response time <15 s and heat-cleaning time of 60 s at 400°C. Compositions based on TiO_2-LiO_2-MgO-V_2O_5 have been used for non-heating-type humidity sensors. A $MgCr_2O_4$-TiO_2 porous material has also shown promise as a humidity sensor.

Anodic spark deposition has been used to prepare porous alumina humidity sensors[77]. This process also involves a reanodization of MgCr films in either borax aqueous or H_2SO_4 solutions to form a MgCr layer at the pore base. These modified films show a high response to water vapor change at -76 to +20°C dewpoint with improved long-term stability.

Porous glasses, which have been used as absorbents, catalysts carriers, and liquid separation membranes, have also been applied to fiber optic moisture sensors for high-temperature applications[78]. Rutgers University has done considerable work in this area with an earlier design using $CoCl_2$ as the sensing element, with a sensitivity of 1% or less. A latter design improved the response time and operated at temperatures up to 300°C.

To prepare porous glass optical fibers, an alkali borosilicate glass preform is made, from which fibers are drawn. These fibers are then heat treated at temperatures between 550 and 600°C to induce the phase separation in the glass fibers to silica and alkaliborate. The borate-rich phase is leached away using diluted mineral acid, leaving a silica-rich porous fiber skeleton. Only a surface layer (the sensing region) of the phase-separated glass is leached so that the inner part of the fiber remains solid. Figures 9 and 10 show the process flow and schematic for these sensors.

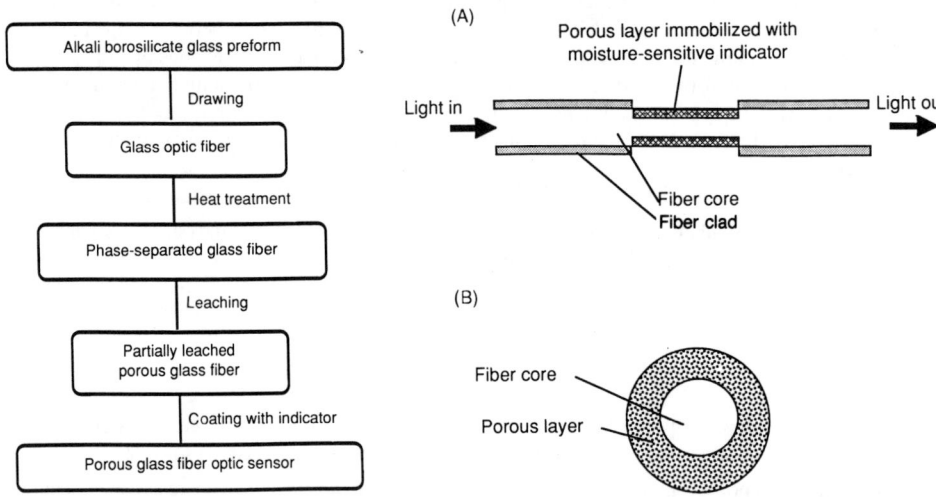

Figure 9 Flow diagram for preparing porous fiber optic sensors.

Figure 10 (A) Schematic of a glass fiber sensing element. (B) Cross section of the sensing segment

References
[1] B. Blake-Coleman and D. Clarke, Method for Selective Filtering of a Fluid Using Porous Piezoelectric Material, Public Health Lab Service Board, U.S. Patent 4904394, Feb. 27 1990
[2,3] J. Saggio-Woyansky, C. Scott, and W. Minnear, "Processing of Porous Ceramics," *Am. Ceram. Soc. Bull.*, Vol. No. 11, 1674-1682, 1992, The American Ceramic Society.
[4] A. Tsuchinari, T. Hokii, O. Shimobayashi, and C. Kanaoka, "Influence of Grain Size and Firing Temperature on the Structure of Porous Magnesia Ceramics," *Journal of the Ceramic Society of Japan*, Int. Edition, Vol. 100, 216-220, 1992.
[5] H. Abe, H. Seki, A. Fukunaga, and M. Egashira, "Preparation and Water Permeation Property of Bimodal Porous Cordierite Ceramics," *Journal of the Ceramic Society of Japan*, Int. Edition,

Vol. 100, 32-36, 1992.
[6] N. Greulich, W. Kiefer, and V. Rehm, Method of Producing Open-Pore Sintered Glass Filters and Product, Schott Glaswerek, U.S. Patent No. 4927442, May 22, 1990.
[7] A. Kawahara, H. Ichinose, S. Furata, and H. Katsuki, "Preparation of Porous Alumina Ceramics by Using the Gelation of Ammonium Alginate," *Journal of the Ceramic Society of Japan*, Int. Edition, Vol. 100, 228-230, 1992.
[8,9] L. Klein and N. Giszpenc, "Sol-gel Processing for Gas Separation Membranes," *Am. Ceram. Soc. Bull.*, Vol. 69, No. 11, pp. 1821-6. 1990. The American Ceramic Society.
[10] L. Klein, C. Yu, R. Woodman, and R. Pavlik, "Microporous Oxides by the Sol-Gel Process: Synthesis and Applications," *Catalysis Today*, Vol. 14, 165-73, 1992.
[11] L. Sheppard, "Corrosion-Resistant Ceramics for Severe Environments," *Am. Ceram. Soc. Bull.*, Vol. 70, No. 7, 1146-1160, 1991, The American Ceramic Society.
[12] R. Orenstein and D. Green, "Thermal Shock Behavior of Open Cell Ceramic Foams," *Center for Advanced Materials Newsletter*, Vol. 5, No.3, pp. 39-41, The Pennsylvania State University, 1991.
[13] L. Minjolle, "Alveolar Ceramic Filters for High-Melting Metals," Ceramiques et Composites, U.S. Patent No. 4921616, May 5, 1990.
[14] A. Sane, A. Gee, and D. Eichermiller, "Filteration of Molten Ferrous Metal", Carborundum Co., U.S. Patent No. 5045111, September 3, 1991.
[15] R. Orenstein and D. Green, "Thermal Shock Behavior of Open Cell Ceramic Foams," *Center for Advanced Materials Newsletter*, Vol. 5, No.3, 39-41, The Pennsylvania State University, 1991.
[16] D. Fain, G. Roettger, and D. White, "Ceramic Membranes for High Temperature Hydrogen Separation," *Proceedings of the Conference on Fossil Energy Materials*, Oak Ridge National Laboratory, 1992.
[17, 18] K. Chan and A. Brownstein, "Ceramic Membranes--Growth Prospects and Opportunities," *Am. Ceram. Soc. Bull.*, Vol. 70, No. 4, 703-707, 1991, The American Ceramic Society.
[19] *Ceramic Materials Technology in Japan*, November 1991, COMLINE News Service, p.6.
[20] T. Flottmann and J. Tretzel, "Micro/Ultrafiltration Membranes with a Fixed Pore Size Formed through Irradiation with Pulsed Lasers and Process for Manufacturing the Same," Akzo NV, U.S. Patent No. 4923608, May 8, 1990.
[21] M. Besland, S. Galaj, and J. Gillot, "Membrane Filter," Societe des Ceramiques Techniques, U.S. Patent No. 4946592, August 7, 1990.
[22,23] *Extended Abstracts*, American Institute of Chemical Engineers 1991 Annual Meeting and Third Topical Conference on Emerging Technologies in Materials, 1991, American Institute of Chemical Engineers.
[24] *High Tech Ceramics News*, February 1991, 4, Business Communications Co.
[25] E. Butler, R. Landham, and M. Thomas, "Composite Membrane," Alcan International Ltd., U.S. Patent No. 4938870, July 3, 1990.
[26] Y. Lin and Anthonie Burggraaf, "Preparation and Characterization of High-Temperature Thermally Stable Alumina Composite Membrane," *J. Am. Ceram. Soc.*, Vol. 74, No. 1, 219-24, 1991, The American Ceramic Society.
[27] G. Fountain and J. Spivey, "Ceramics Research at Research Triangle Institute: Expanding Electronic Processing and Membrane Technologies," *Am. Cer. Soc. Bull.*, Vol. 69, No. 10, 1699-1702, 1990, The American Ceramic Society.
[28] *High Tech Ceramic News,* April 1991, Business Communications Co.

[29][30][31]Y. Lin and A. Burggraaf, "Preparation and Characterization of High-Temperature Thermally Stable Alumina Composite Membrane," *J. Am. Ceram. Soc.*, Vol. 74, No. 1, 219-224, 1991, The American Ceramic Society.

[32,33]J. Kloeppel, "The Many Uses of Molecular Sieves," *Research Horizons*, Fall 1990, 16-18, Georgia Institute of Technology.

[34]"Zinc-Aluminum-Phosphorous-Silicon Oxide Molecular Sieve Composition," UOP, U.S. Patent No. 4935216.

[35,36]D. Fain, G. Roettger, and D. White, "Ceramic Membranes for High Temperature Hydrogen Separation," *Proceedings of the Conference on Fossil Energy Materials*, Oak Ridge National Laboratory, 1992.

[37,38,39,40]M. Alvin, T. Lippert, and J. Lane, "Assessment of Porous Ceramic Materials for Hot Gas Filtration Applications," *Am. Ceram. Soc. Bull.*, Vol. 70, No. 9, 1491-1498, 1991.

[41]J. Zievers, P. Eggerstedt, and E. Zievers, "Porous Ceramics for Gas Filtration," *Am. Ceram. Soc. Bull.*, Vol. 70, No. 1, 108-111, 1991, The American Ceramic Society.

[42]A. Butkus, High-Temperature Filter, Norton Co., U.S. Patent No. 4946487, August 7, 1990.

[43]D. Ciliberti and T. Lippert, "Compact Ceramic Tube Filter Array for High-Temperature Gas Filtration," Electric Power Research Institute, U.S. Patent No. 4904287, February 27, 1990.

[44]*High Tech Ceramic News*, May 1992, Vol.4, No.1, Business Communications Co.

[45,46]Ann M. Thayer, "Catalyst Suppliers Face Changing Industry," *Chemical and Engineering News*, March 9, 1992, 27-49, The American Chemical Society.

[47,48]*Ceramic Materials Technology in Japan*, November 1991, COMLINE News Service, (12,16), 1991.

[49]J. Ward and D. Delaney, "Silica-Alumina Catalyst Containing Phosphorus", Union Oil Co. of California, U.S. Patent No. 5051386, September 24, 1991.

[50]I. Lachman, J. Williams, and K. Zaun, "Phosphate-Containing Ceramic Structures for Catalyst Support and Fluid Filtering," Corning, Inc. U.S. Patent No. 5039644, August 13, 1991.

[51]C. Mauldin and K. Riley, "Titania-Supported Catalysts and Their Preparation for Use in Fischer-Tropsch Synthesis," Exxon Research and Engineering Co., U.S. Patent No. 4992406, February 12, 1991.

[52]F. Luck, "Catalyst for the Selective Reduction of Nitrogen Oxides and Process for the Preparation of the Catalyst," Rhone-Poulenc Chimie, U.S. Patent No. 5037792, August 6, 1991.

[53]H. Hsing, "Al_2O_3 Alkene Isomerization Process and Catalyst," Texas Petrochemicals Corp., U.S. Patent No. 5043523, August 27, 1991.

[54]"Absorption Composition Comprising Zinc Titanate for Removal of Hydrogen Sulfide from Fluid Streams," Phillips Petroleum Co., U.S. Patent No. 5045522, September 3, 1991.

[55]H. Schoennagel, Y. Tsao, and T. Yan, "Thermally Stable Nobel Metal-Containing Zeolite Catalyst," Mobil Corp., U.S. Patent No. 5041401, August 20, 1991.

[56]T. Yamada, M. Funabiki, and K. Kayano, "Exhaust Gas Purification Catalyst for Suppressing Hydrogen Sulfide Evolution, and Process for Production Thereof," N.E. Chemcat Corp., U.S. Patent No. 5039650, August 13, 1991.

[57]Laurel M. Sheppard, "Automotive Performance Accelerates with Ceramics," *Am. Ceram. Soc. Bull.*, Vol.69, No.6, 1012-1013, 1990, The American Ceramic Society.

[58]S. Gulati, "Design Considerations for Diesel Flow-Through Converters," *SP-896: Diesel Particulate Control, Trap, and Filtration Systems*, 89-101, 1991, SAE International.

[59,60]B. Farrauto, J. Adomaitis, J. Tiethof, and J. Mooney, "Reducing Truck Diesel Emissions: A Status Report," *Automotive Engineering*, February 1992, Vol. 100, No.2, 19-23, 1992, SAE International.

[61] K. Baumgard and J. Johnson, "The Effect of Low Sulfur Fuel and a Ceramic Particle Filter on Diesel Exhaust Particle Size Distributions," *SP-896: Diesel Particulate Control, Trap, and Filtration Systems*, 207-215, 1991, SAE International.

[62] N. Kilkarn, "Cellular Ceramic Products Help Curb Diesel Engine Emissions," *Automotive Engineering*, January 1992, Vol. 100, No. 1, 21-25, 1992, SAE International.

[63] N. Khalil and Y. Levendis, "Development of a New Diesel Particulate Control System with Wall-Flow Filters and Reverse Cleaning Regeneration," *SP-896: Diesel Particulate Control, Trap, and Filtration Systems*, 217-231, 1991, SAE International.

[64] H. Bressler and M. Dool, "Filter for Cleaning Exhaust Gases of Diesel Engines," Swiss Aluminum Ltd., U.S. Patent No. 4913712, April 3, 1990.

[65] M. Kuhn, "Soot Burn-Off Filter for Diesel Engines," Daimler-Benz Ag, U.S. Patent No. 4895707, January 23, 1990.

[66] J. Dunnigan and H. Menard, "Filter for Removing Cancer-Causing Compounds from Exhaust Fumes," L'Institut de L'Amiante, U.S. Patent No. 4875910, October 24, 1990.

[67] S. Gulati, D. Lambert, M. Hoffman, and A. Tuteja, "Thermal Durability of a Ceramic Wall-Flow Diesel Filter for Light Duty Vehicles," *SP-896: Diesel Particulate Control, Trap, and Filtration Systems*, 69-79, 1991, SAE International.

[68] K. Pattas and C. Michalopoulou, "Catalytic Activity in the Regeneration of the Ceramic Diesel Particulate Trap," *SP-896: Diesel Particulate Control, Trap, and Filtration Systems*, 123-134, 1991, SAE International.

[69] K. Pattas, Z. Samaras, *et al.*, "Cordierite FIlter Durability with Cerium Fuel Additive: 100000 km of Revenue Service in Athens," *SP-896: Diesel Particulate Control, Trap, and Filtration Systems*, 289-301, 1991, SAE International.

[70] J. Kitagawa, S. Asami, K. Uehara, and T. Hijikata, "Improvement of Pore Size Distribution of Wall Flow Type Diesel Particulate Filter," *SP-896: Diesel Particulate Control, Trap, and Filtration Systems*, pp 81-87, 1991, SAE International.

[71] Y. Goto, T. Abe, T. Sata, and M. Hayashida, "Study on Regeneration of Diesel Particle Trapper by Electrical Self-Heating Type Filter," *SP-896: Diesel Particulate Control, Trap, and Filtration Systems*, 13-23, 1991, SAE International.

[72] M. Barris, E. Wilson, and W. Hough, "A Filter Tube Trap System for Light Truck Applications," *SP-896: Diesel Particulate Control, Trap, and Filtration Systems*, 1-11, 1991, SAE International.

[73] A. Mayer and A. Buck, "Knitted Ceramic Fibers--A New Concept for Particulate Traps," *SP-896: Diesel Particulate Control, Trap, and Filtration Systems*, 103-114, 1991, SAE International.

[74] B. Farrauto, J. Adomaitis, J. Tiethof, and J. Mooney, "Reducing Truck Diesel Emissions: A Status Report," *Automotive Engineering*, February 1992, Vol. 100, No.2, 19-23. 1992. SAE International.

[75] Y. Yeh, T. Tseng, and D. Chang, "Electrical Properties of TiO_2-$K_2Ti_6O_{13}$ Porous Ceramic Humidity Sensor," *J. Am. Ceram. Soc.*, Vol. 73, No. 7, 1992-98, 1990.

[76] C. Cantalini and M. Pelino, "Microstructure and Humidity-Sensitive Characteristics of α-Fe_2O_3 Ceramic Sensor," *J. Am. Ceram. Soc.*, Vol. 75, No. 3, 546-51, 1992.

[77] Z. Chen and M. Jin, and C. Zhen, "Properties of Modified Anodic-Spark-Deposited Alumina Porous Ceramic Films as Humidity Sensors," *J. Am. Ceram. Soc.*, Vol. 74, No. 6, 1325-30, 1991.

[78] J. Ding, M. Shahriari, and G. Siegel, "Fiber Optic Moisture Sensors for High Temperatures," *Am. Ceram. Soc. Bull.*, Vol., 70, No. 9, 1513-1517, 1991.

Section II. Production Methods

PREPARATION OF POROUS MATERIALS BY THE SOL-GEL METHOD

Sumio Sakka, Hiromitsu Kozuka and Tatsuhiko Adachi*
Institute for Chemical Research, Kyoto University
Uji, Kyoto-Fu 611, Japan
*Ube-Nitto Kasei Co., Ltd.
2-1-1 Yabuta-Nishi, Gifu 500, Japan

The sol-gel method is known to produce materials of various microstructures, such as dense glasses and ceramics, inorganic-organic composites, functionally gradient glasses, and porous gels and ceramics. This paper reviews our work on sol-gel preparation of porous silica gels of different pore sizes, their properties, and behavior.

It is shown that the variation of the composition of the starting solution leads to the porous silica bodies of which the average pore diameter is about 2.5 nm, 16 nm and 5000 nm, respectively. The gels with 2.5 nm pores are transparent and are applied to fabrication of transparent discs doped with functional organic molecules. The gels with 5000 nm pores are used as filters and supports of catalysts and enzymes.

With the gels having 16 nm pores, mechanical stability on exposure to solvents of varying surface tension, and changes of porosity, hardness and elastic moduli with heat treatment will be discussed.

INTRODUCTION

The first products encountered in the preparation of materials by the sol-gel method are gels, which consist of oxide backbones (associated particles) and pores. The size of particles and pores shows marked dependence on the composition of the starting solution, the temperature of gelation, the atmosphere during sol-gel conversion, and so on. Accordingly, the sol-gel method is attracting much attention as the method of making porous materials and composite materials. This paper discusses the sol-gel preparation of silica gels of different pore sizes, stability of the resultant porous gels to solvents, mechanical properties of the gels, and application of the gels to preparation of optically active transparent gels. Most of the data referred to in this review article are based on the work in our laboratory.

PREPARATION OF SILICA GELS OF DIFFERENT PORE SIZES

In sol-gel method [1-3], gels of bulk bodies such as disks and rods, fibers and coating films, can be directly prepared without requiring a powder processing step. The problem encountered in preparing bulk bodies is the occurrence of cracks and fracture during the course of drying of wet gel bodies. The capillary force arising in the solution in pores may be the cause for the occurrence of cracks. A number of attempts [4] have been made to reduce the capillary force at the final stage of drying.

We have made three series of silica gel bodies [5,6,7] without occurrence of cracks. Each gel has an average pore size different from each other. The typical compositions of starting solutions, appearance, average pore diameter and porosity of gels are shown in Table 1. The source of SiO_2 is tetramethoxysilane $Si(OCH_3)_4$ for all the samples.

Table 1. Typical Compositions of the Starting Solutions and the Appearance, Average Pore Size and Porosity of Resulting Silica Gel Bodies.

No.	Composition of the starting solution (mole ratio)					Properties of gel			Ref.
	$Si(OCH_3)_4$	H_2O	CH_3OH	$(CH_3)_2NCHO$	HCl NH$_3$OH	Appearance	Average pore size	Porosity	
1	1	4	2		0.4	Transparent	2.5 nm	0.46	5
2	1	12	2	1	5x10^{-4}	Translucent	16 nm	0.73	6
3	1	1.53	2		0.4	Opaque	5-10 μm*	0.6	7

*Size of pores between large, secondary particles. There are finer pores within individual secondary particles.

Gel No. 1 is transparent [5] and so can be employed as optical materials such as lenses and prisms. A more important application of this gel is the fabrication of optically active materials by doping with organic dye molecules. Advanced optical materials showing fluorescence [8], laser oscillation [9], nonlinear optical effect [10], photochemical hole burning [11] and photochromism [12] can be produced from these types of gels. The transparency of this gel is attributed to fine pores which are too small at 2.5 nm to cause significant light scattering.

Gel No. 2 is not transparent but translucent [6] as a result of a pore size of 16 nm, which causes a certain degree of light scattering. The larger pore size compared with Gel No. 1 has been produced by addition of dimethylformamide and the hydrolysis-polycondensation at higher pH brought about by a small amount of ammonia. Gels of this type may be used as catalyst or enzyme supports and microfilters. Another use of this gel is the fabrication of transparent silica glass. On heating to 1050 °C, the gel converts to transparent silica glass without crack formation [13].

Gel No. 3 has much larger spherical particles of about 5 μm in diameter [7] connected to each other as shown in Fig. 1(a). There

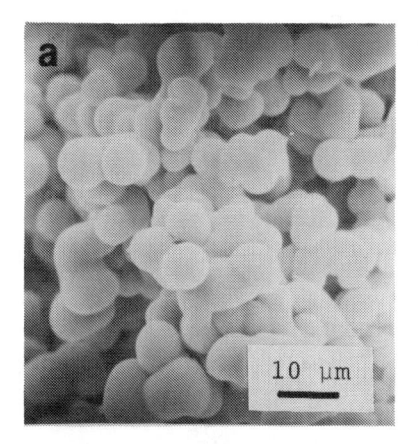

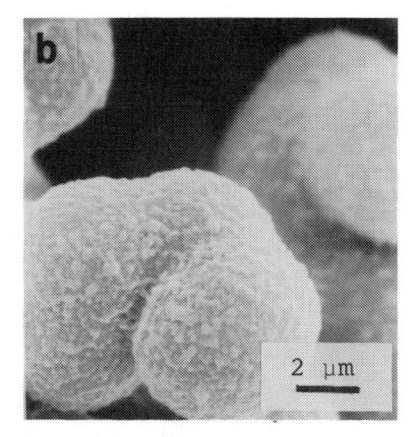

Figure 1 SEM pictures showing the microstructure of Gel No. 3 [7]. (a) Secondary particles connected with each other. (b) Magnified picture showing that each large particle has small pores.

are continuous pores of 5-10 μm in diameter between 5 μm particles. As shown in Fig. 1 (b), each of these large 5 μm particles consists of finer particles and finer pores. This is also shown by the surface measurements which indicate that the surface area ranges from 50 to 200 m^2/g, instead of less than 0.1 m^2/g calculated by assuming that the 5 μm particles are non-porous dense particles [14]. In other words, Gel No. 3 is a double pore system suitable for catalyst and enzyme carriers. In the double pore system, larger pores are used for the rapid transport of the substrate and large surface areas of smaller pores are used for the deposition of a large amount of catalysts and accordingly the efficient catalytic action [15].

On heating to 1300°C, Gel No. 3 becomes glassy in the appearance (opaque with luster), but is still porous. This indicates that a porous silica glass is formed.

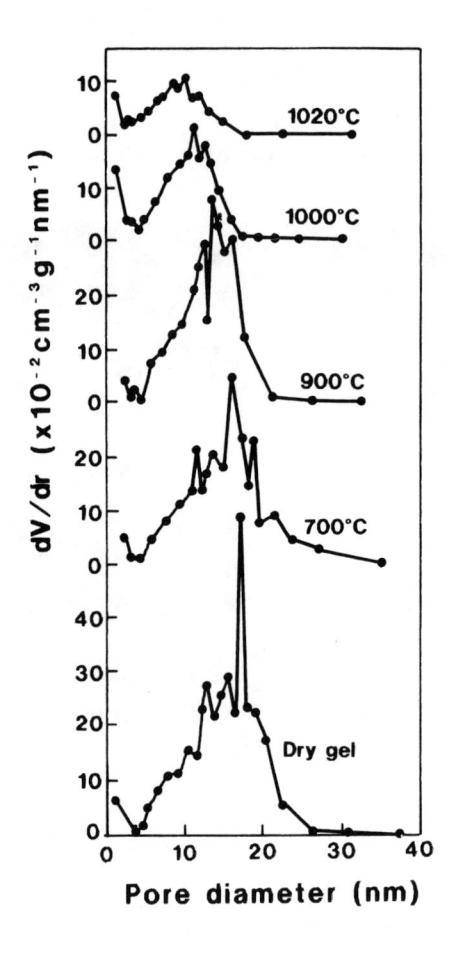

Figure 2 Pore size distributions for Gel No. 2 after heating to various temperatures [16].

BEHAVIOR OF POROUS SILICA GEL BODIES ON HEATING

Gels No. 1, 2 and 3, which have continuous pores of different size range, show different pore behavior on heating. A 6 mm thick and 30 mm diameter disc of Gel No. 1 is fractured, and smaller pieces expand after heating to about 800°C. A 32 mm diameter and 20 cm long rod (cylinder) of Gel No. 2 becomes a dense, transparent silica glass without any fracture and cracks on heating to 1050 °C, shrinking to a rod of 20 mm in diameter and 13 cm in length. As stated in the previous section, a large disc of Gel No. 3 becomes an opaque, glassy, porous material on heating to 1300 °C.

The behavior of pores of Gel No. 2 will be described below. It is interesting to note that a very large change of porosity ranging from 73% to 0% can be obtained. Gel No. 2 has an average pore size of 16 nm and bulk density of 0.62 after drying.

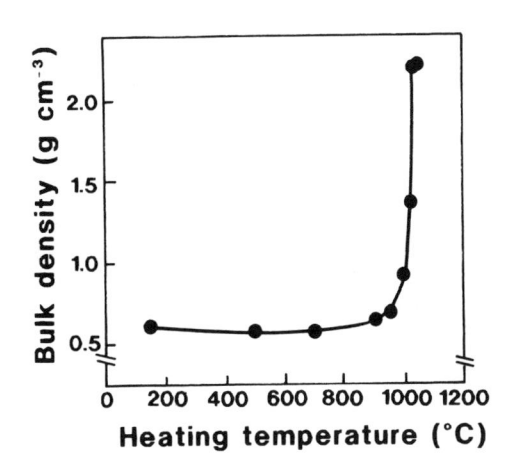

Figure 3 Bulk density of Gel No. 2 as a function of heating temperature [16].

Figure 2 and 3 show the pore size distribution curves and the bulk density for Gel No. 2 heated to the given temperatures [16]. It is seen that heating to temperatures lower than 900 °C does not cause any noticeable change in the average pore size and bulk density. When the gel is heated at temperatures higher than 900 °C, however, the decrease in the average pore size becomes noticeable. The pore size drops from 14 nm to 11 nm from 900 °C to 1000 °C, and drops to 10 nm at 1100 °C. It is noted that the gel heated to 1030 °C has almost no open pores. Heating up to 1050 °C and holding for 2 h gives a transparent and solid glass without pores. In addition, the bulk density shows a rapid increase with heating temperature, reaching a value of 2.2, which coincides with that of silica glass prepared by conventional melting of quartz crystal at high temperatures.

This indicates that it is possible to produce gels with both varying bulk densities and porosities by controlling the heating temperature and time.

STABILITY OF THE GEL TO SOLVENTS

It is known that the bulk samples of porous silica glass prepared by acid-leaching a phase-separated sodium borosilicate glass are easily cracked or fractured during handling. This happens when the porous silica glasses containing continuous pores of several tens of nm stored in water are exposed to air or the dried samples are immersed in solutions. Similar phenomena are observed in sol-gel derived silica gel. It is believed that the cracks or fracture may be caused by the capillary force. More detailed observations on this phenomenon are described below.

Observations were made [17] on dry gel bodies (Gel No. 2 in Table 1), which have a porosity of 73% and an average pore diameter of 16 nm. Discs of 8 mm thickness and 32 mm diameter were cut from the dried gel cylinder and used for the experiment. The discs were dried in vacuum at 100 °C for 12 h and then exposed to vapor of a solvent for 48 h. The discs were then subsequently immersed in a liquid solvent at room temperature at 25 °C. In this test one surface of the exposed disc sample was immersed in the solvent, whereas the other surface was not immersed.

The result of these observations are shown in Table 2 [17]. Some of the examples of crack formation are shown in Fig. 4 [18]. It is seen that when the surface tension of the liquid solvent is small, no cracks occur. When the surface tension is large, however, cracks occur. The occurrence of cracks may be attributed to the tensile stress produced by the capillary force based on the surface tension of the liquid entering the micropores. Assuming that micropores in the gel are represented by capillary tubes of diameter D, the capillary force, ΔP, acting on the capillary walls of the silica skeleton at the menisci is expressed by

Table 2. Observation on the Crack Formation in the Gel Disc 8 mm Thick and 32 mm Diameter due to Exposure to the Vapor and Subsequent Immersion in a Liquid Solvent [17].

Solvent	Solvent Surface tension (dyne cm^{-1})	Boiling point (°C)	Crack formation in the gel Exposure to a solvent vapor	Subsequent immersion in a liquid solvent
Diethyl ether	17.1	34.6	No Crack	No crack
Di-iso-propylamine	20.0*	83.4	No crack	No crack
Acetone	23.7	56.2	No crack	No crack
Benzene	28.9	80.1	No crack	No crack
Dichloroethane	32.2	83.5	No crack	No crack
Dioxane	34.5**	101.3	No crack	No crack
Dimethylformamide	36.8	153	No crack	No crack
Phenol	40.9	181.8	No crack	No crack
Dimethylsulphoxide	43***	189	No crack	No crack
Aniline	44.5	184.7	No crack	No crack
Ethylene glycol	46.5	197.9	A few cracks	Many cracks
Diethylene glycol	48.5***	244.3	Many cracks	Broken into tiny fragments
Monoethanolamine	48.9	180	A few cracks	Many cracks
Water	72.8	100	A few cracks	Many cracks

Surface tension at *16°C, **15°C, ***25°C.

Porous Materials

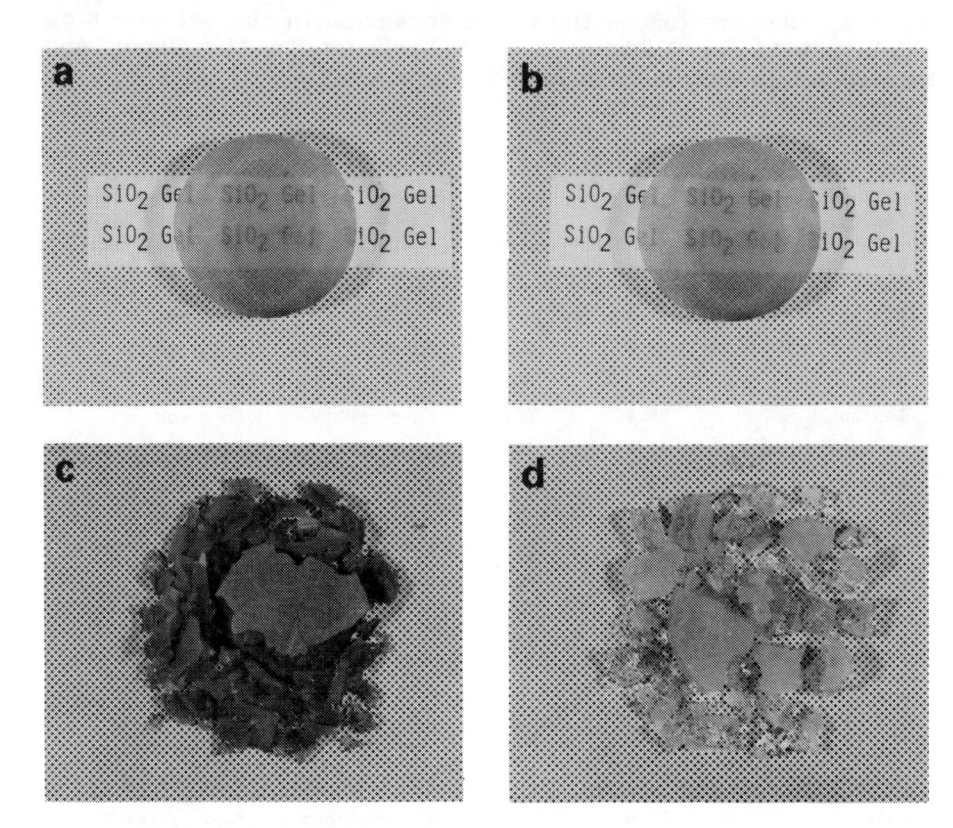

Figure 4 Appearance of the dried gel after being immersed in various solvents and dried [18]. (a) Methanol, (b) dimethylformamide, (c) formamide and (d) water.

$$\Delta P = 4\gamma \cos\theta / D$$

where γ is the surface tension of the liquid filling the micropores and θ is the contact angle between the liquid and the SiO_2 wall of the micropores. The layer in which a liquid is incorporated tends to shrink and undergoes a tensile stress from the dry region, i.e. that part of the gel which is not reached by the liquid. When this stress exceeds the strength of the gel network, cracks will occur.

The above observation tells us the importance of the surface tension of the medium solution in dealing with the brittle porous materials. It has to be remembered that in order to incorporate a liquid or solution into a dried gel for the purpose of stuffing, a solution of low surface tension has to be used or the pore surface has to be chemically modified.

MECHANICAL PROPERTIES OF GELS

Mechanical properties such as strength, hardness, and elastic moduli of gels are very important from both a practical as well as scientific view point. The sol-gel method can produce gels with a large range of porosities by heating the as-prepared gel to different temperatures and for various times. This serves in establishing the mechanical property - porosity relationships.

For some applications, e.g. for application as biomedical material, the dense ceramics and glasses usually are too hard and too rigid. Silica gels will be useful because these gels can be made with hardness and elestic moduli appropriate for a particular application by varying the treatment condition.

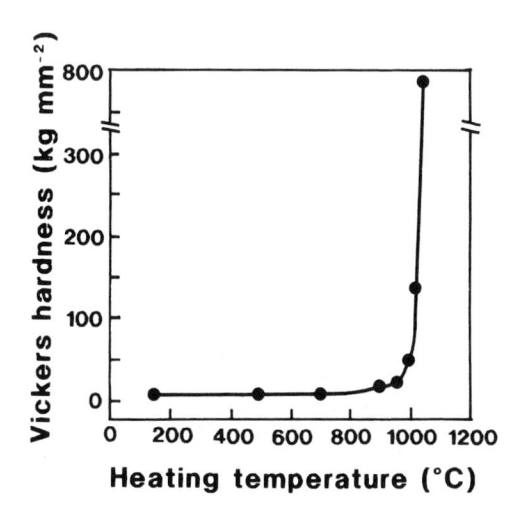

Figure 5 Change of Vickers hardness of Gel No. 2 as a function of heating temperature [19].

Fig. 5 and 6 show the change of Vickers hardness and Young's modulus of Gel No. 2 (see Table 1), respectively, as a function of heating temperature [19]. It is seen that above 900 °C both properties rapidly increase. These rapid increases correspond to the rapid increase of bulk density as shown in Fig. 3 of section 3. The hardness and Young's modulus increased from 4 to 780 kg/mm^2 and from 0.95 to 72.5 GPa, respectively, after densification of a dried gel with 73% porosity [18].

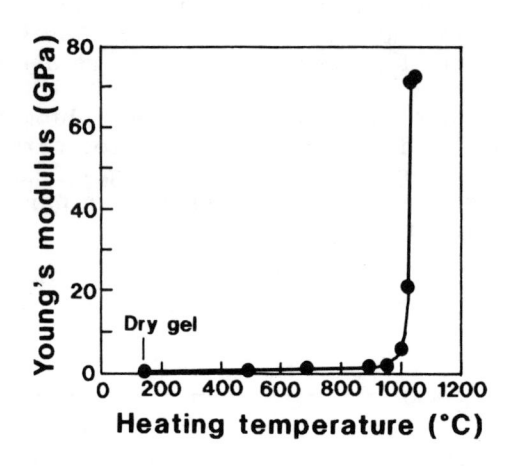

Figure 6 Change of Young's modulus of Gel No. 2 as a function of heating temperature [19].

In order to discuss the Young's modulus-porosity relationship with the present data, Young's modulus was plotted against the porosity for dry gels prepared from different starting solutions (HC-Gel, AM-Gel and Gel), heated gels (H1-Gel and H2-Gel) and foamed glasses, as shown in Fig. 7. In Fig. 7, three curves corresponding to those three series are drawn by dashed lines. For a given porosity, Young's modulus of the heated gels is lower than that of foamed glass, with the dried gels having the lowest value. This indicates that the Young's modulus of the backbone (silica skeleton) of the dried gel is lower than that of the heated gel and foamed glass. The foamed glasses have isolated closed pores and this might result in a higher Young's modulus.

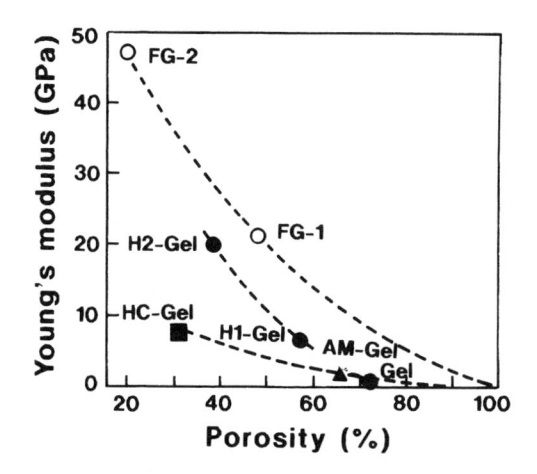

Figure 7 Young's modulus-porosity relationships for various gels and foamed glasses. FG-1 and FG-2: foamed glasses having closed pores. HC-Gel: prepared by acid catalysis without dimethylformamide. AM-Gel: prepared by ammonia catalysis without dimethylformamide. Gel: Gel No. 2 made with dimethylformamide and dried. H1-Gel and H2-Gel: after heating Gel No. 2.

It has been shown [19] that the porous silica samples obtained by heating the silica gel (Gel No. 2) above 700 °C can be regarded as having skeletons of similarly high strength and, therefore, their Young's moduli can be approximately expressed as a function of porosity only. For these samples, the Young's modulus-porosity data fits certain equations, which are applicable to the porous systems having isolated pores in a limited porosity range [19]. It is interesting to see that the equations for isolated pores are valid for the continuous pores.

APPLICATION UTILIZING THE PRESENCE OF PORES IN GELS

The potential applications of sol-gel derived porous gels are listed in Table 3. The gels are suitable for applications for microfiltration, separation and absorption and catalytic reaction, because a wide range of pore sizes and various porosities can be obtained.

In addition to these applications, the gels are used as components for producing new materials. Organic-inorganic composites are prepared by infiltrating or impregnating a gel with monomer solutions and polymerizing monomers in the fine pores of the gel. The organic-inorganic composites are expected to be light-weight,

Table 3. Potential Applications of Sol-Gel Derived Porous Gels.

Effect	Application
Microfiltration	Microfilter
Separation	Separation of mixed gas molecules
Absorption	Water cleaning
Catalytic reaction	Catalyst support
Components for making new materials	
(1) Organic-inorganic composites	Light weight optical parts
(2) Graded index lenses	Microlenses
(3) Porous TiO_2 coating film	Photocatalyst Solar cell
(4) Anitireflecting coating	Coating film for increasing damage threshold of laser optical system

transparent, and less brittle than silica gel.

Graded index lenses [20,21] are prepared by ion exchanging, stuffing or destuffing the wet or dry gel with index-changing chemical species, and therefore producing a refractive index gradient.

TiO_2 coating films are used as a highly efficient photocatalyst based on the photoelectrochemical effect [22], and as a solar cell for electrical power generation [23]. This is partially attributed to the presence of a large active surface area due to the structure of the film having fine pores.

Antireflecting coating films have been obtained by etching a silica gel coating film with hydrofluoric acid. This produces tapered pores so that the refractive index is larger in the direction of the glass substrate and becomes smaller until it coincides with that of the substrate at the interface [24].

CONCLUSION

Porous gels are used for microfiltration, separation of gas molecules, absorption of contaminants to clear waste water, and catalytic reaction as catalyst support. Also the gels can be used to produce new materials because of the presence of micropores. Properties of porous SiO_2 gels related to thses applications have been presented.

REFERENCES

1. S. Sakka, The Science of the Sol-Gel Method, Agne-Shofu-Sha, Tokyo, 1988, pp. 201 (in Japanese).
2. S. Sakka, "Gel method for making glass", in Treatise on Materials Science and Technology, 22, Glass III, ed. M. Tomozawa and R.H. Doremus, Academic Press, New York, 1982, p.129-167.
3. C.J. Brinker and G.W. Scherer, The Sol-Gel Science, Academic Press, San Diego, 1990, pp.908.
4. S. Sakka, "Formation of particles in sol-gel process", KONA: Powder Science and Technology in Japan No.7, Hosokawa Micron Company, Osaka, 1989, p.106-118.
5. S. Sakka, K. Aoki and H. Kozuka, "Sol-gel synthesis of silica disc as applied to supports of organic molecules with optical functions", submitted to J. Mater. Sci.
6. T. Adachi and S. Sakka, "The role of N,N-dimethylformamide, a DCCA, in the formation of silica gel monoliths by sol-gel method", J. Non-Cryst. Solids, **99**, 118-128 (1988).
7. H. Kozuka and S. Sakka, "Formation of particulate opaque gels from highly acidic solutions of tetramethoxysilane", Chem. Mater., **1**, 398-404 (1989).
8. D. Avnir, D. Levy and R. Reisfeld, "The nature of the silica cage as reflected by spectral changes and enhanced photostability of trapped rhodamine 6G", J. Phys. Chem., **88**, 5956-59 (1984).
9. R. Reisfeld, "Theory and application of spectroscopically active glasses prepared by the sol-gel method", Proc. SPIE, vol. 1328, Sol-Gel Optics, 1990, p.29-39.
10. M. Nakamura, H. Nasu and K. Kamiya, "Preparation of organic dye-doped SiO_2 gels by the sol-gel process and evaluation of their optical non-linearity", J. Non-Cryst. Solids, **135**, 1-7 (1991).
11. T. Tani, A. Makishima, H. Namikawa and K. Arai, "Photochemical hole burning study of 1,4-dihydroxyanthraquinone in amorphous silica", J. Appl. Phys., **58**, 3559-65 (1985).

12. D. Levy, S. Einhorn and D. Avnir, "Applications of the sol-gel process for the preparation of photochromic information-recording materials: synthesis, properties, mechanisms", J. Non-Cryst. Solids, **113**, 137-145 (1989).
13. T. Adachi and S. Sakka, "Preparation of monolithic silica gel and glass by the sol-gel method using N,N-dimethylformamide", J. Mater. Sci., **22**, 4407-10 (1987).
14. J. Yamaguchi, H. Kozuka and S. Sakka, "Structural change of gels prepared from highly acidic tetramethoxysilane solutions in the drying process", paper presented at Foundation Memorial Symposium of Advanced Materials Science and Engineering Society, Tokyo, March, 1989.
15. S. Sakka, "Formation and chemical properties of fine pores of porous ceramics", Ceramics (Japan), **23**, 702-706 (1988) [in Japanese].
16. T. Adachi and S. Sakka, "Microstructure changes in sol-gel derived silica gel monolith with heating as revealed by the crack formation on immersion in water, J. Ceram. Soc. Japan, **97**, 203-207 (1989) [in Japanese].
17. S. Sakka and T. Adachi, "Stability of sol-gel derived porous silica monoliths to solvents", J. Mater. Sci., **25**, 3408-14 (1990).
18. S. Sakka and T. Adachi, "Stability of solutions, gels and glasses in the sol-gel glass synthesis, J. Non-Cryst. Solids, **102**, 263-268 (1988).
19. T Adachi and S. Sakka, "Dependence of the elastic moduli of porous silica gel prepared by the sol-gel method on heat treatment", J. Mater. Sci., **25**, 4732-37 (1990).
20. M. Yamane, H. Kawazoe, A. Yasumori and T. Takahashi, "Gradient-index glass rods of $PbO-K_2O-B_2O_3-SiO_2$ system prepared by the sol-gel process", J. Non-Cryst. Solids, **100**, 506-510 (1988).
21. S. Konishi, K. Shingyouchi and A. Makishima, "R-GRIN glass rods prepared by a sol-gel method", J. Non-Cryst. Solids, **100**, 511-513 (1988).
22. T. Yoko, A. Yuasa, K. Kamiya and S. Sakka, "Sol-gel derived TiO_2 film semiconductor electrode for photocleavage of water", J. Electrochem. Soc., **138**, 2279-85 (1991).
23. B. O'Regan and M. Grätzel, "A low-cost, high-efficiency solar cell based on dye-sensitized colloidal TiO_2 films", Nature, **353**, 737-740 (1991).
24. B.E. Yoldas and D.P. Partlow, "Formation of broad band antireflective coatings on fused silica for high power laser applications", Thin solid Films, **129**, 1-14 (1985).

GEL PROCESSING ROUTES TO POROUS CERAMICS

A. Atkinson, AEA Technology, Building 429, Harwell Laboratory, Didcot, OX11 0RA, U.K.

Gel processing routes are particularly attractive for the production of porous ceramics as monoliths, coatings and powders for a wide variety of applications. The pore volume, size and size distribution are all controllable by manipulation of chemical and physical interactions between submicron porous particles. The method has been pursued for many years and processing routes have been developed to a variety of porous ceramics; some examples of which are described here. The processes cover aqueous sol-gel, gel precipitation and intercalation of layered structures. The examples given have application as chromatographic media, catalyst supports, adsorbents and gas sensors.

INTRODUCTION

In recent years there has been a great increase in the use of chemical routes for the synthesis of ceramic materials, and one class of such routes that has received particular attention is gel processing. The term "gel processing" encompasses processing methods in which a fluid precursor is converted into a solid gel and subsequently, by heat treatment, to the desired ceramic. The most widely studied gel processing route is known as sol-gel [1,2] and has two main variants; one based on the hydrolysis of metal alkoxides in alcoholic solution and the second, less popular, involving suspensions in aqueous media. In this paper we consider only the aqueous sol-gel method. Another type of aqueous gel processing considered here is gel precipitation in which metal hydroxides produce a gel on precipitation in the presence of a gel-forming organic polymer.

Gel processing routes are well suited to the production of porous oxide ceramics for a wide range of applications; including chromatographic

© 1993, UKAEA.

media, catalyst supports, adsorbents for effluent treatment, membranes and sensors. For this reason they have been investigated and applied at our laboratory for many years. In this paper several examples are presented to illustrate the breadth of applications and the control over porosity and pore-size distribution that can be achieved by these methods.

POROUS CERAMICS BY AQUEOUS SOL-GEL

In the aqueous sol-gel process the liquid-like sol is a dispersion of solid "particles" in an electrolyte and the suspension is stabilised electrostatically by repulsive electrical forces between the charged particles. The particles may be true solids (e.g of oxide, hydrated oxide or hydroxide) submicron in size, or they may be large polynuclear ions grown from solution. If the interactions between particles are strongly repulsive, the particles are unagglomerated (type I sols) and the sol is gelled by loss of water, the gel will be dense and heat treatment will produce a ceramic with little, or no, porosity. If, on the other hand, the particles are aggregated into open clusters in the dispersion (type II sols), or they are gelled by changing the interparticle forces (e.g. by changing the pH) then the primary particles cannot pack together efficiently and the gel will have low density and lead to a porous ceramic. An alternative method is to modify the forces which arise during drying of the gel so that they do not cause particle rearrangement and thereby preserve an open structure. This can be done by exchanging the aqueous pore fluid with one of lower surface tension (e.g. an alcohol).

The effect of aggregation of primary particles on the density of gels and ceramics of various oxides is illustrated in Table 1. The aggregated sols give porous gels which can be heat treated to give porous ceramics because the low co-ordination number of particles in the low density gels inhibits densification by sintering.

The aggregated sols have been used in a wide variety of catalyst applications (e.g. alumina supports for precious metal catalysts in vehicle exhaust treatment) in which their high surface area, porosity and resistance to densification are valuable attributes [3,4].

Table 1. The Effect of State of Aggregation in the Sol upon the Density of Gels and Ceramics Made from those Sols by Heating to 1000 °C [3] (Densities in g cm^{-3}).

Oxide	Sol	Gel	Ceramic	Theoretical
Titania	aggregated	1.49	3.73	4.26
	non-aggregated	2.59	4.11	4.26
Silica	aggregated	0.46	0.53	2.2
	non-aggregated	1.83	2.01	2.2
Ceria	aggregated	2.0	2.3	7.1
	non-aggregated	4.0	6.2	7.1

The control of pore size and pore size distribution offered by manipulation of sol aggregation and gel drying has been exploited in porous materials for high performance liquid chromatography. These materials are produced as small spheres (e.g. with diameters of 5 or 10 microns) in silica or alumina under the registered trademark "Spherisorb". Their internal pore structure is characterised by having a small size and very narrow distribution (Figure 1) as a result of the controlled agglomeration of uniformly sized primary particles.

The pore size distribution can also be manipulated by the incorporation of water soluble organic polymers into the sol. Figure 2 illustrates how the addition of 1% of polyvinyl alcohol (PVA) to a boehmite sol eliminates large pores from the resulting alumina ceramic and produces a much narrower pore size distribution [5].

GEL PRECIPITATION

In gel precipitation a hydoxide gel is formed by precipitation of hydoxides from aqueous solution in the presence of a gel-forming polymer. The purpose of the polymer is to control the precipitation, by interactions between the polymer and metal ions in solution, so that a rigid gel is produced rather than a loose precipitate. The process was originally developed in the nuclear industry for the preparation of oxide nuclear fuel particles. The attraction of the method in this

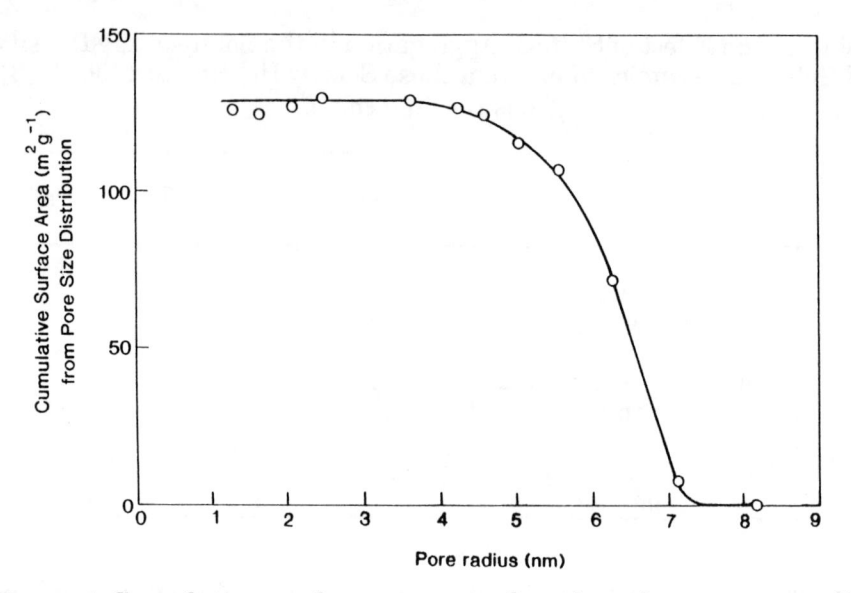

Figure 1 Cumulative surface area as a function of pore entry radius from gas adsorption/desorption isotherms on a commercial alumina chromatographic material produced by the sol-gel route.

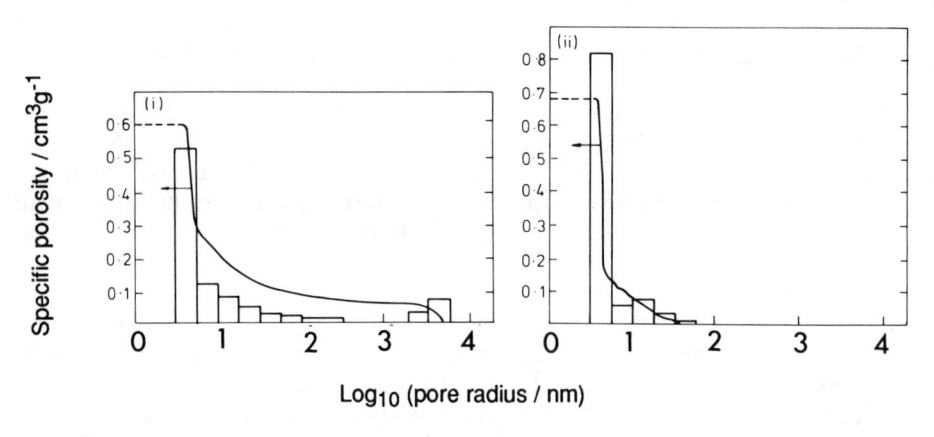

Figure 2 Cumulative pore entry radius distributions from mercury intrusion porosimetry on alumina ceramics (calcined at 700 °C) produced by the sol gel process (i) with no additive and (ii) with 1.0 wt% PVA [5].

application is that spherical oxide particles (about 1mm in diameter) can be produced and, being a liquid-based, the process is dust-free. The method is suitable for producing particles of either dense or porous hydroxide (or hydrous oxide) gels or oxide ceramics.

A typical example of gel precipitation is given by Ramsay et al. [6] for the preparation of spherical particles of porous hydrous zirconia gels. In that process a zirconium nitrate solution containing 4% polyacrylamide (the gel-forming polymer) was fed through a vibrating jet to give spherical droplets approximately 1 mm in diameter. The droplets passed through ammonia vapour, which caused initial gelation, and then through ammonium hydroxide solution, to complete the precipitation. The porosity of the gel spheres can be controlled by the way in which they are dried. For example, the zirconia gel spheres had a fractional porosity of 0.31 and mean pore radius of 2 nm when dried in air. However, if the water was exchanged for butanol before drying the fractional porosity was increased to 0.90 and the mean pore radius to 22 nm.

The internal structure of the gels has been investigated using neutron scattering [6]. Small angle neutron scattering allows the structure to be studied over different characteristic length scales determined by the scattering vector, Q; the smaller the scattering angle the larger the characteristic distance scale that is probed. Scattering data for hydrous zirconia gels prepared by gel precipitation and then dried either in air or after exchange of water by butanol are shown in Figure 3. At larger scattering vectors both types of gel give a scattering intensity proportional to Q^{-4}. This scattering is from the primary hydrous oxide particles and is consistent with their size being approximately 30 nm in both cases. The difference in the form of the scattering curves at lower Q indicates that the two gels are different on the scale of many primary particles. In the case of the gel which had been butanol exchanged the observed dependence of $Q^{-2.75}$ is close to that ($Q^{-2.5}$) produced by scattering from fractal clusters of particles formed by diffusion limited aggregation of the primary particles. The scattering from the air dried sample, on the other hand, indicates a closer-packed arrangement of primary particles having higher nearest neighbour coordination number. These observations show that air drying the water-saturated gel results in rearrangement and dense packing of the primary particles under the large surface tension forces of the water/air interface. The lower surface tensions of butanol/water and butanol/air are insufficient to move the primary particles significantly and the open structure of the original wet gel is preserved.

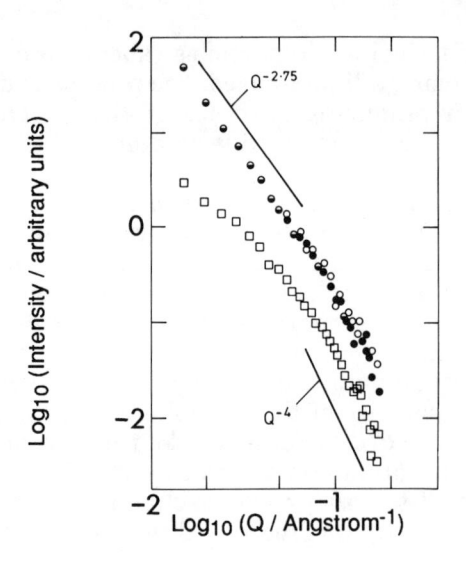

Figure 3 Small angle neutron scattering data from gels of hydrous zirconia produced by the gel precipitation route: (a) as prepared wet gel, o, (b) wet gel dried by butanol displacement and then reimmersed in water, •, and (c) wet gel dried in air and reimmersed in water, □. Q is the scattering vector, $4\pi \sin\theta / \lambda$ [6].

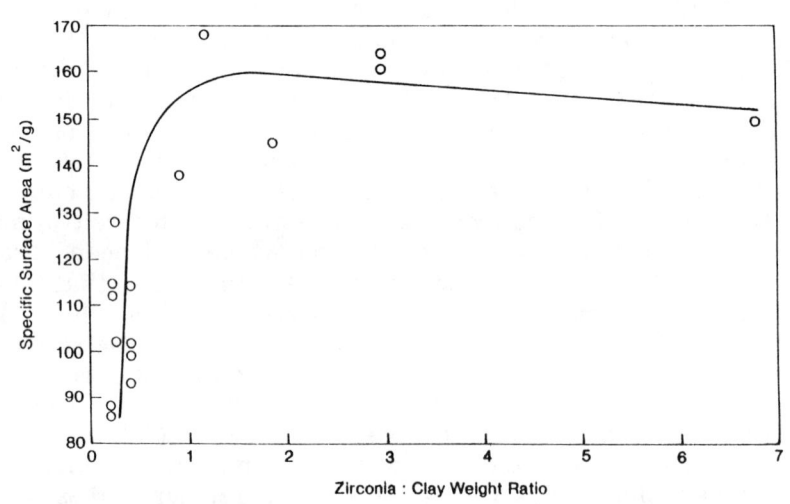

Figure 4 The effect on surface area (measured by nitrogen adsorption) of montmorillonite clay caused by ion exchange using zirconia sol at various clay to zirconia ratios and subsequent calcination [7].

PILLARING LAYERED COMPOUNDS WITH SOLS

Porous materials can also by produced by opening up the layers in layered compounds. For example, clay minerals such as montmorillonite are composed of alumino-silicate layers held together by interstitial water molecules and exchangeable mono- and di-valent cations. Positively charged sol particles may be regarded as giant cations which have the possibility of entering the interlayer regions by exchanging with the original cations and, because of their larger size, open up the layers to generate porous material. If the layers are propped open the process is known as pillaring, but if the opening is too large the layers may delaminate completely and generate a "house of cards" structure. Pillaring often generates micropores (< 2 nm) whereas the porosity in delaminated material is in the mesopore region.

We have used zirconia sols to open the layers in clays containing montmorillonite [7]. (This is not strictly a gel processing method because a macroscopic gel is not necessarily produced.) The sols used had a relatively wide range of particle size and therefore the effect on the clay was to create a partially pillared and partially delaminated porous material. Figure 4 shows the effect of zirconia sol exchange on the specific surface area of the clay. Such materials have potential applications as supports in catalysis and adsorbents in effluent treatment.

THIN FILM GAS SENSORS

Gas sensors are required to detect small quantities of gases in air. Typical gases of interest are CO, H_2, CH_4, NH_3 and the devices need to be inexpensive to enter the market. Many different sensing principles have been investigated as a basis for these devices; such as field effect transistors, surface wave transducers, ionic conductors and semiconducting oxides. Here we describe sensors based on semiconducting tin oxide produced as porous thin films by the sol-gel route.

Tin oxide is an n-type semiconductor and oxygen will adsorb onto its surface to form ions on the surface. Abstraction of the electrons from the tin oxide causes a high resistivity layer depleted in charge carriers to form around the tin oxide crystal grains. A sample of this type will therefore show a high resistance in the presence of air. Reducing gases adsorb preferentially onto tin oxide in the presence of air and are also oxidised by the adsorbed oxygen (i.e. tin oxide is a strong oxidation catalyst). The combined adsorption and reaction removes some of the adsorbed oxygen ions and the resistance of the polycrystalline tin oxide is decreased. This is the qualitative basis of the sensing action of tin oxide.

Tin oxide gas sensors are usually produced by screen printing a paste of powdered tin oxide slurry (thick film technology). We have developed a sol-gel process for producing thin films of porous, polycrystalline tin oxide that is cheap and easy to carry out and gives sensors that are more reproducible and more robust than the thick film sensors [8]. The response of a thin film tin oxide sensor, produced using the sol-gel route, to 1% of CO or CH_4 in air is shown in Figure 5.

CONCLUSIONS

Gel processing in aqueous media is a powerful and flexible method for the production of porous ceramic oxides. By manipulation of aggregation in the sol, gelation, drying and heat treatment it is possible to obtain materials with fractional porosities ranging from almost zero to 90%. The same variables also permit control of pore radius, from 2 to 100nm, and pore size distribution. Furthermore the gelation process allows control over the external shape of the porous material. Thus gel processing has been used to produce monodisperse porous spheres having diameters ranging from 5 microns to several mm and to porous films for sensors and catalytic coatings.

ACKNOWLEDGEMENT

The author is grateful to all colleagues, past and present, whose work is referred to in this article.

REFERENCES

1   C.J. Brinker and G.W. Scherer, *Sol-gel Science*, Academic Press, London and San Diego, 1990.
2   D.L. Segal, *Chemical Synthesis of Advanced Ceramic Materials*, Cambridge University Press, 1989.
3   R.L. Nelson, J.D.F. Ramsay, J.L. Woodhead, J.A. Cairns and J.A.A. Crossley, *The coating of metals with ceramic oxides via colloidal intermediates*, Thin Solid Films, 81, 329-337, 1981.

4   A. Atkinson, J. DiSanza, A. Dyer, N. Jorgensen and D.L. Segal, *Aqueous sol-gel processing routes to supported metal catalysts*, Mat. Res. Soc. Extended Abstract, EA 24, 229-232, 1990.

5   K. Kunze and D.L. Segal, *Modification of the pore structure of sol-gel-derived ceramic oxide powders by water-soluble additives*, Colloids and Surfaces, 58, 327-337, 1991.

6   J.D.F Ramsay, P.J. Russell and S.W. Swanton, *Gel-precipitated oxide gels with controlled porosity - determination of structure by small angle neutron scattering and adsorption isotherm measurements*, pp 257-265, in Characterisation of Porous Solids II, Eds. F. Rodriguez et al., Elsevier Science Publishers, Amsterdam, 1991.

7   D.L. Segal and J. DiSanza, *Delaminated bentonite: zirconia sols*, UK Patent Application P91 05983.2, 1991.

8   A. Atkinson and P.T. Moseley, *Thin film electroceramics*, accepted for publication in Applied Surface Science, 1992.

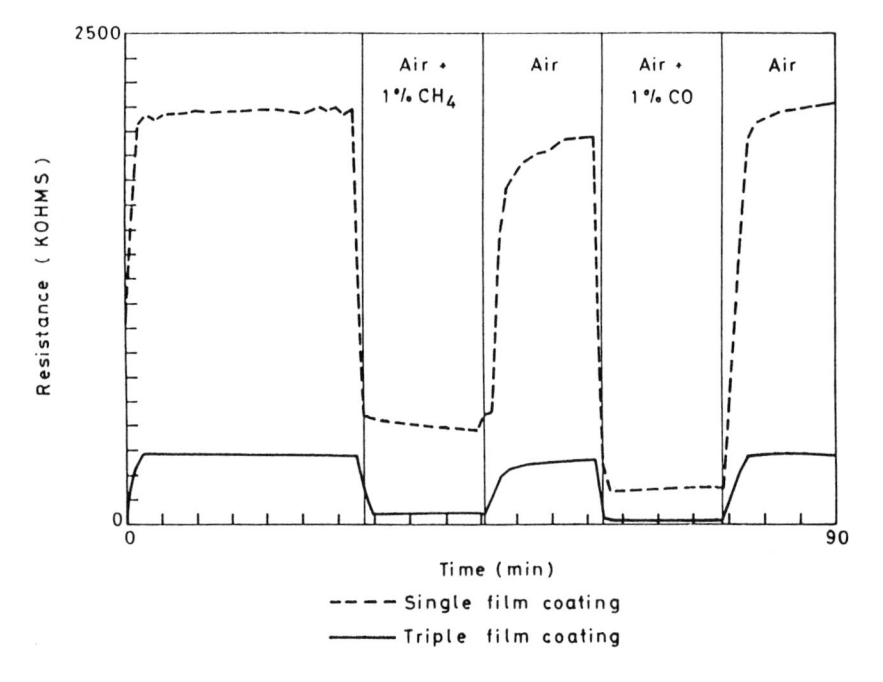

Figure 5 Response of resistive thin film tin oxide gas sensors, made by an aqueous sol-gel route, to CH_4 and CO in air [8].

MACROPORE STRUCTURE DESIGN OF SOL-GEL DERIVED SILICA BY SPINODAL DECOMPOSITION

Kazuki NAKANISHI, Hironori KAJI and Naohiro SOGA
Department of Industrial Chemistry, Faculty of Engineering,
Kyoto University, Sakyo-ku, Kyoto 606-01, JAPAN

INTRODUCTION

Various methods to control morphologies of multicomponent materials which involve phase separation processes are now widely utilized, especially in the field of 'polymer alloying'[1]. An extremely low diffusion rate of entangled polymer molecules makes it easy to 'freeze', both physically and chemically, a transient structure developed by a phase separation process as a permanent material morphology. Recently, a similar concept of "freezing of a transient structure of a phase separation by an inorganic sol-gel transition" has been successfully applied to several variations of alkoxy-derived silica sol-gel systems[2-4]. Gels with well-defined interconnected pores and skeletons in a length scale up to micrometers have been obtained for a variety of coexisting components in the sol-gel reactions. This paper, employing three typical systems as examples, focuses on the correspondence between the constituent of the systems and the pore characteristics of respective resultant gels. A qualitative explanation based on the Flory-Huggins type phase relation of a polymerizing system is also given.

EXPERIMENTAL

Gel Preparation

Low-Water Polar Solvent System

Appropriate amounts of tetramethoxysilane (TMOS), 1 mol dm^{-3} aqueous nitric acid solution and formamide, as listed in table 1, were mixed together at 0°C for hydrolysis. After stirring for 1 min, the resultant transparent solution was transferred into a container, and kept at 40°C under tightly sealed condition for gelation and aging for 1 day. An as-aged gel was then immersed in 1 mol dm^{-3} aqueous nitric acid solution for a few hours, finally dried at 60°C for 2 days.

High-Water Systems Containing Water-Soluble Organic Polymer

Poly(acrylic acid) (HPAA) or poly(ethylene oxide) (PEO) was used as the polymer component. TMOS or tetraethoxysilane (TEOS) was hydrolyzed with an aqueous solution containing appropriate amounts of organic polymer and nitric acid at room temperature. The resultant transparent solution was allowed to gel in a sealed container at 40 or 80°C, then aged until any more spontaneous contraction of the wet gel could not be observed. The aged gel was immersed in an equi-volume mixture of ethanol and distilled water in order to extract the components in macropores, and then soaked in an aqueous solution of nitric acid. Drying of the gel piece was performed at 60 or 80°C for a few days.

Characterization

Gel morphologies larger than 20 nm were examined by a scanning electron microscope (SEM; S-510, Hitachi Co.) using flat fractured surfaces of heat-treated gels. Pore size distributions were determined by mercury porosimetry (PORESIZER-9310; Micromeritics Corp.) and nitrogen adsorption methods (ASAP-2000; Micromeritics Corp.). The distribution of mesopores (pore diameter smaller than 20 nm and larger than 1.7 nm) was determined according to the BJH method[6]. When the cumulative pore volume data were obtained from both of these techniques, they were combined to represent a continuous distribution curve.

RESULTS AND DISCUSSION

General Features of Morphology Formation

The free energy change of mixing in a binary polymer-containing system can be expressed as follows[7];

$$\Delta G \approx RT \left[(\phi_1 / P_1) \ln \phi_1 + (\phi_2 / P_2) \ln \phi_2 + \chi_{12} \phi_1 \phi_2 \right] \qquad (1)$$

where ϕ_i and P_i (i = 1, 2) respectively denote the volume fraction and degree of polymerization of the component i, and χ_{12} the interaction parameter, R the gas constant, T the temperature. The first two terms in the bracket, both exhibit negative values, represent contributions of mixing entropy and their absolute values decrease when the corresponding component polymerizes, i.e. P_i increases. If the value of χ_{12} is positive, which implies a repulsive interaction between the components, a polymerizing system would exhibit an upward convex binodal line moving toward higher temperature in a composition-temperature diagram. Thus, with a progress of polymerization reaction, a starting composition is thrust into the two-phase region, and an initially homogeneous solution starts to decompose into conjugate phases.

In a gel-forming polymerization, if one can induce a phase separation concurrently with the sol-gel transition, the transient structure of the phase separation process would be 'frozen' in a gel network. Especially in the case where the spinodal decomposition occurs, an interconnected structure with well-defined domain size can be obtained. The laser-light scattering measurements of the gelling silica solutions has shown that the morphology of a dried gel exactly reflected that formed by the spinodal decomposition in a gelling solution[3]. Generally, earlier initiation of a phase separation relative to the sol-gel transition results in a coarser morphology.

Low-Water System Containing Formamide

Figure 1 schematically shows the phase relation expected for the polymerizing solution prepared by hydrolyzing TMOS with understoichiometric amount of acidic water under coexistence of formamide. Because of an incomplete hydrolysis, substantial fraction of terminal groups attached to the growing silica polymers are alkoxy groups, which lowers the polarity of the polymer. On the other hand, the solvent components, methanol, formamide and acidic water, are highly polar. Thus, the solution is likely to decompose into silica-rich and solvent-rich phases. Since all the starting components are low-molecular weight species, the corresponding binodal line is assumed to be nearly symmetrical. With a progress of polymerization, it raises and the critical composition, where the binodal and spinodal lines coincide with each other, shifts toward the solvent-rich side. Near the critical composition, a formation of interconnected phases by spinodal decomposition is favorable. When a gel morphology is formed in this composition range, the volume fraction of solvent-rich phase tends to be higher than that of the silica-rich phase due to the asymmetry of the binodal line.

The pore size distributions of gel samples heat-treated at 300°C for 2 h are shown in Fig. 2. A very small change of formamide concentration is enough to alter the morphology from mesoporous to macroporous. Mesopores are formed by a microscopic phase separation in a tightly crosslinked gel network. As the concentration of formamide increases, however, a spinodal decomposition initiates in an earlier stage of the sol-gel transition where the mobility of gel network is higher. As a result, the macropores are formed in addition to the mesopores with their average size and volume fraction increasing and their distribution sharpening dramatically with formamide concentration. Further increase of formamide results in a sedimentation of silica-rich phase because the phase separation initiates much earlier than the sol-gel transition. The strong dependence of gelation time on the starting composition severely restricts the composition range in which the interconnected structure can be obtained in the Formamide-TMOS system. An alternative use of trifunctional alkoxide, methyltrimethoxysilane, in place of TMOS extends the composition range to some extent[5].

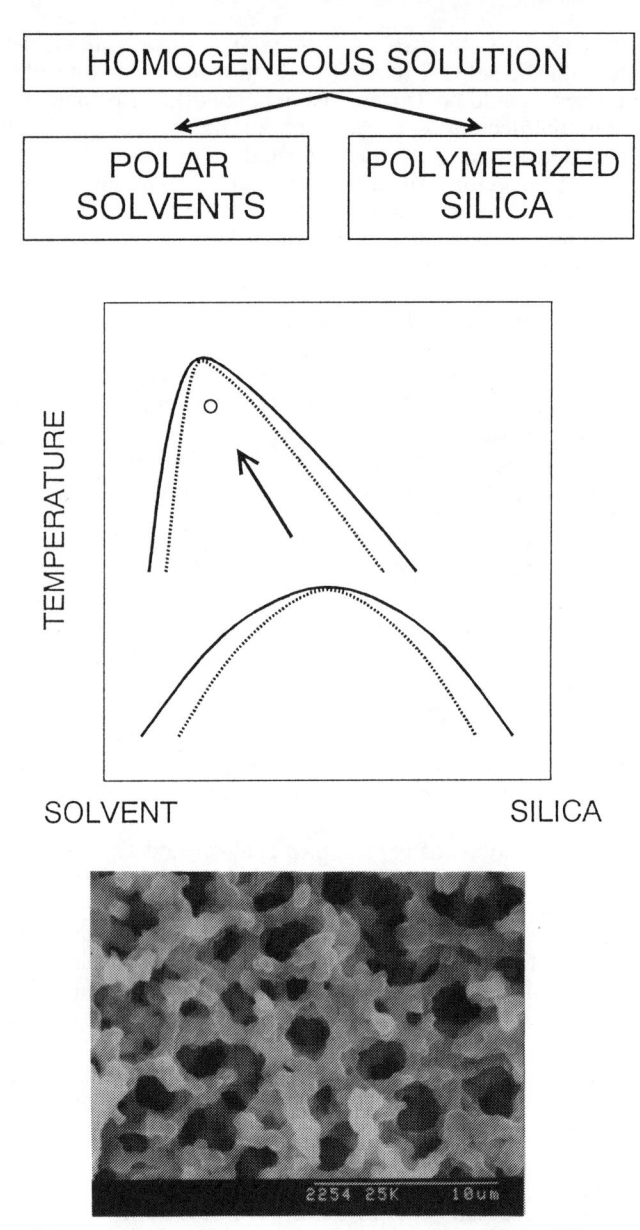

Figure 1 Schematic phase diagram of polymerizing Formamide–Silica system and SEM photograph of a typical morphology of a heat-treated gel.

Table 1. Starting Compositions of Formamide–TMOS System. (Unit:g)

Sample	Formamide	H_2O^{*1}	60%HNO_3^{*1}	TMOS
F–1.0[*2]	3.05	1.746	0.203	10.3
F–1.6[*2]	4.88	1.746	0.203	10.3
F–2.0[*2]	6.10	1.746	0.203	10.3
F–2.2[*2]	6.71	1.746	0.203	10.3
F–2.8[*2]	8.53	1.746	0.203	10.3

*1 : Mole ratio of total H_2O to TMOS is fixed at 1.5.
*2 : The figure after F– denotes the mole ratio of Formamide to TMOS.

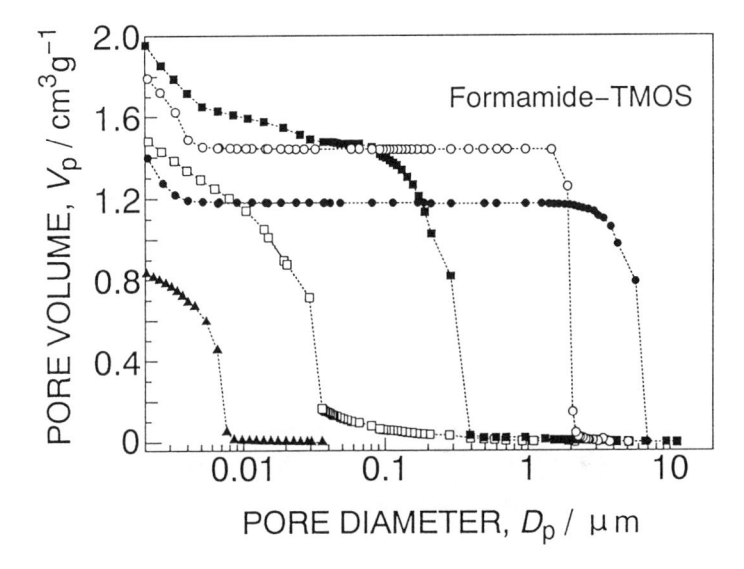

Figure 2 Pore size distribution of gel samples of Formamide–TMOS system heat–treated at 300°C for 2h.
▲:F–1.0, □:F–1.6, ■:F–2.0, ○:F–2.2, ●:F–2.8.

Porous Materials

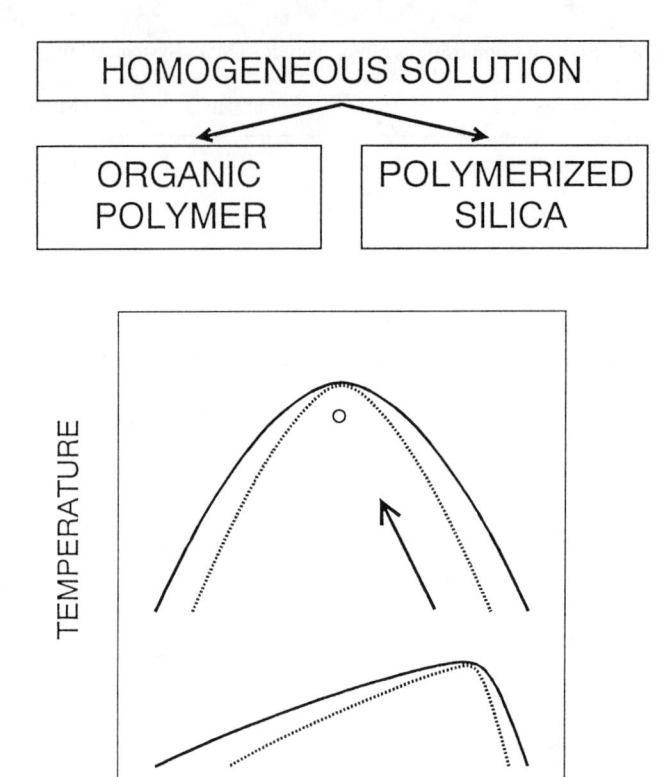

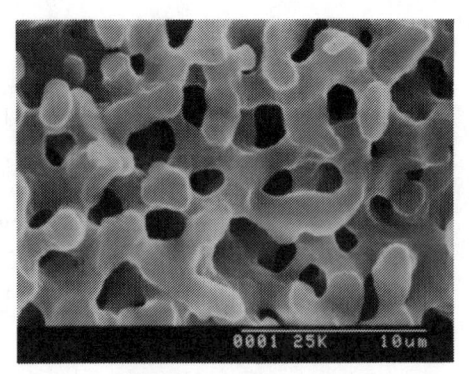

Figure 3 Schematic phase diagram of polymerizing HPAA–Silica system and SEM photograph of a typical morphology of a heat-treated gel.

High-Water System Containing Poly(acrylic acid)

The system containing HPAA decomposes into HPAA-rich and silica-rich phases. Because both polymerized silica and HPAA have high affinity with alcohol and water, the phase relation among HPAA - silica - solvent corresponds to that of polymer A - polymer B - good solvent. Since a good solvent works only to dilute the repulsive interaction between polymeric species[8], we can use a quasi two component diagram with HPAA-rich and silica-rich end members, schematically shown in Fig. 3, for discussion. As the polymerization of silica proceeds, initially asymmetrical binodal line raises and gradually becomes symmetrical. Thus, the interconnected structure developed by the spinodal decomposition tend to retain comparable volume fractions of conjugate phases. Consequently, the amount and composition of the solvent phase have minimum influence on the pore volume but major influence on the pore size, which strongly depends on the mutual solubility between HPAA and silica.

The pore size distribution curves of gels heat-treated at 1000°C for 2 h are shown in Fig. 4 and respective starting compositions are listed in table 2. The nanometer-range pores are completely diminished due to sintering for these samples. The obtainable pore size range covers nearly two orders of magnitude, and the macropore volume measures between 0.3 and 0.4 cm^{3}g^{-1}, which corresponds to about 40 to 50 % of volume fraction. The morphology gave a smooth interconnected appearance when the pore size was larger than about 0.1 µm. Samples with pores below this range consisted of interconnected particulate skeletons. Since the pore size depends only weakly on HPAA concentration, variations of molecular weight, additional components and reaction temperature have to be introduced in order to design the materials in a varied pore size range[9,10].

Table 2. Starting Compositions of HPAA-TMOS, TEOS System. (Unit:g)

Sample	HPAA (M.W.)	H$_2$O	60%HNO$_3$	Alkoxide	Additive
PA9E48[*1]	0.40 (90 000)	5.0	0.51	TEOS 6.51	None
PA9E44[*2]	0.40 (90 000)	4.0	0.40	TEOS 6.51	EtOH[*3] 2.38
PA9E30F[*1]	0.30 (90 000)	3.0	0.30	TEOS 6.51	FA[*4] 2.28
PA9E35F[*1]	0.35 (90 000)	3.0	0.30	TEOS 6.51	FA[*4] 2.28
PA25E2F[*1]	0.20 (250 000)	5.0	0.51	TEOS 6.51	FA[*4] 5.70
PA25M3[*2]	0.30 (250 000)	5.0	None	TMOS 5.15	None

*1: Gelled at 80°C. *2: Gelled at 40°C.
*3: EtOH denotes ethanol.
*4: FA denotes formamide.

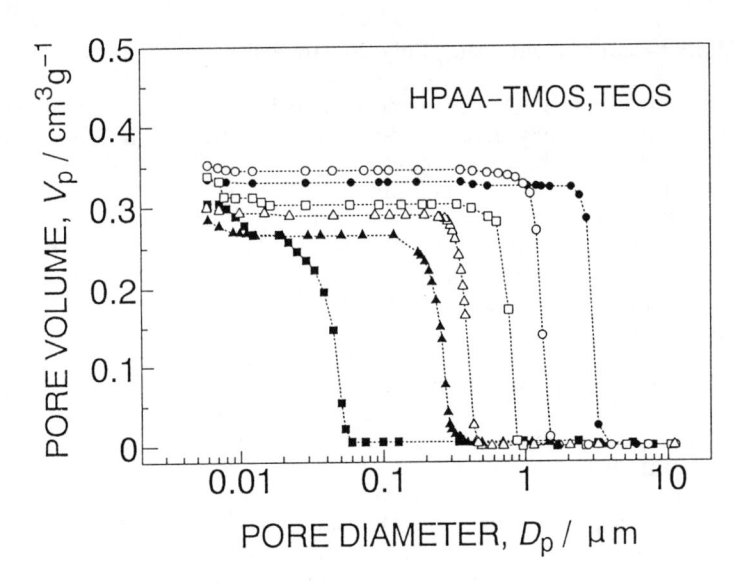

Figure 4 Pore size distribution of gel samples of HPAA–TMOS,TEOS system heat-treated at 1000°C for 2h.
▲ :PA9E30F, △ :PA9E35F, ■ :PA25M3, □ :PA25E2F,
● :PA9E44, ○ :PA9E48.

High-Water System Containing Poly(ethylene oxide)

Since PEO strongly adsorbs the surfaces of colloidal silica especially at a low pH region[11], the coverage of silica oligomers with PEO chains is expected to increase with the progress of polymerization. A thermal analysis proved that the proportion of silica distributed in the PEO-rich phase increased with PEO concentration and that most of PEO and silica existed in the same phase when an appropriate amount of PEO coexists. Thus, the phase relation can be explained by assuming a quasi two component diagram with PEO-rich and solvent-rich end members. PEO exhibits a lower critical solution temperature when dissolved in water, due to the instability of the hydrogen bond between them at high temperature, and the bottom of binodal line locates at quite low concentration of PEO when the molecular weight is higher than few tens thousand[12]. Hence, the spinodal decomposition is expected to take place in an asymmetrical miscibility gap similar to that of the Formamide–TMOS system. However, even if an interconnected structure developed near the critical composition where PEO concentration is low, the morphology cannot be 'frozen' unless enough concentration of highly polymerized silica is associated with the PEO-rich phase. With an increase of PEO concentration, a phase separation occurs in a later stage of polymerization reaction, where silica polymers with increased concentration and molecular weight are associated with PEO molecules. As a result, the structure freezing by a sol-gel transition becomes easier at a higher volume fraction of PEO-rich phase.

The pore size distribution curves of gels heat-treated at 600°C for 2 h are shown in Fig. 5. The pore size is mainly governed by PEO concentration, which affects the relation between the initiation of phase separation and the sol-gel transition as described above. The water concentration had a much weaker influence on the pore size but affected the pore volume according to the phase relation between PEO and water. In the typical samples shown in the figure, the pore volumes exhibit intermediate values between those of HPAA and Formamide systems.

Table 3. Starting Compositions of PEO-TEOS System. (Unit:g)

Sample	PEO (M.W.)	H_2O	60%HNO_3	TEOS
PO1069[*1]	0.69 (100 000)	8.0	0.81	6.51
PO1076[*1]	0.76 (100 000)	8.0	0.81	6.51
PO1084[*1]	0.84 (100 000)	8.0	0.81	6.51

*1: Gelled at 40°C.

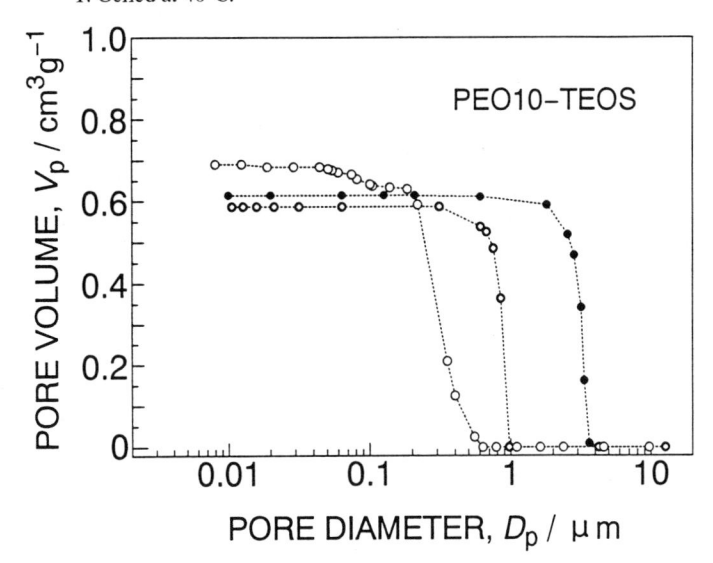

Figure 5 Pore size distribution of gel samples of PEO10-TEOS system heat-treated at 600°C for 2h.
● :PO1069, ◐ :PO1076, ○ :PO1084.

CONCLUSION

The macropore structure of silica gels can be controlled by changing coexisting components and other reaction parameters in a sol-gel reaction. The pore volume is largely determined by the phase relation of polymerizing system, while the pore size is determined by the relation between the kinetics of the phase separation and the sol-gel transition. The effects of incorporated organic polymers are quite different depending on the degree and nature of their interactions between polymerizing silica.

ACKNOWLEDGEMENTS

The authors thank Ms. M. Yamamoto and Mr. T. Kawaguchi, Advanced Glass R&D Center, Asahi Glass Company Ltd., for performing the mercury porosimetry measurements. A financial support by a Grant-in-Aid for Scientific Research on Priority Areas, New Functionality Materials Design Preparation and Control from the Ministry of Education, Science and Culture, Japan is gratefully acknowledged.

REFERENCES

[1] D.R. Paul and S. Newman, ed., *Polymer Blends, I & II*, Academic Press, N.Y., (1978).
[2] K. Nakanishi and N. Soga, "Phase separation in gelling silica-organic polymer solution: Systems containing poly(sodium styrenesulfonate)", *J. Am. Ceram. Soc.*, **74**, 2518-30(1991).
[3] K. Nakanishi and N. Soga, "Phase separation in silica sol-gel system containing polyacrylic acid I. Gel formation behavior and effect of solvent composition", *J. Non-Cryst. Solids*, **139**, 1-13(1992).
[4] K. Nakanishi and N. Soga, "Phase separation in silica sol-gel system containing polyacrylic acid II. Effects of molecular weight and temperature, *J. Non-Cryst. Solids*, **139**, 14-24(1992).
[5] H. Kaji, K. Nakanishi and N. Soga, "Preparation of porous silica gels with micrometer-range interconnected structures from trifunctional and tetrafunctional silanes", Proc. Intn'l. Conf. Science and Technology of New Glasses, eds. S. Sakka and N. Soga, 93-98(1991).
[6] E.P. Barrett, L.S. Joyner and P.P. Halenda, "The determination of pore volume and area distributions in porous substances. I. Computations from nitrogen isotherms", *J. Am. Chem. Soc.*, **73**(1951), 373-380.
[7] P.J. Flory, *Principles of Polymer Chemistry*, Cornell University Press, Ithaca, N.Y., (1971).
[8] R.L. Scott, "The thermodynamics of high polymer solutions. V. Phase equilibria in the ternary system: Polymer1 - Polymer2 - Solvent", *J. Chem. Phys.*, **17**, 279-284(1949).
[9] K. Nakanishi and N. Soga, "Phase separation in silica sol-gel system containing polyacrylic acid. III. Effect of catalytic condition", *J. Non-Cryst. Solids*, **142**, 36-44(1992).
[10] K. Nakanishi and N. Soga, "Phase separation in silica sol-gel system containing polyacrylic acid. IV. Effect of chemical additives", *J. Non-Cryst. Solids*, **142**, 44-54(1992).
[11] J. Rubio and J.A. Kitchener, "The mechanism of adsorption of poly(ethylene oxide) flocculant on silica", *J. Coll. Interf. Sci.*, **57**, 132-142(1976).
[12] S. Saeki, N. Kuwahara, M. Nakata and M. Kaneko, "Upper and lower critical solution temperatures in poly(ethylene glycol) solutions", *Polymer*, **17**, 685-689(1976).

POROUS MATERIALS PREPARED FROM SILICA GEL BY HYDROTHERMAL HOT-PRESSING

Kazumichi Yanagisawa, Mamoru Nishioka, Koji Ioku, and Nakamichi Yamasaki,
Research Laboratory of Hydrothermal Chemistry, Faculty of Science,
Kochi University, 2-5-1 Akebono-cho, Kochi-shi 780, JAPAN

Porous materials in which pores with uniform diameter were homogeneously distributed, were produced by hydrothermal hot-pressing from a commercial silica gel consisting of fine spherical particles with an average diameter of 18 nm. The ceramics prepared at 300°C and 25 MPa for 1 hour were porous with low bulk density (0.88 g/cm^3) and high pore volume (0.59 cm^3/g). Pore size was uniform, about 45 nm in diameter. In spite of high porosity, they had high compressive strength (70 MPa). A high compressive pressure decreased the pore volume and the pore diameter, and increased the compressive strength. The increase in the reaction temperature slightly decreased the pore volume, and increased the pore diameter. The IR absorption of Si-OH bond decreased with an increase in the reaction temperature. It was considered that the dehydration followed by the condensation of Si-OH bond proceeded the linkage of the silica gel particles.

INTRODUCTION

The hydrothermal hot-pressing technique was designed to solidify inorganic powders at low temperatures below 350°C [1]. The idea for this technique came from the fact that sedimentary rocks are naturally produced from loose particles occurring in sediment by a chemical process termed lithification that reduces the original porosity by compaction and cementation [2]. In the hydrothermal hot-pressing technique, compression of a powder from outside an autoclave accelerates the compaction of the powder, and the bonding of the particles proceeds under hydrothermal conditions. Many kinds of powders such as amorphous and crystalline silica, calcium carbonate, calcium hydroxyapatite, titania and zirconia, are solidified by this technique [3]. The addition of alkaline solution accelerates the solidification of silica powders mixed with the other compounds. Thus, this technique is also useful for solidifying industrial wastes such as radioactive wastes from nuclear power plants [4-6], and sludge ashes [7].

The densification method of powders by hydrothermal hot-pressing is considered to be one of the sintering methods. It is similar to hot-pressing in the presence of liquid phase. In this technique, the liquid phase is water and/or solution under hydrothermal conditions. If the liquid phase remains in grain boundaries after sintering, the resulting material must be porous. In this paper, the hydrothermal hot-pressing technique was used to produce porous materials from a commercial silica gel.

Concerning porous silica materials, the following substances have been produced: aerogel and xerogel derived from the sol-gel process of alkoxides [8-10]; porous glass produced by segregation of glass [8,9]; sintered glass [11]; and so on. Though each of these materials has attractive properties, they have certain processing disadvantages. Alkoxides are expensive and the process for the production of porous glass and sintered glass is complicated. The authors showed recently that spherical silica particles prepared by hydrolysis of tetraethyl orthosilicate were transformed to a densified body by the hydrothermal hot-pressing technique without any additive [12,13]. This result means that the silica ceramics prepared by this technique will have high purity, depending on the purity of the starting material. The simplicity of the process for the hydrothermal hot-pressing technique and the low cost of the starting material make this method cost-effective. The production process under relatively mild hydrothermal conditions might give hydrophilicity to the products. This paper describes the effect of hydrothermal hot-pressing conditions on properties of the products, such as pore size distribution.

EXPERIMENTAL PROCEDURE

The apparatus for hydrothermal hot-pressing is shown in Fig.1(a). The autoclave shown in Fig.1(b) is a cylinder made of steel with a cylindrical chamber, 3.5 cm in diameter. A starting material, including water, is compressed through pistons from above and below by an electric oil pump equipped with an accumulator. The pistons are designed with a cavity (10.5 cm^3, each), into which water contained the starting powder is released during hydrothermal hot-pressing. Gland packing made of Teflon prevents leakage.

The starting material was the commercial silica gel (Nipsil VN3) produced by Nippon Silica Industrial Co. LTD., Japan. Its composition is shown in Table 1. The gel consisted of fine spherical particles with the average diameter of 18 nm (Fig.2). The specific surface area measured by BET method was 160 m^2/g. The amount of absorbed water was 9.1 wt%. After the silica gel (30 g) contained absorbed water was mixed with water (3 g), it was placed in the autoclave and compressed at 10-50 MPa. The autoclave was heated to a desired temperature from 100 to 300°C by an induction heater at a rate of about 5°C/min. Reaction time at the temperature was 10-180 min. During the heating treatment, the compressive pressure was

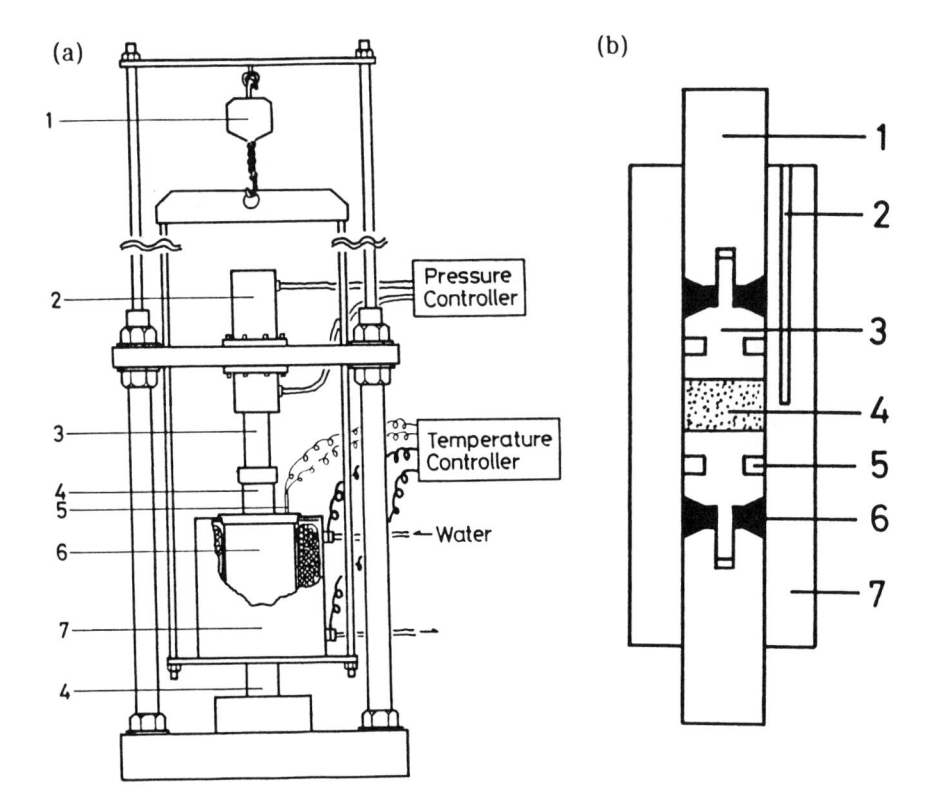

Figure 1 (a) Apparatus and (b) autoclave for hydrothermal hot-pressing.
(a) 1,hoist for adjustment of furnace position; 2,pump; 3,ram; 4,push rod of autoclave; 5,thermocouple; 6,autoclave; 7,induction furnace.
(b) 1,push rod; 2,well for thermocouple; 3,piston; 4,sample; 5,water storage; 6,gland packing; 7,cylinder.

Table 1. Composition of the Starting Silica Gel (wt%)

Oxide	Amount(wt%)	Oxide	Amount(wt%)	Oxide	Amount(wt%)
SiO_2	98.63	TiO_2	0.10	MgO	0.01
Al_2O_3	0.66	Fe_2O_3	0.04	SO_4^{2-}	0.10
Na_2O	0.43	CaO	0.02	Cl^-	0.01

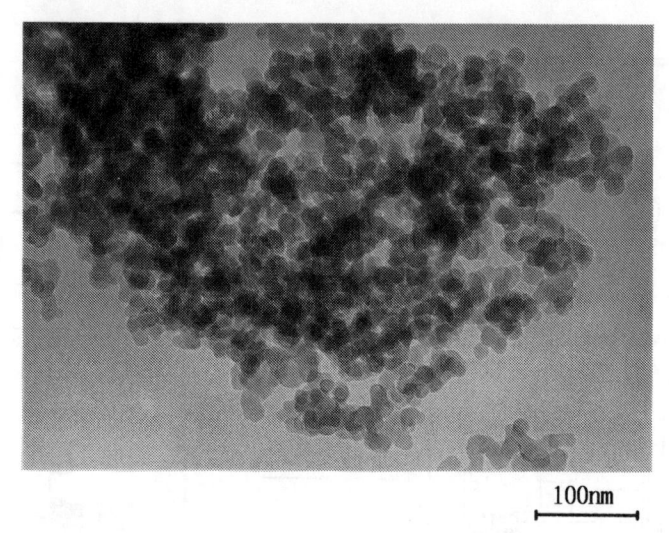

Figure 2 Photograph of starting silica gel particles.

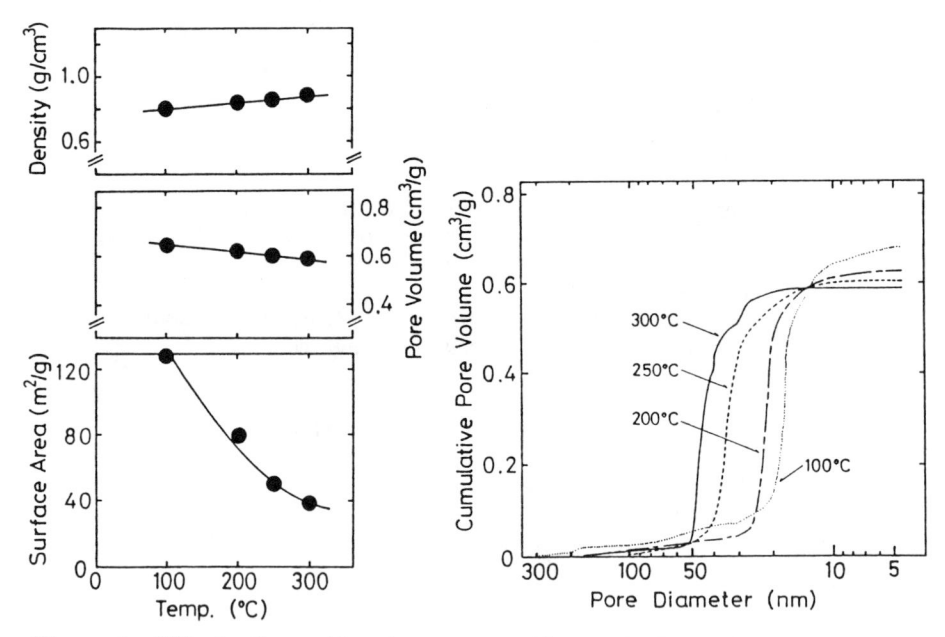

Figure 3 Effect of reaction temp. on properties of the ceramics produced at 25 MPa for 60 min.

Figure 4 Pore size distribution of the ceramics shown in Fig.3.

kept constant.

Porous ceramics obtained were dried at 120°C, and then their properties were evaluated. Bulk density was calculated from their apparent volume and weight. Compressive strength and Vickers hardness were measured by uniaxial compression tests at room temperature and by the micro-Vickers indentation method with 0.98 N loading for 15 s, respectively. Specific surface area was measured by BET method using N_2 gas as an absorbent. Pore volume and distribution of pore diameter were measured by mercury intrusion porosimetry. The ceramics were identified by X-ray diffraction and IR absorption. Their fractured surfaces were observed by a scanning electron microscope.

RESULTS AND DISCUSSION

Effects of Reaction Temperature

Ceramics were produced at various temperatures at a constant compressive pressure and reaction time of 25 MPa and for 60 min, respectively. Figure 3 shows the effect of reaction temperature on bulk density, pore volume and specific surface area. The increase in reaction temperature slightly increased the density and decreased the pore volume. The surface area was remarkably decreased. The fact of the large decrease in the surface area and the small decrease in the pore volume suggested that the pore diameter was increased by hydrothermal hot-pressing at higher temperatures. In fact, the pore diameter measured by mercury intrusion porosimetry (Fig.4) was increased with an increase in reaction temperature. These results are similar to Iler's summary [14] which says that hydrothermal treatment of silica gel does not change the pore volume even though the surface area decreases, and the pore diameter increases by a movement of the gel from the surface of wider capillaries to fill in the smaller capillaries or pores. During hydrothermal hot-pressing, shrinkage of the silica gel was observed. The shrinkage corresponded to the increase in the density and the decrease in pore volume. The pore diameter of the ceramics produced at each temperature was uniform, about 20 nm at 100°C and 45 nm at 300°C.

According to X-ray diffraction, the ceramics produced in this study were amorphous without any crystalline compounds. The IR spectra of the ceramics are shown in Fig.5. The absorptions at 950 and 800 cm^{-1} were assigned to Si-OH and Si-O, respectively [15,16]. The intensity of the absorption assigned to Si-OH was decreased with an increase in reaction temperature. The ceramics produced at above 250°C were strong enough to cut and polish, but those produced at lower temperatures were friable. Thus, mechanical properties of the ceramics also improved with an increase in reaction temperature. These results suggested that the linkage of the silica gel particles proceeded by the dehydration followed by the condensation of Si-OH bond.

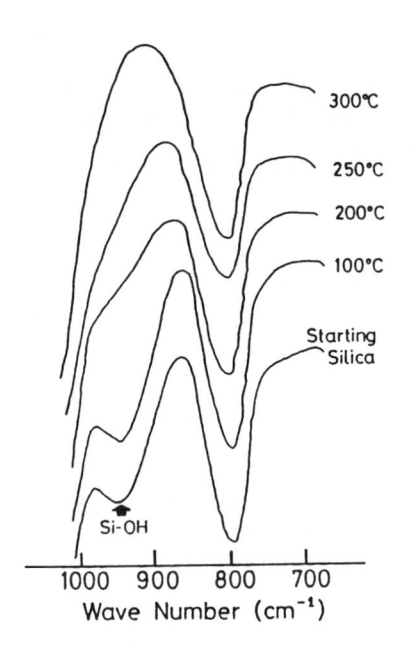

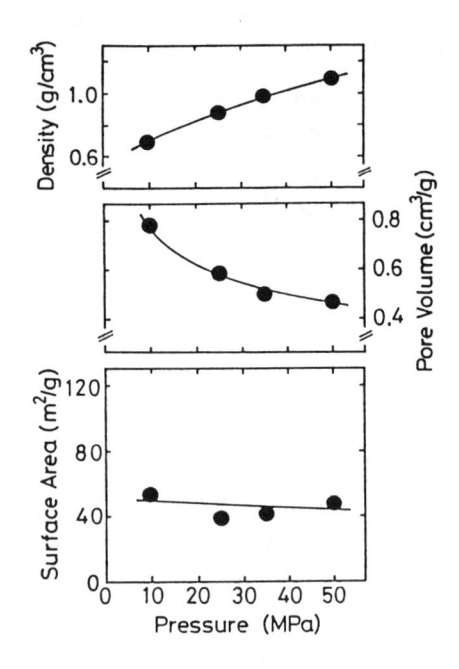

Figure 5 IR spectra of the ceramics shown in Fig.3.

Figure 6 Effect of compressive pressure on properties of the ceramics produced at 300℃ for 60 min.

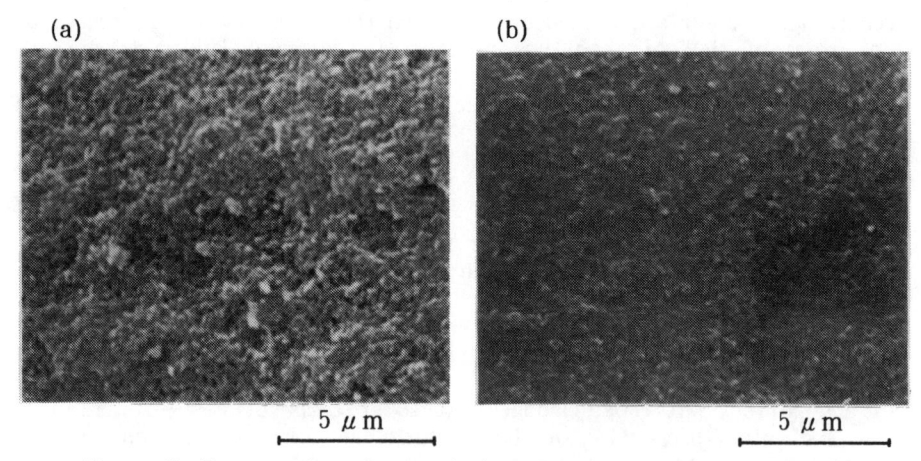

Figure 7 Fractured surfaces of the ceramics produced at (a) 10 MPa and (b) 50 MPa.

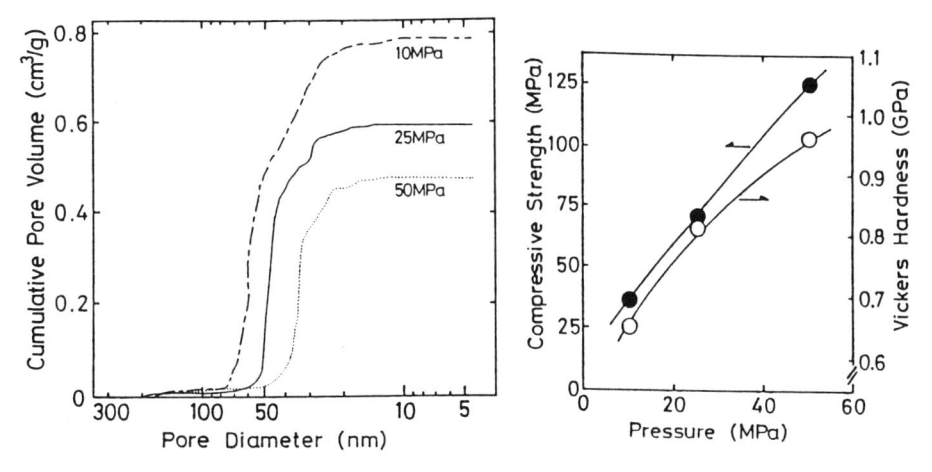

Figure 8 Pore size distribution of the ceramics shown in Fig.6.

Figure 9 Mechanical properties of the ceramics shown in Fig.6.

Effects of Compressive Pressure

Ceramics were produced at various pressures at a constant temperature and reaction time of 300°C and 60 min, respectively. The effect of compressive pressure on the properties of the ceramics is shown in Fig.6. The bulk density was remarkably increased and the pore volume decreased with an increase in compressive pressure. The change in the surface area was small. The compressive pressure accelerated the densification of the starting silica gel powder. The observation of the fractured surface of the ceramics (Fig.7) also showed that small pores were homogeneously distributed in both samples and the densification proceeded at higher pressure. The distribution of the pore diameter is shown in Fig.8. The average pore diameter was decreased from 60 nm to 35 nm with an increase in compressive pressure from 10 MPa to 50 MPa, and it was uniform even when the compressive pressure changed.

The mechanical properties also improved by hydrothermal hot-pressing at high pressure (Fig.9). The silica ceramics produced from spherical silica particles by hydrothermal hot-pressing at 250°C and 220 MPa had high mechanical strength (compressive strength = 230 MPa, Vickers hardness = 2.1 GPa). It is expected that a high compressive pressure may increase mechanical strength.

Effects of Reaction Time

Ceramics were produced at 25 MPa and 250°C while the reaction time was

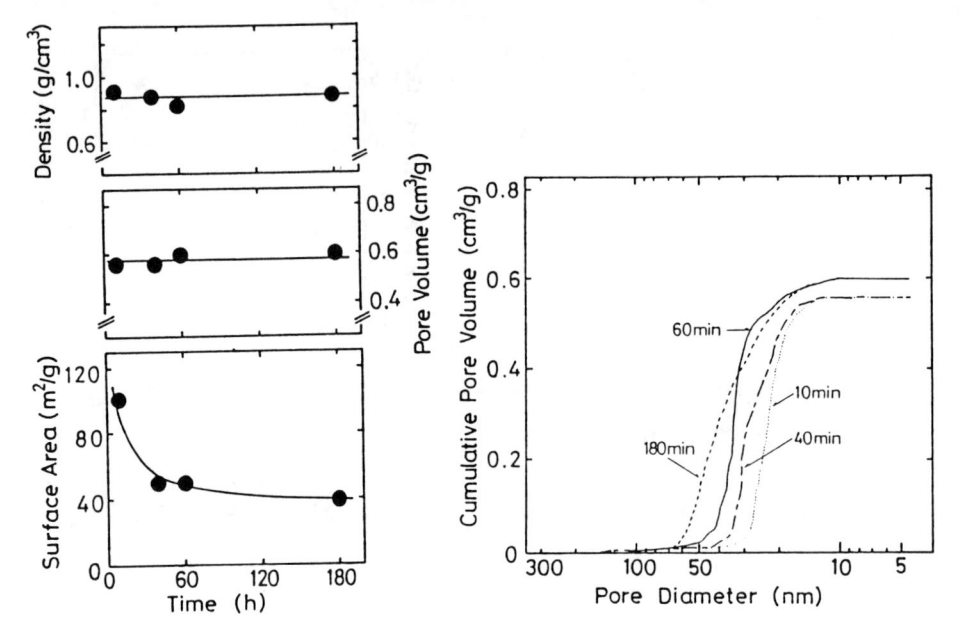

Figure 10 Effect of reaction time on properties of the ceramics produced at 250°C and 25 MPa.

Figure 11 Pore size distribution of the ceramics shown in Fig.10.

changed. The effect of reaction time on the properties of the ceramics is shown in Fig.10. The density and the pore volume did not change, which suggested that the densification proceeded while heating to 250 °C and did not continue after holding the temperature at 250°C. The surface area was significantly decreased in a short time. The increase in pore diameter (Fig.11) with an increase in reaction time showed that the hydrothermal treatment of the silica gel proceeded. The ceramics produced with long reaction time of 180 min had a large pore size distribution. The long reaction time reduced the homogeneity in pore size.

CONCLUSION

The commercial silica gel consisting of fine spherical particles with an average diameter of 18 nm was solidified by the hydrothermal hot-pressing technique. Porous silica materials with high porosity were successfully produced. In the ceramics, pores with uniform diameter were homogeneously distributed. For example, the ceramics produced at 300°C and 25 MPa for 60 min had high porosity (pore volume = 0.59 cm^3/g) with homogeneous pore diameter (about 45 nm), and low bulk density (0.88 g/cm^3). In spite of the high porosity, they had high mechanical properties (compressive strength

= 70 MPa, Vickers hardness = 0.83 GPa). The increase in reaction temperature slightly decreased the pore volume, and increased the pore size. The compression at high pressure accelerated the densification of the silica gel, decreased the pore volume and the pore diameter, and increased the mechanical strength. The homogeneity in pore size was reduced after long reaction times. The decrease of Si-OH absorption in IR spectra at higher temperature suggested that the linkage of the silica gel particles proceeded by dehydration followed by condensation of Si-OH bond.

ACKNOWLEDGMENT

The authors wish to express their thanks to Nippon Silica Industrial Co. LTD. for offering the silica gel.

REFERENCES

[1] N.Yamasaki, K.Yanagisawa, M.Nishioka, and S.Kanahara, "A Hydrothermal Hot-Pressing Method: Apparatus and Application," J. Mater. Sci. Letters, 5, 355 (1986).
[2] G.M.Friedman and J.E.Sanders, "How Sediments Become Sedimentary Rocks," pp.144-163, "Principles of Sedimentology," John Wiley & Sons, New York (1978).
[3] N.Yamasaki, K.Yanagisawa, and M.Nishioka, "From Synthetic Rocks Produced under Hydrothermal Conditions to Immobilization of Radioactive Wastes," New Ceramics, 2 (10), 81 (1989) (in Japanese).
[4] K.Yanagisawa, M.Nishioka, and N.Yamasaki, "Immobilization of Radioactive Waste by Hydrothermal Hot-Pressing," Bull. Am. Ceram. Soc., 64 (12), 1563 (1985).
[5] K.Yanagisawa, M.Nishioka, and N.Yamasaki, "Immobilization of Radioactive Wastes in Hydrothermal Synthetic Rock, (III) Properties of Waste Form Containing Simulated High-Level Radioactive Waste," J. Nucl. Sci. Technol., 23 (6), 550 (1986).
[6] N.Yamasaki, K.Yanagisawa, K.Kinoshita, and T.Kashiwai, "Solidification of Waste Containing Sodium Borate by Hydrothermal Hot-Pressing" J. At. Energy Soc. Jpn., 30 (8), 714 (1988) (in Japanese).
[7] M.Nishioka, K.Yanagisawa, and N.Yamasaki, "Soloidification of Sludge Ash by Hydrothermal Hot-Pressing," Res. J. water Pollution Control Federation, 62 (7), 926 (1990).
[8] K.Kamiya, "Porous Silica and Its Application," pp313-325, "Application Technology of High-Purity Silica," Ed. by T.Kagami and A. Hayashi, Shi-emu-shi, Tokyo (1991) (in Japanese).
[9] S.Sakka. "Formation and Chemical Properties of Fine Pores of Porous Ceramics," Ceram. Jpn. (Bull. Ceram. Soc. Jpn.), 23 (8), 702 (1988) (in Japanese).
[10] M.Nogami, "Xerogel," pp.45-53, A.Makishima, "Synthesis of Porous Glass," pp.88-96, "Development and Applications of Porous Ceramics II," Ed. by M.Hattori and S.Yamanaka, Shi-emu-shi, Tokyo (1991)

(in Japanese).
[11] E.M.Rabinovich, D.W.Johnson,JR., J.B.MacChesney, and E.M.Vogel, "Preparation of High-Silica Glasses from Colloidal Gels: I, Preparation for Sintering and Properties of Sintered Glasses," J. Am. Cram. Soc., 66 (10), 683 (1983).
[12] K.Yanagisawa, M.Nishioka, K.Ioku, and N.Yamasaki, "Neck Formation of Spherical Silica Particles by Hydrothermal Hot Pressing," J. Mater. Sci. Letters, 10, 7 (1991).
[13] K.Yanagisawa, M.Nishioka, K.Ioku, and N.Yamasaki, "Low Temperature Sintering of Spherical Silica Powder by Hydrothermal Hot-Pressing," Seramikkusu Ronbunshi (J. Ceram. Soc. Jpn.), 99 (1), 59 (1991) (in Japanese).
[14] R.K.Iler, "Wet Gel Treatment," pp.528-549, "The Chemistry of Silica," John Wiley & Sons, New York (1979).
[15] K.Kamiya, S.Sakka, and M.Mizutani, "Prepararion of Glass Fibers and Transparent Silica Glass from Silicon Tetraethoxide," Yogko-Kyokai-shi (J. Ceram. Soc. Jpn.), 86 (11), 552 (1978) (in Japanese).
[16] Y.Azuma, Y.Tajima. N.Oshima, and K.Suehiro, "Synthesis of Spherical Silica Particles and Their Thermal Behavior," ibid., 94 (6), 559 (1986) (in Japanese).

PREPARATION OF LOW-DENSITY AEROGELS AT AMBIENT PRESSURE FOR THERMAL INSULATION

Douglas M. Smith [1], Ravindra Deshpande [1], C. Jeffrey Brinker [1,2]
[1] UNM/NSF Center for Micro-Engineered Ceramics, Albuquerque NM 87131 USA. Tel: (505) 277-2861, Fax: (505) 277-1024
[2] Sandia National Laboratories, Albuquerque NM 87185.

Low density ceramic aerogels have numerous properties which suggest a number of applications such as ultra high efficiency thermal insulation. However, the commercial viability of these materials has been limited by their high cost associated with drying at supercritical pressures, low stability to water vapor, and low mechanical strength. Normally, critical point drying is employed to lower the surface tension and hence, capillary pressure, of the pore fluid to essentially zero before drying. However, the University of New Mexico has developed a process to control capillary pressure and gel matrix strength by employing a series of aging and pore chemistry modification steps such that the gel shrinkage is minimal during rapid drying at <u>ambient pressure</u>. The total processing time from gelation to the final dried product is less than 48 hours. The properties (density, surface area, pore size, SAXS) of aerogel monoliths prepared from base-catalyzed silica gels using this technique, CO_2 critical point drying, and supercritical ethanol drying are compared. An additional advantage of this approach is that the final gels are <u>hydrophobic</u>. Densities in the range of 0.15 to 0.3 g/cm^3 with pore sizes less than 100 nm are routinely made. Thermal conductivity is on the order of 0.02 W/mK at room temperature and atmospheric pressure.

To the extent authorized under the laws of the United States of America, all copyright interests in this publication are the property of The American Ceramic Society. Any duplication, reproduction, or republication of this publication or any part thereof, without the express written consent of The American Ceramic Society or fee paid to the Copyright Clearance Center, is prohibited.

INTRODUCTION

There is significant interest in producing new generations of high performance thermal insulation as energy costs rise and environmental constraints become more severe. This problem has become more problematic as replacement of chlorinated fluorocarbons (CFC's) is mandated. These are the blowing agents of choice for most foam insulation because of their low thermal conductivity and desirable foam blowing properties. For high performance insulation, the following properties are desired; 1) thermal conductivity less than air (k~0.02 W/mK), 2) nonflammable, 3) small effect of temperature on thermal conductivity, 4) environmental stability, and 5) low cost.

Aerogels are extremely porous materials representing the lower end of the density spectrum of man-made materials. Densities as low as 0.003 g/cm^3 have been reported for silica aerogels [1]. These fascinating materials have numerous unique properties as a result of their ultra-fine microstructure. Aerogels with the porosities in the range of 85% to 98% may be transparent and translucent. They exhibit strong Rayleigh scattering in blue region and very weak in red region. In the infrared region of electromagnetic spectrum, the radiation is strongly attenuated by absorption leading to application in radioactive diode systems, which effectively transmit solar radiation but prevent thermal IR leakage [2].

Silica aerogels with their thermal conductivities as low as 0.02 W/mK (0.01 W/mK when evacuated) find application in superinsulation systems [3]. Fricke [3] has described several aerogel applications based on their insulating properties. They include reduction of heat losses through windows, energy-effective greenhouses, translucent insulation for solar collectors, and solar ponds for long-term energy storage. Fricke [3] has also discussed the mechanism for thermal transport through aerogel tiles and the effects of density and gas pressure on their thermal conductivity.

Since aerogels are made by sol-gel processing, their microstructure may be tailored for a specific application. Precursors, including metal alkoxides, colloidal suspensions, and a combination of both under several mechanisms of gelation may be used to synthesize gels [4]. By varying aging conditions such as time, temperature, pH, and pore fluid, the parent wet gel micro-structure may be altered [4,5,6,7]. Considerable shrinkage and reduction in surface area is usually observed during drying, which may be a result of condensation of the groups on the pore surface and/or surface tension induced collapse. The capillary pressure generated during drying is related to the pore fluid surface tension, γ_{LV}, and contact angle, θ, between the fluid meniscus and pore wall as follows,

$$P_c = -(2\,\gamma_{LV} \cos\theta)/a \tag{1}$$

where a is the pore radius. For the submicron capillaries in silica gels, very large stresses are developed during drying. For aerogel synthesis under supercritical conditions, γ_{LV} is essentially zero.

Aerogels are not new materials. They were first reported by Kistler over 60 years ago [8]. However recent advances in sol-gel processing technology along with increasing environmental concern caused by alternate insulation material manufacturing regenerated interest in energy conservation and thermal insulation applications. The most common method of making aerogels involves directly removing the pore fluid above its critical point (for ethanol T_c=243 °C, P_c=63 bar). This avoids liquid-vapor menisci and therefore, capillary stresses, during drying and essentially preserves the wet gel structure. An alternate low temperature method involves replacing the pore fluid with liquid CO_2 and then removing CO_2 above its critical point (T_c = 31 °C, P_c = 73 bar). Brinker and Scherer [4] and Fricke and Caps [2] have reviewed research on aerogels made from a variety of oxides.

Although aerogels exhibit unique properties, they suffer from <u>drawbacks for commercialization</u>. The major disadvantage is their <u>high processing cost</u>. High pressure associated with supercritical drying requires large autoclaves and increases capital and operating costs. For the high temperature process, significant changes in gel structure occur due to the accelerated rates of aging whereas the lengthy low temperature CO_2 exchange is limited to certain pore fluids since they must be miscible in liquid CO_2.

Additionally, moisture condensation in the pore network leads to loss of insulating properties and capillary stress induced collapse.

In this study we explore an alternative process to reduce aerogel cost by designing a route at ambient pressure. We use a series of aging and pore surface chemistry modification steps to alter the gel matrix strength and for capillary pressure management.

EXPERIMENTAL PROCEDURE

Silica gel samples were prepared by a two-step acid/base catalyzed hydrolysis of tetraethylorthosilicate (TEOS) as described in previous studies [5,6,7]. These samples were washed with aprotic solvents to remove unreacted monomer. The details of aging, surface chemical modification, and subsequent drying are described elsewhere [9]. Several different aging treatment were used to tailor the porosity in the resulting aerogels. These gels were typically dried at 70 °C and 1 bar and are denoted as ambient pressure aerogels (APA). Similar wet gels were washed with ethanol and supercritically dried (high temperature aerogel) or washed with ethanol followed by liquid carbon dioxide, and supercritically dried (low temperature aerogel). Also a xerogel was made by drying the wet gel at 70 °C without surface treatment.

N_2 sorption at 77 K measured with automated volumetric analyzers was used to study pore structure of the dried gels. Samples were outgassed under vacuum at 373 K for two hours. BET theory was used to obtain the surface area and a single condensation point was used to find the pore volume. Measuring aerogel structure using either adsorption or mercury porosimetry is problematic because of their fragile nature. Therefore, total pore volume was determined using the skeletal density (helium) and the bulk density measured by mercury displacement at ambient pressure. Contact angles for solvent/surface modification combinations were found by measuring the capillary rise rate (Figure 1) versus time (wetting fluids) and by measuring the pressure required to force fluids into the pores (nonwetting fluids). Wet gel mechanical strength was measured using a 3-point bend apparatus (Figure 2).

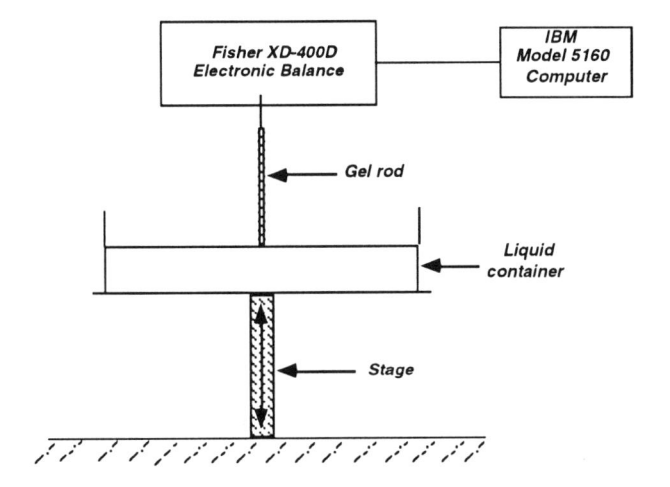

Figure 1 Gravimetric capillary rise apparatus.

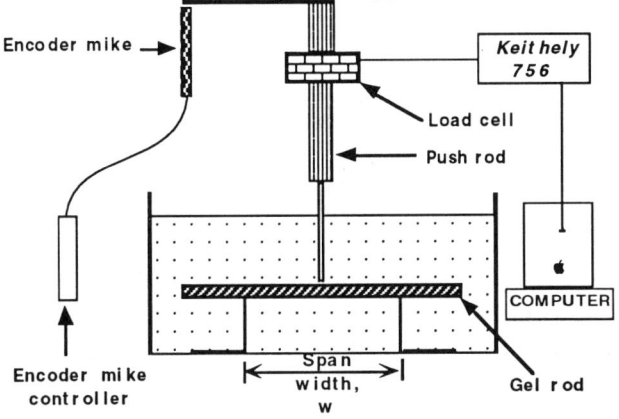

Figure 2 Wet gel mechanical strength measurement apparatus.

RESULTS AND DISCUSSION

Thermal insulation properties of aerogels are a direct result of their ultra-fine microstructure and low densities. We have previously presented a micro-structural comparison of our ambient pressure (APA) aerogels with aerogels made by low and high temperature supercritical fluid extraction routes as well as a xerogel made by

Porous Materials 75

evaporation [9]. Pore size distributions obtained by nitrogen condensation/BJH analysis indicated a maximum pore size of less than 100 nm. The bulk density of the supercritical and ambient pressure aerogels was the same (~0.3 g/cm^3). More recently, we have made a series of ambient pressure aerogels with bulk density between 0.15 and 0.3 g/cm^3. SAXS experiments were performed on the ambient pressure and low temperature aerogels. The curves are virtually identical over the entire length scale probed (10-1000 Å).

In order to provide a better understanding of the mechanisms associated with drying in the APA process, continuous measurement of sample size and weight were made using a combined fiber optic probe/gravimetric apparatus. Plots of shrinkage versus the fraction of pore fluid evaporated is presented in Figure 3 for APA gels dried from two different aprotic solvents of varying surface tension. Both gels exhibited shrinkage much less than gels which were not surface modified. Interestingly enough, the gel dried from the lower surface tension solvent actually expanded during the final stage of drying. Depending upon the aging and surface modification scheme employed in the APA process, the sample length usually returned to within 1% of the original length. For the higher surface tension fluid, the sample fractured during the rapid drying conditions employed.

Several modifications were made in the ambient pressure aerogel synthesis procedure to achieve lower density with more efficient use of industrially acceptable solvents. Total sample shrinkage and bulk density as a function of surface modifying agent quantity is presented in Table 1. For all samples, the same amount of solvent was used to disperse the surface modifying agent. It is clear from the table that a minimum quantity of surface modifying agent is necessary to dry silica gel at ambient pressure without shrinkage. Using a smaller amount of surface modifying agent results in considerable shrinkage and an increase in the bulk density. Predictions based on the surface area of a wet gel and extent of reaction of the surface modifying agent with the silica surface suggests a minimum quantity of surface modifying agent is required to change the contact angle sufficiently to reduce capillary pressure induced collapse during drying. In order to understand the relationship between surface modification and solvent,

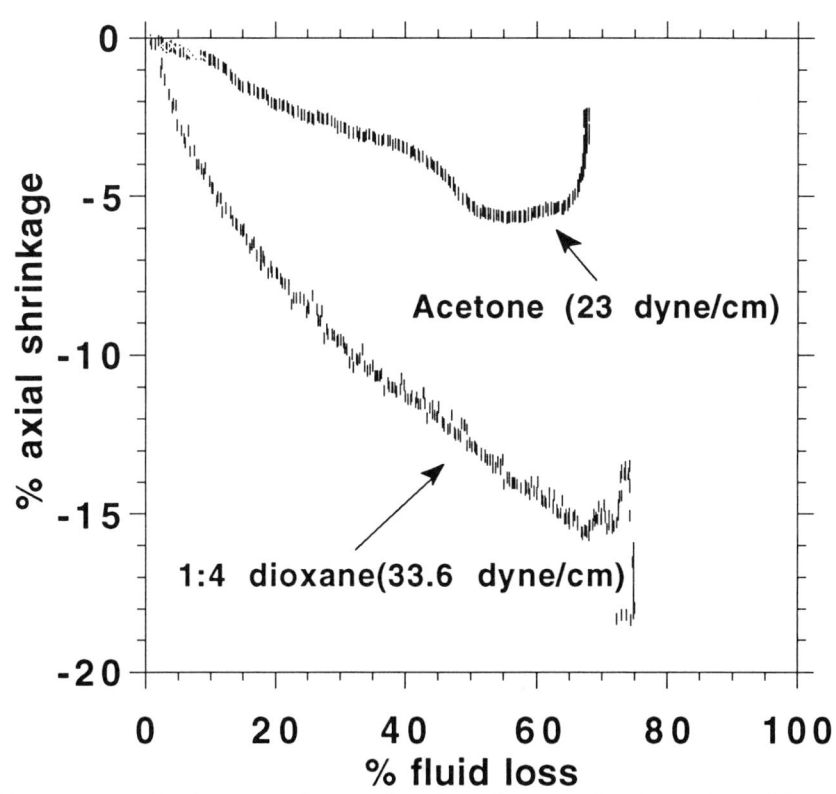

Figure 3 Variation of sample size during drying of ambient pressure aerogels.

Table 4 Shrinkage and bulk density for APA synthesized with varying quantity of surface modifying agent

Relative quantity of surface modification	% linear shrinkage	Bulk density g/cm^3
1	~ 1	0.202
0.75	~ 1	0.179
0.5	~ 1	0.195
0.25	~ 25 and cracking	0.415
0.1	~ 40 and cracking	0.493

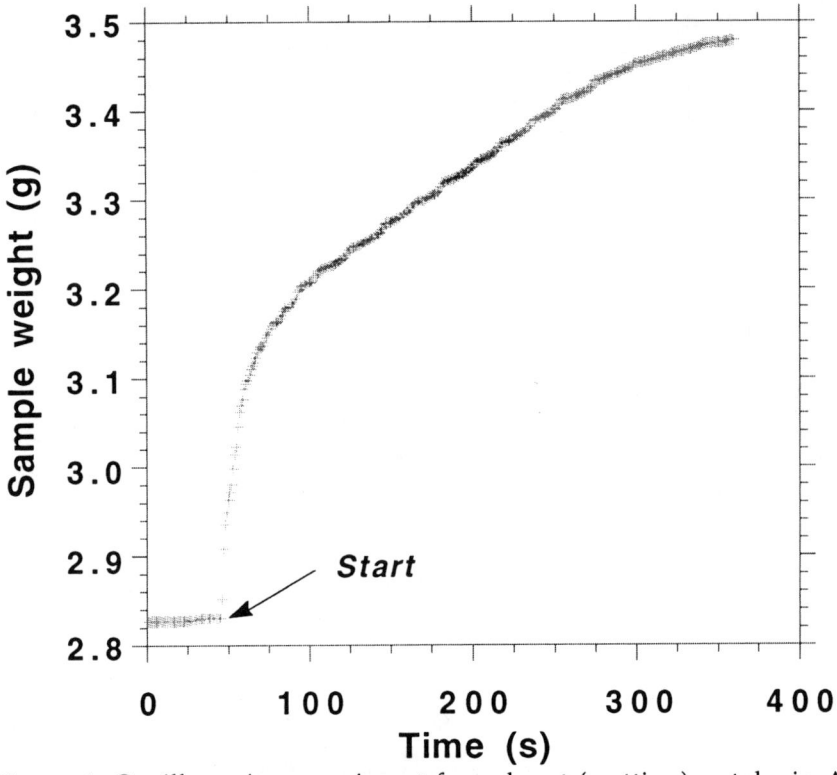

Figure 4 Capillary rise experiment for solvent (wetting) uptake in APA.

capillary rise experiments have been performed on a range of APA samples. Figure 4 shows a typical curve. For samples which did not show significant shrinkage during drying, contact angles on the order of 70-80° were found. The detailed study of the relationships between contact angle, pore size, mechanical strength, surface modification, pore fluid, and drying is currently underway.

Attempts were made to lower APA density by diluting the sol. Table 2 shows the results of TEOS stock solution dilution with ethanol on elastic modulus and modulus of rupture of the wet gels. Elastic modulus decreases with increased dilution due to a decrease in Si-O-Si bonds per unit volume of the gel. The bulk density of the resulting APA's is a result of the density of the wet gel and the extent of capillary-induced collapse. Wet gels with progressively lower density (and therefore, lower elastic modulus) are expected to undergo an increasing extent of capillary stress induced collapse. As shown in Table 2, the bulk density of the most dilute sample shows a increase in density due to a greater extent of collapse of the relatively complient structure.

As a result of the surface modification used during the APA processing, the final aerogels are hydrophobic (see Figure 5). This avoids a major problem associated with the use of aerogels which is collapse due to atmospheric moisture. Samples have been submerged in water for over 1 year with no loss of their hydrophobic nature. Previously, we have noted that pressures of greater than 10,000 psia are required to force water into APA's [9].

Figure 5   Demonstration of hydrophobic nature of APA.

Table 2 Effect of dilution on density and mechanical properties.

Dilution	Elastic modulus (MPa)	Modulus of rupture (MPa)	density g/cm^3
100 : 0	.225	.037	0.1899
90 : 10	.166	.043	0.2112
80 : 20	.084	.043	0.2020
60 : 40	.035	.021	0.2150
40 : 60	.0075	did not break	0.3413

CONCLUSIONS

Low density silica aerogels were prepared via sol-gel processing at ambient pressure. The hydrophobic nature of ambient pressure aerogels gave them structural stability against humidity. Thermal conductivity of the APA's was ~0.02 W/mK at room temperature and pressure.

ACKNOWLEDGEMENTS

This work has been supported by the UNM/NSF Center for Micro-Engineered Ceramics which is a collaborative effort of the National Science Foundation (CDR-8803512), Los Alamos and Sandia National Laboratories, and the ceramics industry.

REFERENCES

[1] L.W. Hrubesh, " Aerogels : World's Lightest Solids", Chemistry and Industry, 24 (1990) 824.

[2] J. Fricke and R. Caps, " Aerogels-A Fascinating Class of Porous Solids ", in Ultrastructure Processing of Advanced Ceramics: ed. J.D. Mackenzie and D.R. Ulrich (Wiley, New York, 1988).

[3] J. Fricke, " Thermal Insulation Materials by Sol-Gel Process ", in Sol-Gel Technology for Thin Films, Fibers, Preforms, Electronics and Speciality Shapes: Ed. L.C. Klein (Noyse Publications, Park Ridge NJ, 1988).

[4] C.J. Brinker and G.W. Scherer, Sol-Gel Science: The Physics and Chemistry of Sol-Gel Processing, (Academic Press, San Diego, 1990).

[5] P.J. Davis, C.J. Brinker and D.M. Smith, " Pore Structure Evolution of Silica Gels during Aging/Drying I. Temporal and Thermal Aging ", Journal of Non-Crystalline Solids, 142 (1992) 189.

[6] P.J. Davis, C.J. Brinker, D.M. Smith and R.A. Assink, " Pore Structure Evolution of Silica Gels during Aging/Drying II. Effect of Pore Fluids ", Journal of Non-Crystalline Solids, 142 (1992) 197.

[7] R. Deshpande, D.W. Hua, C.J. Brinker and D.M. Smith, " Pore Structure Evolution of Silica Gels during Aging/Drying III. Effect of Pore Fluid Surface Tension ", Journal of Non-Crystalline Solids, 144 (1992) 32.

[8] S.S. Kistler, " Coherent Expanded Aerogela and Jellies ", Nature, 127 (1931) 741.

[9] R. Deshpande, D.M. Smith and C.J. Brinker, "Preparation of High Porosity Xerogels by Surface Chemical Modification", U.S. Patent Application, 1992.

CERAMIC MEMBRANES AND APPLICATION TO THE RECOVERY OF SOY SAUCE

Nobuyuki Maebashi
New Ceramic Reseach Division, Basic Reseach Laboratory, TOTO LTD.
2-8-1 Honson, Chigasaki-shi, Kanagawa 253, Japan

INTRODUCTION

Ceramic membranes have exellent uniform and narrow pore size distributions and better thermal, mechanical and chemical properties than polymeric membranes. Therefore, in recent years, ceramic membranes have been applied in the food and pharmaceutical industries, as well as in waste water treatments and many other fields.

Recently we have developed two types of ceramic membranes; a disk type microfiltration membrane*, and a multi-channel tubular type ultrafiltration membrane. This paper reports the structures and physical properties of the membranes, and describes one application of the rotating disk type membrane filtration system.

MANUFACTURING PROCESS

Figure 1 shows the scanning electron micrograph of the microfiltration membrane cross section, and Fig.2 shows the process flow diagram of the membranes.

The membranes consist of 3 or 4 layers; porous support (body), intermediate layer, microfiltration layer, and ultrafiltration layer.

To obtain a high chemical resistance the support is formed by pressing or extrusion of high purity raw materials. The outer layers on the support are formed by coating with suspended particles. The pore size distribution is controlled by controlling the particle size distribution of the raw materials and the sintering condition.

*The rotating disk type membrane filtration system is developed in cooperation with the Hitachi Plant Engineering & Construction Co.,Ltd.

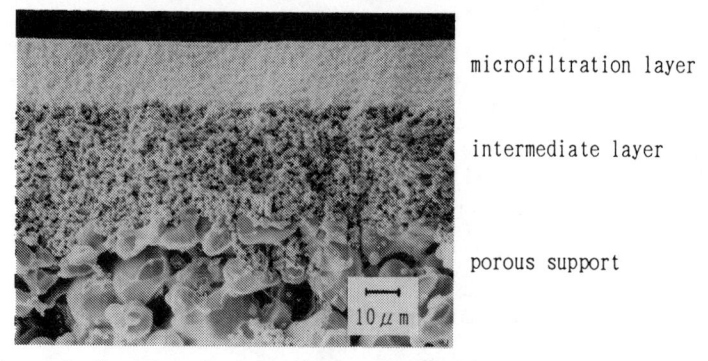

Figure 1 Scanning electron micrograph of membrane cross section

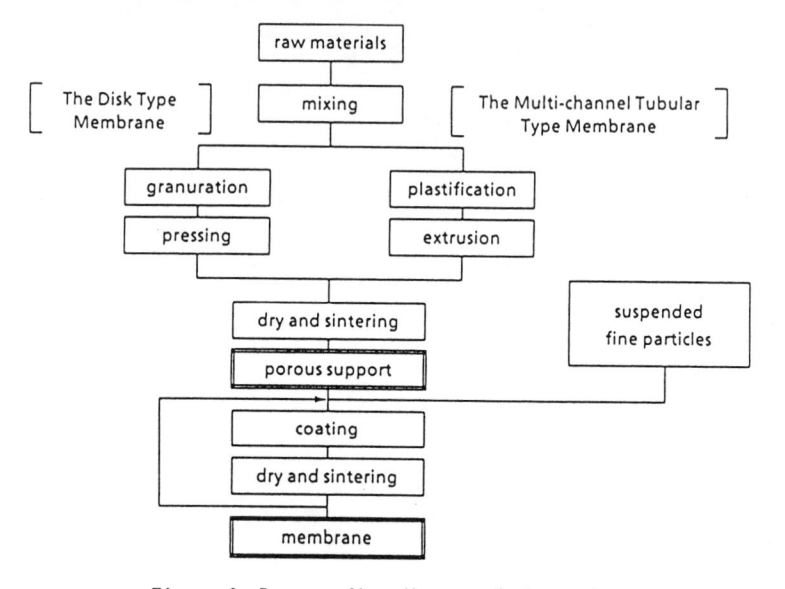

Figure 2 Process flow diagram of the membranes

STRUCTURES AND PHYSICAL PROPERTIES OF THE MEMBRANES

Figure 3 shows TOTO Ceramic Membranes, and Table 1 shows the physical properties. The disk type membrane has a microfiltration layer on the surface of the body, and the multi-channel tubular type has an ultrafiltration layer on the inside of the tubes.

The pore sizes of the intermediate layer and the microfiltration layer were measured by means of the bubble point method. The pore size of the intermediate layer is 1 μm, and that of the microfiltration layer is 0.1 μm. The ultrafiltration membranes have six types of molecular weight cut-off values ranging from 1000 to 550000. These were measured using PEG(polyethylene glycol) and dextran solutions.

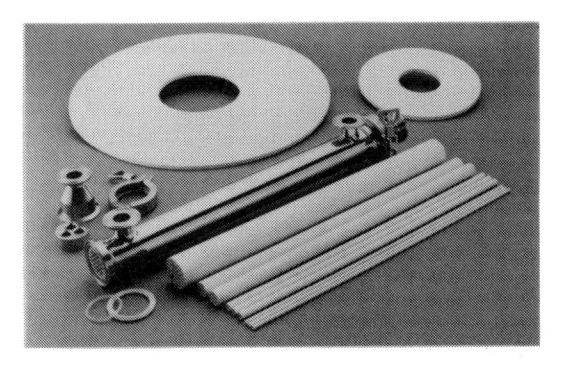

Figure 3   TOTO Ceramic Membranes

Table 1. Physical Properties of TOTO Ceramic Membranes

	Microfiltration(MF) membrane	Ultrafiltration(UF) membrane
Pore sizes	Nominal pore diameters 0.1 μm 1 μm	Nominal molecular weight cut-off values 2000 10000 20000 55000 200000 550000
Material Support layer MF layer UF layer	Al_2O_3 Al_2O_3 ZrO_2	Al_2O_3 ZrO_2 ZrO_2
Dimensions of elements (mm)	disk O.Dϕ230-I.Dϕ100 O.Dϕ500-I.Dϕ200	multi-channel tubular ϕ20-L 500-I.Dϕ4- 7channels ϕ30-L 500-I.Dϕ4-19channels ϕ30-L1000-I.Dϕ4-19channels
Temperature resistance	steam sterilization up to 121°C	
Chemical resistance	use at pH 1~14	resistance for solvents

The Multi-channel Tubular Type Membrane

Figure 4 shows schematically the cross-flow filtration using a tubular type membrane.

Porous Materials

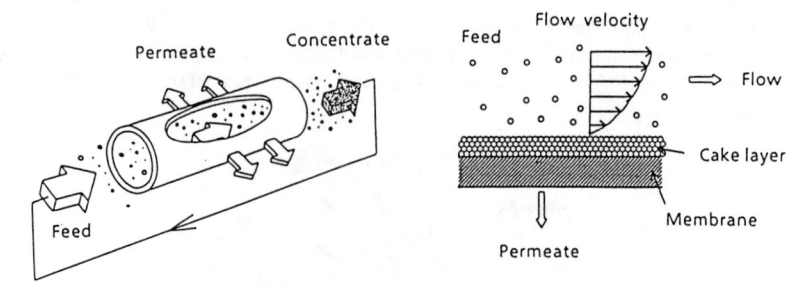

Figure 4 Cross-flow filtration using a tubular type membrane

In conventional dead-end-flow filtration, rejected particles accumulate on the filter surface and follow the permeate flux decline. Therefore the cross-flow filtration is used to limit filter-cake growth. Using the tubular type membranes, the cross-flow filtration employs a high fluid circulation rate tangential to the membrane surface, by using a circulation pump.

Generally the multi-channel tubular type membrane filtration system is suitable for purifications or separations of low concentration fluids.

The Disk Type Membrane

Figure 5 shows schematically the cross-flow filtration using the disk type membrane.

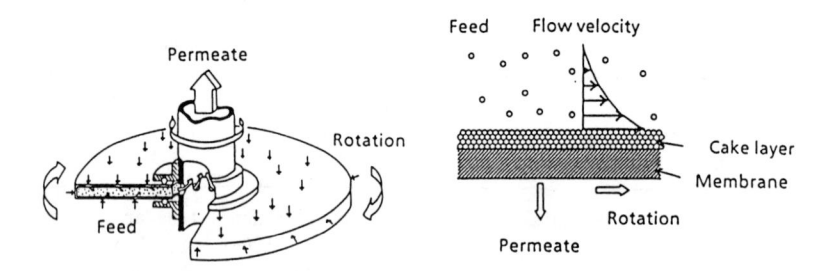

Figure 5 Cross-flow filtration using the disk type membrane

With high suspended solids concentration or viscous fluid, it is difficult to circulate the fluid with a high flow velocity, because of choking of channels or insufficient capacity of the circulation pump. The fluid flows on the disk type membrane surface by rotating of the membrane in the fluid. The rotation limits the filter-cake growth. Therefore, the rotating disk type membrane filtration system needs a lower power pump and motor to feed the fluid to the vessel and rotate the membranes, compared with the tubular type membrane filtration system.

APPLICATION

Figure 6 shows the conventional flow diagram for the production of soy sauce. In the conventional process the sediment contains about 10 wt% of produced soy sauce, and the sediment is scrapped as the residue with diatomaceous earth.

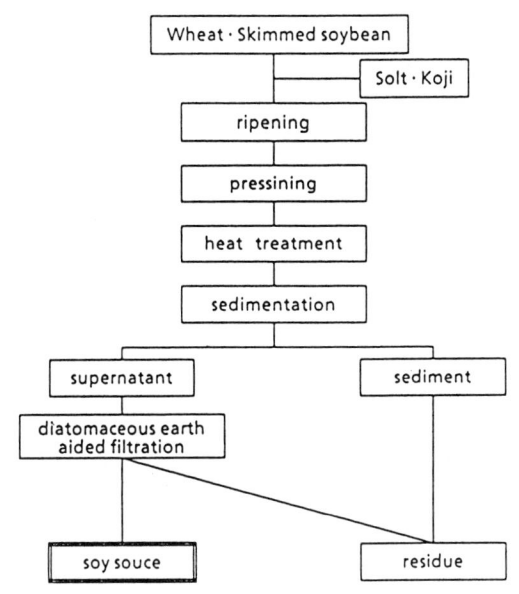

Figure 6 Conventional process flow diagram of soy sauce

The rotating disk type membrame separation system was applied to concentrate the sediment. The purpose of this operation was to reduce the amount of residue and increase the yield of soy sauce.

Fluxes were measured at different transmembrane pressures, rotational frequencies of membranes and soy sauce temperature 298 K, using the 7 membranes module with 0.1 μm pore diameters and 500 mm outer diameters, as shown in Fig. 7 and 8. As rotational frequencies were increased up to 400 rpm, the fluxes increased at each transmembrane pressures. But as the rotational frequencies were increased more than 400 rpm the fluxes increased gently or declined.

The permeate in the support layers tends to move toward the outer wall of the membranes by a centrifugal acceleration caused by increasing of the rotational frequency. Therefore, the fluxes showed a minimal increase or decreased. Those results show that there was an optimum rotational frequency for each transmembrane pressure.

Figure 7 The rotating disk type membrane module

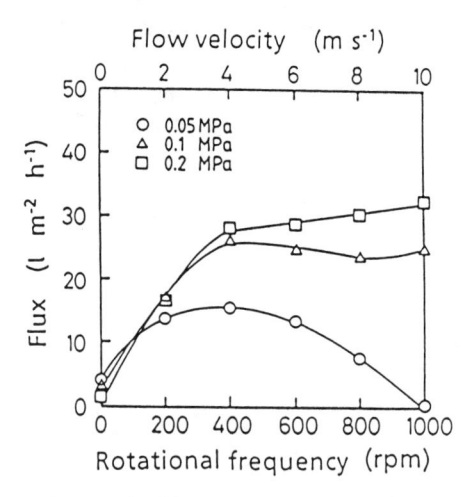

Figure 8 Fluxes measured at different transmembrane pressures and rotational frequencies

The sediment was concentrated at 0.2 MPa transmembrane pressure and 600 rpm rotational frequency. The result is shown in Fig. 9, compared to several other studies[1~4], using tubular type membrane filtration systems.

In these studies there are two types of concentrations; the concentrations of soy sauces before the sedimentation and the concentrations of sediments.
In Fig. 9, starting concentration ratios of the soy sauces are shown as 1, and those of the sediments are shown as 10.

It was found that the yield of soy sauce and the permeate flux using the rotating disk type membrane filtration system were higher than the yields and fluxes obtained with the tubular type membrane filtration systems.
This result suggests that the rotating disk type membrane filtration system is suitable for high ratio concentrations of fluids or filtrations of high suspended solids concentration fluids.

CONCLUSIONS

We have developed two types of ceramic membranes: the disk type microfiltration membrane, and the multi-channel tubular type ultrafiltration membrane.

The rotating disk type membrame filtration system was applied to the concentration of the sediment from heat treated and sedimented soy sauce. The yield of soy sauce and the flux were higher than the yields and fluxes obtained in other studies using tubular type membrane filtration systems.

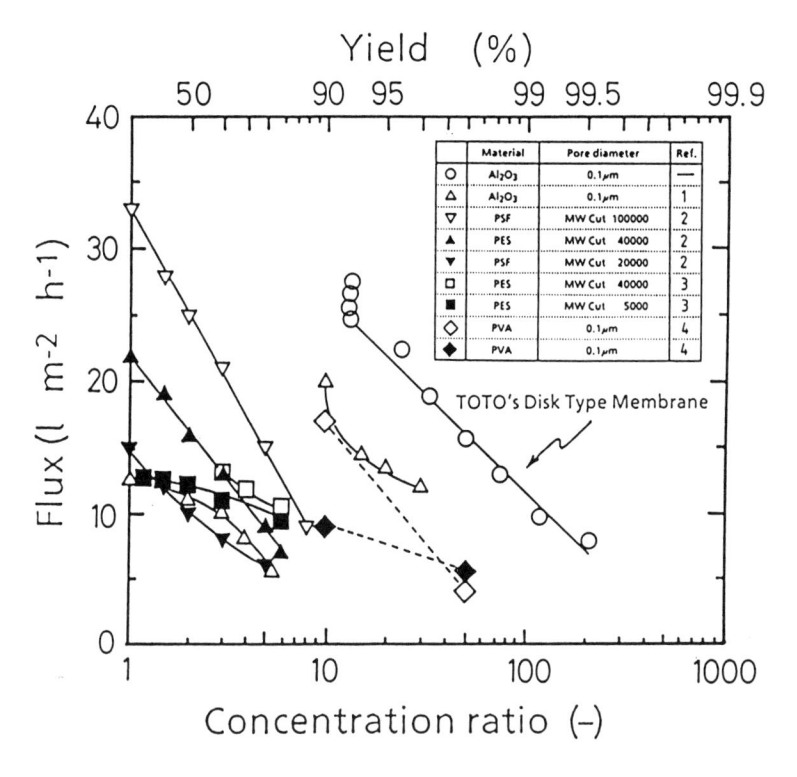

Figure 9 Results of concentrations of soy sauce

It was found that the rotating disk type membrane filtration system was suitable for high ratio concentrations of fluids or filtrations of high suspended solids concentration fluids.

REFERENCES

1 H. Hasegawa, "Applications of Ceramic Membranes in the Food Industries"; *Preprint of 2nd Membrane Research Circle of Food autumn meeting*, (1990).
2 Y. Okusawa, and U. Eguchi, "Development of Membrane Utilization Techniques for Seasoning Liquid. 1"; *Research Report of Food Industrial Research Institute of Saitama Prefecture*, 1986, 103-109(1987).
3 K. Kubo, and S. Hayashida, "Utilization Techniques for Polymer Membranes. 2 Filtration Tests on Soy Sauce Using Membranes"; *Research Report of Industrial Research Institute of Nagasaki Prefecture*, 1987, 97-104(1987).
4 O. Kusudo, Y. Hamamoto, and T. Miyake, "Application of Membrane to Soy Sauce. 2 Filtration of Soy Sauce Precipitates by SF Filter"; *Journal of Japan Soy Souce Laboratory*, 7[3]108-111(1981).

DEVELOPMENTAL STUDIES ON POROUS ALUMINA CERAMICS

P.K.TANTRY, S.N.MISRA, & A.L.SHASHI MOHAN
R&D CENTRE, GRINDWELL NORTON LTD.,
BANGALORE-560049, INDIA

Porous ceramics exhibit a unique combination of physical properties (such as strength, thermal stability, uniform pore structure), and chemical properties (such as inertness, non-wettability by molten metals). Therefore, these materials are being considered as candidates for applications in materials conservation (metal filtration, quality metal castings), energy conservation (thermal insulation, low-mass kiln furniture) and pollution control (supports for catalytic converters for auto emissions).

One of the methods for manufacturing porous ceramics involves: preparation of stable ceramic slurries, impregnation of these slurries into desired polymeric preforms, and drying and firing of impregnated preforms.

In the present work, laboratory experiments have been carried out to prepare high alumina foams (containing 90-95% Al_2O_3) by using polyurethane (PU) foams, of desired pore structure and size, as preforms. Process conditions have been optimized with regard to composition of the alumina slurries, impregnation of the slurries into the PU foams and subsequent firing of the same to obtain the corresponding ceramic foam. Effects of various sintering aids and bond contents on the physical properties of the alumina foams have been investigated.

Details of the experiments and their results are presented in this paper.

INTRODUCTION

Porous or low density (5 to 15% of theoretical density) ceramics possess excellent physical properties such as uniform pore size and shape, mechanical strength, and thermal stability. In addition, most of these materials are chemically stable and inert. Therefore, low density ceramics are gaining importance as candidates for applications in:

* Metallurgy - for metal filtration
* Ceramics - as low mass kiln furniture
* Pollution control - as catalyst carriers
* Furnaces - for thermal insulation
* Aerospace - as insulating tiles for space shuttles

Porous materials are also used in agriculture, aquaculture, food processing, bio-technology and many other areas.

Depending on the method of fabrication, it is possible to generate porous ceramics of at least 4 distinct macro-structures such as:

* Tangled fibre networks
* "Particulate" networks
* Closed cell structures
* Open cell structures

The present work is mainly concerned with the development of open cell macro-structures of alumina ceramics for subsequent applications as metal filters.

Generally, open cell macro-structures are produced by coating the connecting strut network of a reticulated, foamed polymer with a ceramic. The reticulated polymer substrate is itself manufactured by uniformly foaming a thermoset polymer and then dissolving the thin cell walls. The substrate is coated with a ceramic powder and, after drying, is pyrolysed at moderate temperatures. After pyrolysis, the powder coating is densified by sintering.

In the present work, extensive experiments were carried out with regard to optimizing :

* ceramic slurry compositions

* impregnation of the slurries into desired polymer substrate to achieve uniform coating of cell walls.

* drying and sintering cycles to achieve the final ceramic foam.

Further, the resultant ceramic foams were characterised for:

* Density
* Flexural strength
* Coefficient of thermal expansion.

The experimental details and the results thereof are described in subsequent sections.

EXPERIMENTAL DETAILS

Thermal Studies of PU Foams

Reticulated PU foam of 12 kg/m^3 density and 13 pores/cm (35 ppli) was used in all the experiments. A known mass of the foam was heated for a desired interval of time at different temperatures and corresponding weight loss as a function of temperature is shown in Fig. 1.

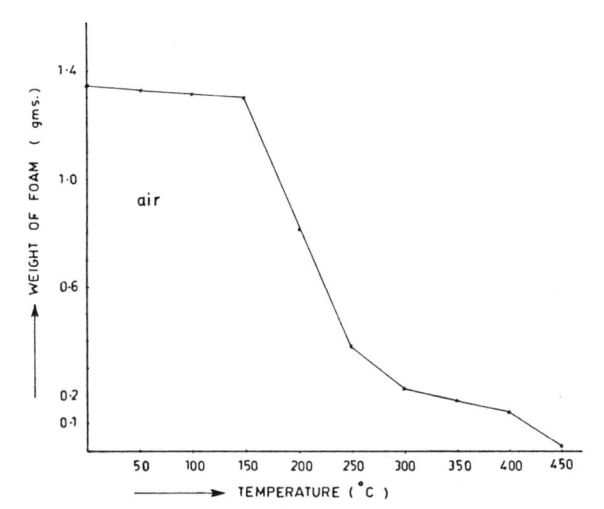

Figure 1 Weight of PU Foam VS Temperature

It may be noted that thermal decomposition of the PU foam starts at around 150°C and nearly 80% of the material is lost within 250°C. In fact, almost the entire foam is pyrolyzed at 450°C.

Slurry Compositions

Aqueous slurries of a fine powder of calcined alumina (99% pure and 5 microns average particle size) was used as the principle raw material. The formulations were optimized using different additives such as clay, calcium and zinc salts, etc. Generally the slurries contained 60 to 70% solids (w/w).

The viscosity of the different slurries was determined at room temperature using a Brookfield viscometer (model RVT).

Slurry Coating of PU Foam

The above aqueous slurry was impregnated into the reticulated PU foam by the following methods:

* The slurry was held in a container and the foam was dipped slowly into it. The foam was impregnated by capillary action.

* The foam was dipped into the slurry and the container evacuated so that all the trapped air was removed and the foam was filled with the ceramic slurry.

* The foam was squeezed and dipped into the slurry so that it acted as a sponge and was impregnated by the slurry during its expansion to original shape and size.

The impregnated foam was passed between two MS rollers, the gap between which could be adjusted as desired. By adjusting the gap between rollers, desired amounts of excess slurry were squeezed out from the impregnated slurry so that different green densities could be achieved.

Drying of Slurry Coated Foams

Both conventional (hot air) oven drying and microwave drying techniques were attempted. It was found that for a given sample the time taken in the conventional oven was about 7-8 hours whereas in the microwave oven the drying step was completed within 1/2 hour. However, all the experiments were carried out in the conventional oven due to paucity of working space in the microwave oven.

Densification of Ceramic Macro-Structures

The dried foam with the solids was heated as per predetermined time-temperature cycles. Based on the thermal decomposition characteristics of the foam (as described earlier) the initial heating rate from room temperture up to 500°C was maintained at around 1°/minute. Subsequently, the temperature was raised to 1400-1500°C in 5-8 hours time. At the final temperature, the material was soaked for 4-8 hours and allowed to cool in the furnace.

Physical Characterization of Fired Sample (Ceramic Foam)

The ceramic foams obtained as above were characterized for:

* dimensional density and porosity
* flexural strength at room temperature (3 point bending)
* coefficient of thermal expansion (RT - 1000°C)

RESULTS & DISCUSSIONS

Viscosity Studies on Alumina Slurries

Figure 2 represents a plot of viscosity (25°C) of alumina slurries containing clay as an additive.

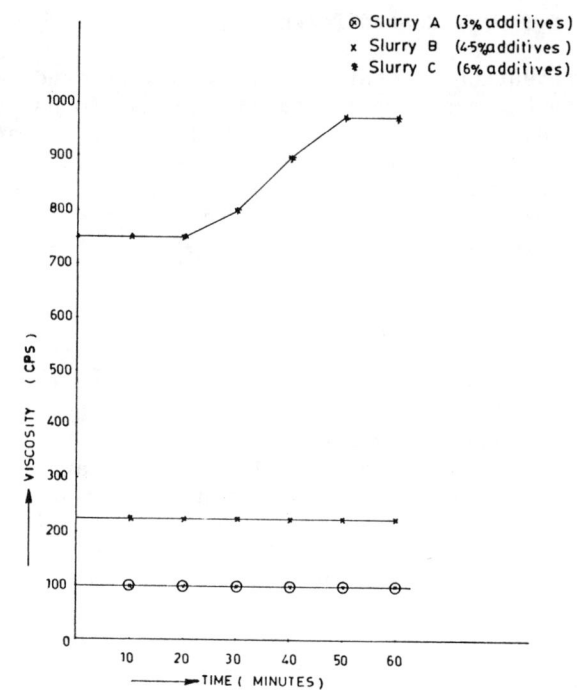

Figure 2 Viscosity VS Time curve for Al_2O_3-Clay slurries

It is observed that (a) viscosity increases with increasing clay content and (b) at 6% clay concentration, the viscosity increases with ageing of the slurry. Perhaps, the ageing phenomenon may be attributed to the swelling of the clay in water at higher clay concentrations.

In Fig. 3, a plot of viscosity Vs time is depicted for slurries containing CaO, SiO_2 and ZnO in different proportions. Here the viscosity appears to decrease with increased ageing, perhaps due to increased deflocculation in presence of cationic species.

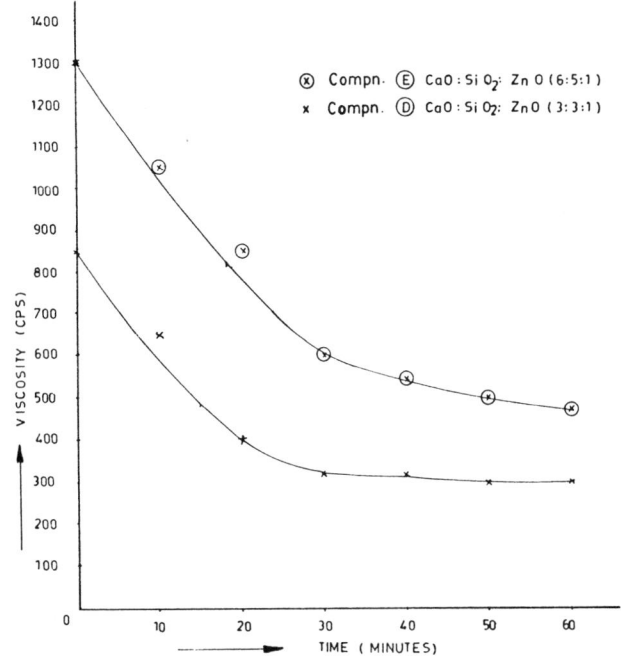

Figure 3 Viscosity VS Time curve for Al$_2$O$_3$ slurries containing CaO, ZnO and SiO$_2$

Optical Microscopic Observations of Fired Samples

Preliminary microscropic examination showed that the thermal decomposition of the impregnated PU foam was topotactic retaining the orginal morphology of the substrate. Also, shrinkage during firing was less than 5% by volume. Thus, it was possible to obtain near-net-shape ceramic foams by the present process.

Flexural Strength (MOR) Studies

Three point bending flexural strength (MOR) data are presented in Fig.4 for the different alumina foam samples.

Porous Materials

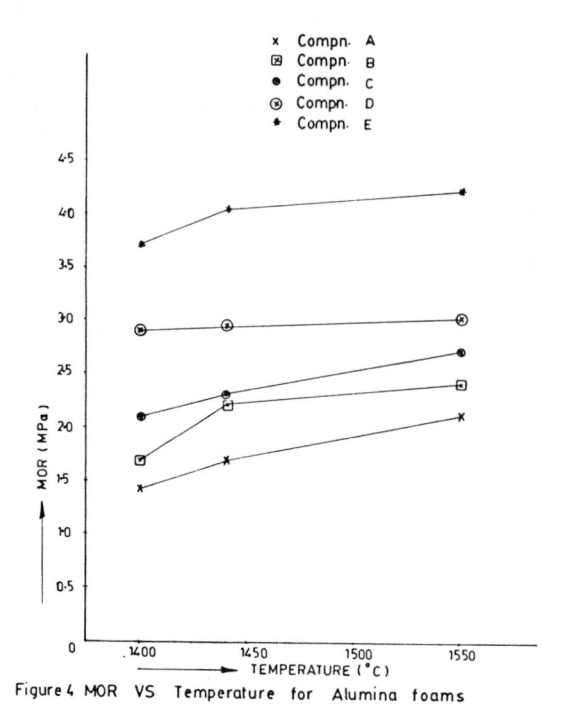

Figure 4 MOR VS Temperature for Alumina foams

The following general observations could be made:

* With increasing clay content, the MOR values show an increasing trend although the differences are not very large.

* Samples containing CaO, SiO_2 and ZnO as additives, have much higher MOR than the clay containing samples.

The above MOR data refer to samples with a fired density of 550 kg/m^3. Thus, the observed trends in MOR values can be ascribed to an increased degree of sintering in the case of clay containing samples. However, in the samples containing CaO etc. the primary bonding mechanism appears to be the formation of vitreous silicate phases, which in turn enhance the MOR values.

The MOR values, however, seem to be independent of the temperature of firing (1400°C, 1440°C and 1550°C) thus indicating that the sintering/ vitreous bond formation is more or less complete at 1400°C.

Figure 5 shows a plot of MOR vs density for samples containing 3% clay as additives. These samples were fired at 1440°C. Here, it may be observed that the MOR values vary linearly with fired density as expected.

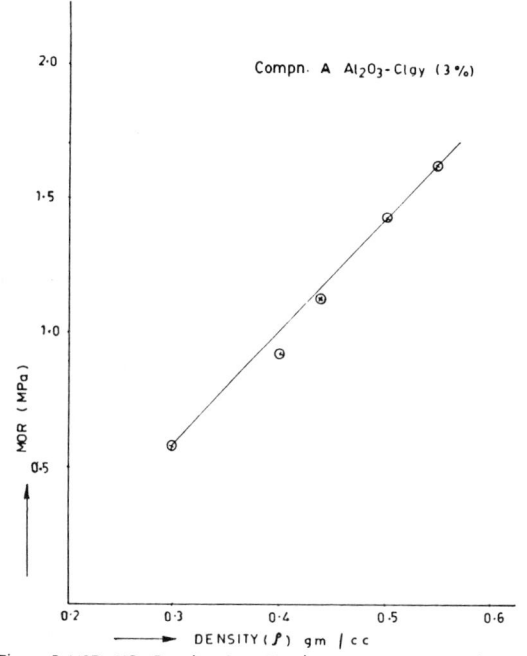

Figure 5 MOR VS Density for Alumina foam   Sample fired at 1440°C

Thermal Expansion Studies

Coefficient of thermal expansion (α) of the above samples between 25 and 1000°C are shown in Table 1 below:

TABLE 1. Thermal expansion measurment of ceramic foams of different compositions (fired at 1440°C)

Composition	(α) Coefficient of Thermal expansion /°C
A	8.3×10^{-6}
B	8.37×10^{-6}
C	8.5×10^{-6}
D	8.41×10^{-6}
E	8.63×10^{-6}

It may be seen that values vary between 8.3 and 8.6, which is nearly the same as that for α-alumina. In other words, the effect of additives on the thermal expansion of the alumina foams is not very significant.

CONCLUSIONS

From the data and explanations described in the foregoing sections, the following general conclusions can be drawn:
a. Thermal decomposition of the alumina impregnated PU foams has been found to be topotactic retaining the morphology of the substrate.
b. The alumina foams produced by the present techniques have been found to have appreciable flexural strengh (about 1.5 to 4.0 MPa)
c. Irrespective of the history of the sample the coefficient of thermal expansion of the alumina foams investigated here coincide with the α value for pure alumina.

The above foams have been tested for filtration of molten aluminium in a major primary smelting industry. The filtration was quite satisfactory. More work is in progress to develop similar materials for filtration of both ferrous and non-ferrous metals.

REFERENCES

(1). W.H.Sutton and D.Apelian, "Filtration of high temperature vacuum-melted alloys", TMS-AIME 112th Ann., Mtg., Atlanta, G.A., March 6-10, 1983.
(2). 'RETICEL' Catalogue of Hi-Tech Ceramics, Inc., Alfred, New York.
(3). E.J.A. Pope and A.Almazan, "Caged Alumina particles in reticular foam", J.Am. Ceram.Soc., 74 (10, 2715-17, 1991
(4). F.F.Lange and K.T.Miller, "Open-Cell, Low Density Ceramics Fabricated from Reticulated Polymer Substrates", Adv. Ceram. Mat., 2(4), 827-831, 1987.
(5). A.Slosanczyk, "Highly Porous Hydroxyapatite Material", Powder Metallurgy Internaional, 21(4), 24-25, 1989.

MICROSTRUCTURE EVOLUTION DURING FABRICATION OF A POROUS CERAMIC FILTER

Kan-Sen Chou, H. C. Liu and K. L. Chang
Department of Chemical Engineering, National Tsing Hua University, Hsinchu, Taiwan 30043 ROC

A simple fabrication process using porous sponges as a pore former to make large pore ceramic filters is reported. The evolution of its microstructure was of special emphasis. Our results showed that the porosity of the final product depended mainly on the particle loading in the sponge during soaking in the slurry. Both the structures of the sponge (its porosity) and the slurry concentrations affect our loading data. The characteristics, i.e. porosity and pore sizes, of the product were determined after it was loaded. The most significant shrinkage of the product occurred above 800°C during the firing step. Permeability of these filters with respect to water and nitrogen was also reported.

INTRODUCTION

Porous ceramic filters have found many applications in different industries. Its continuous expansion in recent years is due to its inherent chemical, physical, mechanical and microbiological stability and improved fabrication technique. Depending on their pore sizes, these filters can be used in conventional filtration, microfiltration, ultrafiltration and reverse osmosis. Filters with large pores (over 100 micron) will be the subject of this report. Uses for these types of filters include, for example, the removal of particles from melting metals, or hot process gases; the recovery of biological products; the structural support for microfiltration and ultrafiltration membranes, etc. [1-5].

In spite of its wide application, however, there are very few articles in the literature discussing the fabrication of these large pore filters, except patents [6]. Pore formers, either organic or inorganic additives, are extensively used in the manufacturing process explained in these patents. Examples include polymer particles, wax, calcium carbonate, calcium chloride etc., which either burn off or decompose to produce gaseous products during the firing process leaving behind pores of various sizes.

The objective of this study was to investigate the mechanism of a fabrication process for making ceramic filters with large pores. We hope to understand the factors that control the porosity and pore sizes in the products. The microstructural evolution during pore formation will be specifically followed. It is hoped that such an understanding of the relation between processing, microstructure, and properties will be beneficial to the production of these filters.

EXPERIMENTAL

Pottery clay (Yamada, No. 640) was used as our raw powder. According to x-ray diffraction (XRD) studies, its main constituents were kaolinite and silica (as quartz), with also a small amount of anorthite (sodium low). Chemical analysis showed 67.1 wt% of silica, 19.7 wt% alumina, 1.9 wt% sodium oxide, 4.8 wt% potassium oxide and 4.4 wt% loss on ignition. We used sodium carbonate at 41 mg/g clay as defloculant to help to disperse this clay in water. Slurries of different concentrations (in terms of weight percentage) were prepared and used in this work. The viscosity of the slurry was measured in an Ostwald type viscosimeter.

Two kinds of polyurethane (PU) sponges were used in this work. They differed in porosity (A: 99.5% versus B: 97.8%) and structure, which both influence the ability to absorb particles from the slurry. The sponges were first cut into appropriate sizes and then were squeezed several times and immersed in the slurry for about 2 hours. After removing from the slurry, the samples were first air-dried at room temperature for one day before being oven-dried at 110°C. These pre-forms were then fired at 1200°C for 12 hours. During the heat-up period, the samples were held at 400 and 800°C for 2-3 hours to allow for the burn-off of sponge and the dehydroxylation of kaolinite.

The porosities of the products were analyzed by the Archimedian method and the pore sizes were determined from enlarged photographs. The usable area of a sample refers to its area of uniform size and shape, which was an average from five to seven pieces of sample.

RESULTS AND DISCUSSION

Sponge Loading and Slurry Concentration

The relationship between sponge loading of clay particles and slurry concentration (wt%) for both types of sponge are shown in Fig.1. The quantity of particles that can enter a sponge depends on: (1) the suction force from a squeezed sponge; (2) the concentration of the slurry; and (3) the ability of the sponge to retain these particles (against gravity force). The first and the third factor are related to the structure of sponge. A more open and porous sponge (i.e. type A in our case) exerts less suction force towards slurry particles. The sponge loading was therefore smaller. However, as the slurry concentration increased to 60wt%, the slurry became very viscous. The particles might then have difficulty entering the dense sponge in the first place. Sponge loading decreased accordingly as shown in Fig. 1.

Another problem related to the forming process is the shape uniformity of the final product. Our experiments showed that the viscosity of the slurry played an important role here. The more viscous the slurry, the

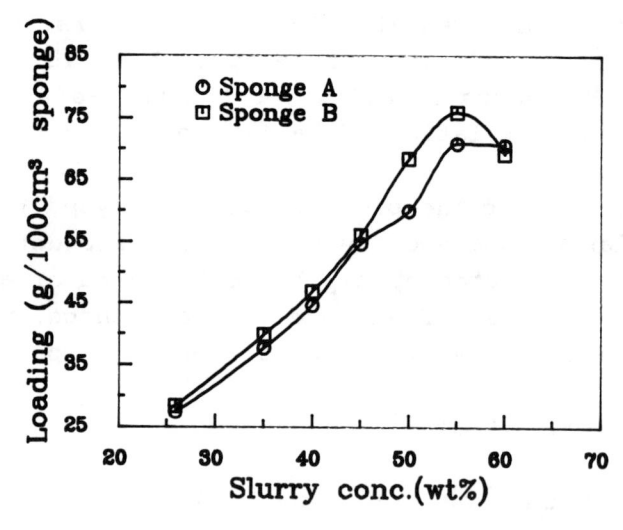

Figure 1 Relationship between particle loading and slurry concentration

more serious was this problem. One can imagine when a squeezed sponge tries to recover to its original shape, its movement will be retarded by the viscosity of slurry. Shown in Table 1 are some data from a series of runs. The correlation between usable area of a product and the viscosity of the slurry is very obvious. When the slurry concentration was more than 45 wt%, its viscosity showed an abrupt increase. This would then limit its use in real practices.

Table 1. Slurry Concentration, Viscosity and Usable Area

slurry conc.	viscosity	usable area
26 wt%	1.7 cp	~ 80%
35	2.0	78
40	4.8	76
45	15.5	70
50	1290.0	60

Product Property and Sponge Loading

Shown in Fig. 2 are representative pictures of our products and its cross section. The pore structure seems homogeneous. Figures 3 and 4 show the effects of sponge loading on the porosity and average pore size of fired products. The data fell on the same curves and no clear difference was noted between these two sponges. It therefore suggested that the sponge structure did not have any effect on the fired product once it was loaded. In other words, the shrinkage of objects and forming of microstructures was not influenced by sponges. Higher loading leads to smaller porosity and smaller pore sizes.

Figure 2 Representative pictures of product and its cross section

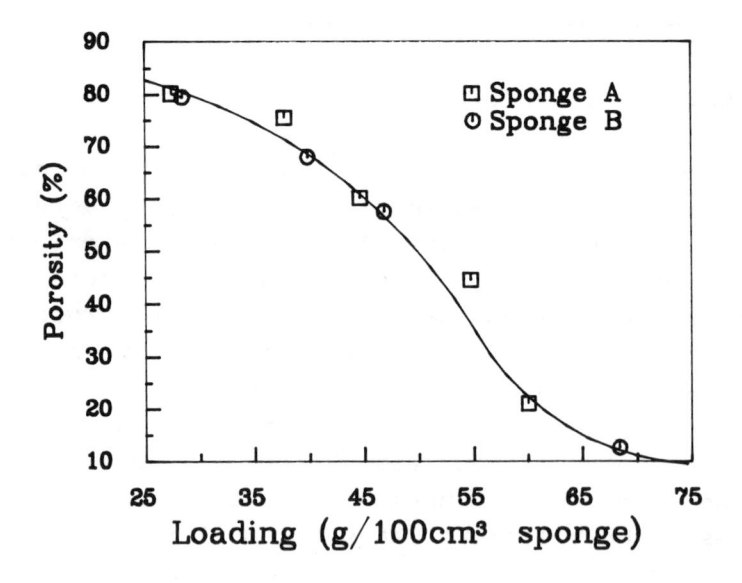

Figure 3 Product porosity versus sponge loading

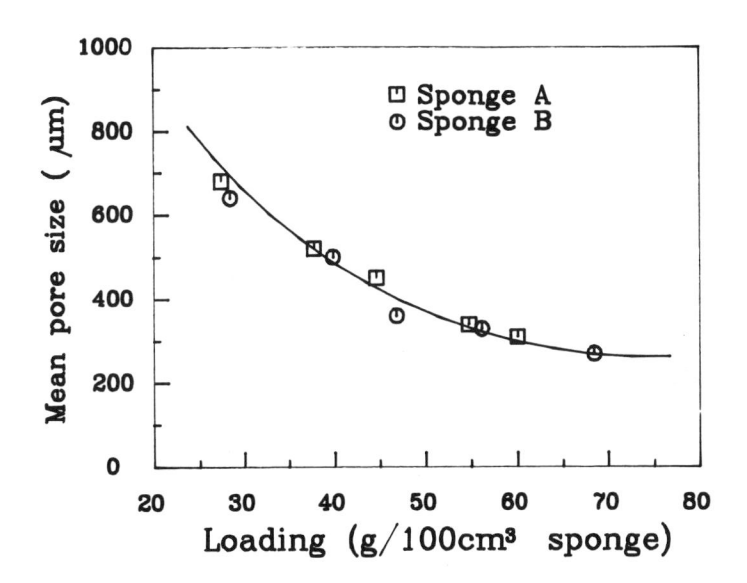

Figure 4 Dependence of mean pore size on sponge loading

Changes in pore size and porosity of samples during the firing process are shown in Fig. 5. The pore structure remained basically unchanged below 800°C. Significant shrinkage occurred above 800°C. The slight increase in porosity below 800°C was due to density changes in the particles. Densities recorded for the original particle and those calcined at 800°C and 1200°C were 2.61, 2.81 and 2.37 g/cm^3 respectively. The sponge burned off around 400°C according to DTA study. Since the shrinkage of the sample did not start at this temperature, the sponge did not show any influence on the final structure of products.

Finally, the products were tested for permeability with respect to water and nitrogen. The permeability was calculated by the following equation:

$$B = \mu \ (Q/A) \ (- \Delta P/L) \qquad (1)$$

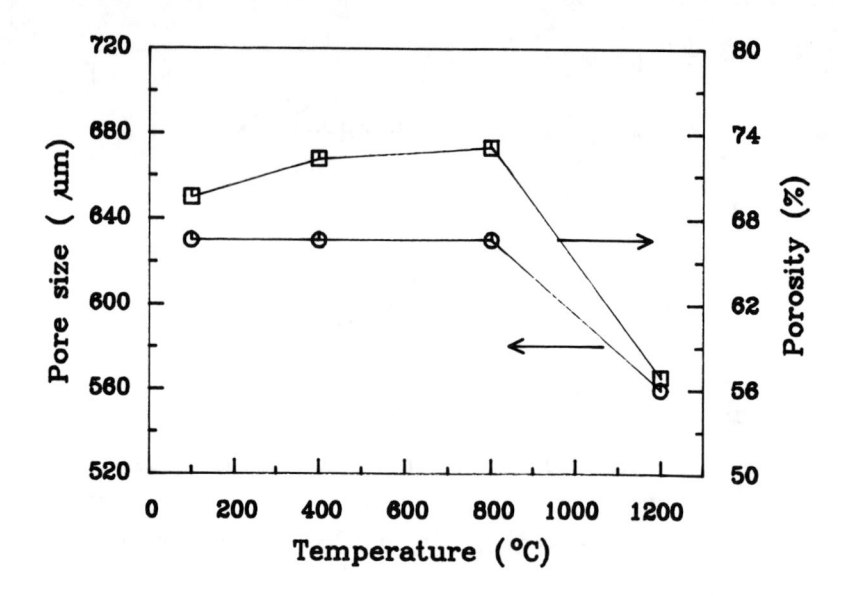

Figure 5 Changes in pore size and porosity during firing

where μ is the fluid viscosity, Q is the volumetric flow rate, A is the cross area for flow, Δ P is the pressure drop across the porous filter and L is the thickness of the filter. It therefore has the unit of cm^2. The agreement between these two fluids was reasonable.

Table 2. Permeability with Respect to Water and Nitrogen for Several Filters

filter porosity	for water	for nitrogen
0.45	0.71	0.83
0.58	1.3	1.4
0.68	6.0	4.7

CONCLUSION

Based on the above results and discussion, we can thus conclude:

(1) The sponge, being the host to particles, also provided the primary suction force during the soaking period. A more open and porous sponge would in general generate smaller suction force than a dense one. It therefore resulted in a smaller solid loading.
(2) Once the particle loading in the sponge was finished, the properties, such as porosity and average pore size, of fired products were fixed. In other words, the structure of the sponge did not have any influence after the soaking step.
(3) It is not appropriate to use a slurry with concentrations of ceramics higher than 45% due to a sharp increase in viscosity. A viscous slurry is detrimental to the formation of uniform products.

ACKNOWLEDGEMENT

The authors wish to thank National Science Council of ROC for financial support of this project. (Grant No. NSC 81-0402-E007-600)

REFERENCES

1. K. Yamada, T. Watanabe, K. Fukuda, T. Kawaragi and T. Tashiro, Removal of Non-metallic Inclusion by Ceramic Filter, Trans. Iron & Steel Inst. Japan, 27, 873-877, 1987.
2. J. Charpin, P. Bergez, F. Valin, H. Barnier, A. Maurel and Jm. Martinet, Inorganic Membranes: Preparation, Characterization, Specific Applications, Industrial Ceramics, 8[1], 23-27, 1988.

3. L. Cot, Ch. Guizard and A. Larbot, Novel Ceramic Material for Liquid Separation Process: Present and Perspective Applications in Microfiltration and Ultrafiltration, Industrial Ceramics, 8[3], 143-148, 1988.
4. J. Charpin, A.J. Burggraff and L. Cot, A Survey of Ceramic Membranes for Separation in Liquid and Gaseous Media, Industrial Ceramics, 11[2], 84-89, 1991.
5. J.F. Zievers, P. Eggerstedt and E.C. Zievers, Porous Ceramics for Gas Filtration, Am. Cer. Soc. Bull., 70[1], 108-111, 1991.
6. ROC Patents No. 117789, 1989; No. 128012, 1990; No. 155696, No. 164212, 1991. All on the manufacture of porous ceramic objects.

PRODUCTION OF POROUS ALUMINA BY HOT ISOSTATIC PRESSING

A. Kuzjukēvičs and K. Ishizaki
Nagaoka University of Technology, Nagaoka, Niigata 940-21, Japan

Production of porous alumina ceramics by capsule-free hot isostatic pressing of bimodally distributed alumina powders with a large size difference (>10) was investigated. The influence of the particle size fraction of bimodally distributed alumina powders on open and closed porosity and Vickers hardness was studied.

Hot isostatic pressing (HIP) at 1600°C provided sintering of mainly fine-grained powder. A small loss in open porosity (about 4%) is observed after increasing the fine powder weight fraction up to 0.4. At this weight fraction, the Vickers hardness increases also by 4X.

INTRODUCTION

For many years open pore materials have been used for many industrial purposes [1,2]. Among the different types of ceramics, alumina is one of the most widely used porous ceramics, especially as various filters [3].

There are different methods of manufacturing porous ceramics, e.g. by using burnout additives [4, 5], sublimable compounds [5], etc. Recently a method of obtaining porous materials was reported [6], where as-pressed powder samples were capsule-free sintered by hot isostatic pressing (HIP). As a result, improvement of mechanical properties with adequate porosity was observed [7]. The controlling factors in the production of porous materials using HIP are particle size of starting powder, pressure of cold isostatic pressing, and pressure and temperature of HIP [8].

In this study, HIP production of porous alumina ceramics and control of

---

To the extent authorized under the laws of the United States of America, all copyright interests in this publication are the property of The American Ceramic Society. Any duplication, reproduction, or republication of this publication or any part thereof, without the express written consent of The American Ceramic Society or fee paid to the Copyright Clearance Center, is prohibited.

porosity are investigated by sintering alumina powder with a bimodal particle size distribution having distinct coarse and fine particle fractions and a large size difference (>10). It has been already shown that fine-grained alumina powder can be used to enhance the densification of a low-reactivity, coarse powder [9]. So far as to the best knowledge of the authors, there is only one patent [10] on manufacturing porous alumina ceramics using the fine- and coarse-grained powder mixture. However, this method also requires a pore forming agent.

EXPERIMENTAL PROCEDURE

To make bimodally distributed alumina powders, a fine-grained powder (Sumitomo Chemical Co., Ltd., average particle size 0.36 μm) and a coarse-grained powder (Taiheiyo Landam Co., Ltd., average particle size 5 μm) with three different weight ratios (0.15 fine + 0.85 coarse, 0.25 fine + 0.75 coarse, 0.40 fine + 0.60 coarse) were mixed for 24 hours in ethanol by a ball milling machine with alumina balls in a plastic pot. Powder slurries were dried in a vacuum furnace at 80°C for 5 hours and then crushed. Two grams of the resultant powders were uniaxially pressed at a pressure of 10 MPa to make pellets of 20 mm in diameter and about 3 mm high. These pellets were then cold isostatically pressed at 20 MPa.

Samples were hot isostatically pressed (Professor O_2-HIP, Kobe Steel, Ltd.) for 1 hour under 50 and 150 MPa in argon with 1% oxygen at 1600 to 1700°C. Porosity of sintered samples was determined by the Archimedes method, details are given elsewhere [6]. Hardness was measured by Vickers indentation (Akashi model AVK-A) with 9.8 N indentation load and 10 s holding time. Average value of Vickers hardness was calculated from measurements of 10 points.

RESULTS

Changes in the amount of open and closed porosity with weight fraction of the fine-grained powder at different HIP pressures and temperatures are shown in Fig. 1. Open porosity of samples sintered at 1600°C and normal pressure (0.1 MPa in Fig. 1) exponentially decreases from 21% for the coarse-grained powder to 1% for the fine-grained powder. At the same time closed porosity increases from 0.2 to about 6% for the mixture with 0.4 of fine powder fraction and then slightly decreases to 4% for the fine-grained powder.

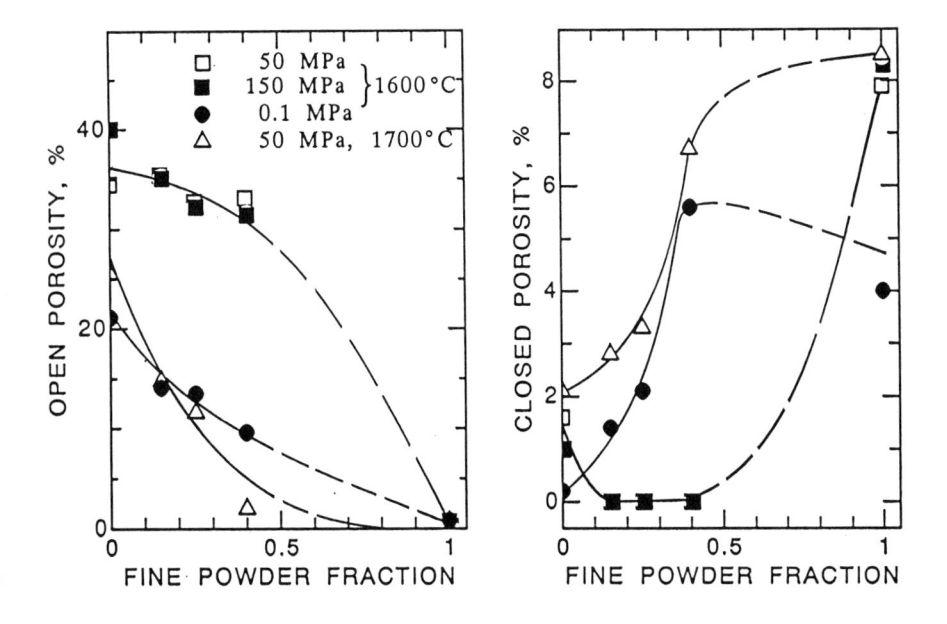

Figure 1 Relationships between open and closed porosity and fine powder weight fraction of bimodally distributed alumina powder.

Increasing sintering pressure to 50 MPa, at which the transition from normal to HIP sintering takes place, has remarkable effects on open and closed porosity. For mixtures with fine powder fraction ≤0.4, the open porosity increases to about 33-35% with a closed pore level ≤0.5%.

Practically no changes were observed for fine-grained powder, except the amount of closed pores increased to about 8%. A further increase of HIP pressure to 150 MPa does not affect the porosity. HIP sintering at 1700°C led to a decrease in the amount of open pores and an increase of closed porosity. The standard deviation of porosity was ±1.5% or less based on three samples.

Figure 2 shows changes of the Vickers hardness with the amount of the fine-grained powder for samples sintered at 1600°C. A steep increase of Vickers hardness is observed for samples with a fine powder fraction of 0.15. The value of Vickers hardness of coarse-grained sample sintered at

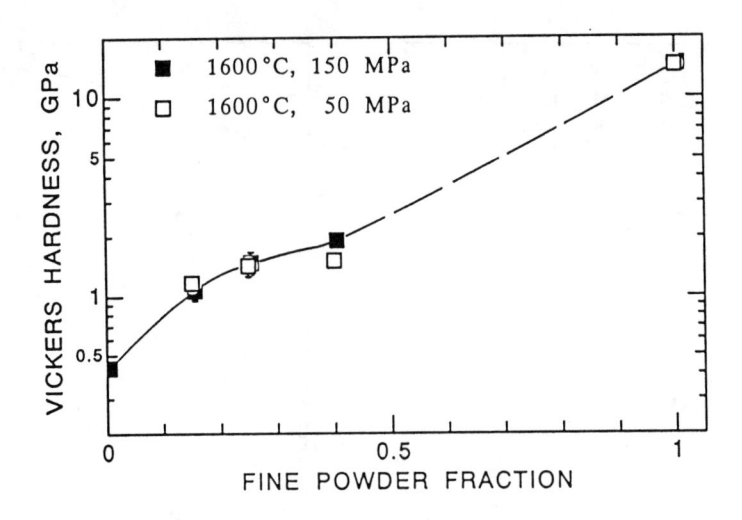

Figure 2 Relationship between Vickers hardness and fine powder weight fraction.

1600°C under 150 MPa is 0.43 GPa and increases to 1.9 GPa for a sample with a fine-grained powder fraction of 0.4, more than a factor of four times.

DISCUSSIONS

As it was shown, fine-grained alumina powder (particle size <0.5 μm) can be used to enhance the densification of a low-reactivity, coarse powder [9]. When heated, fine particles begin to sinter first, and the compact densifies until coarse particles come into contact. Further densification is controlled or inhibited by coarse particles. Fine particles fill up interstices formed by the coarse-grained particles during uniaxial pressing. At normal pressure, open porosity of samples with fine powder fraction ≤0.4 almost linearly decreases with the amount of the fine powder. With an increase in the sintering pressure this dependence becomes weaker. As shown in Fig. 3 the ratio of porosity of samples sintered at high and normal pressures increases when the fine powder fraction approaches 0.4. It is suggested that during HIPing subjected to sintering powder's part is transferred from interstices to contact areas between coarse grains. As a result necks between coarse particles become bigger and the strength of ceramic increases while the loss in open porosity is small.

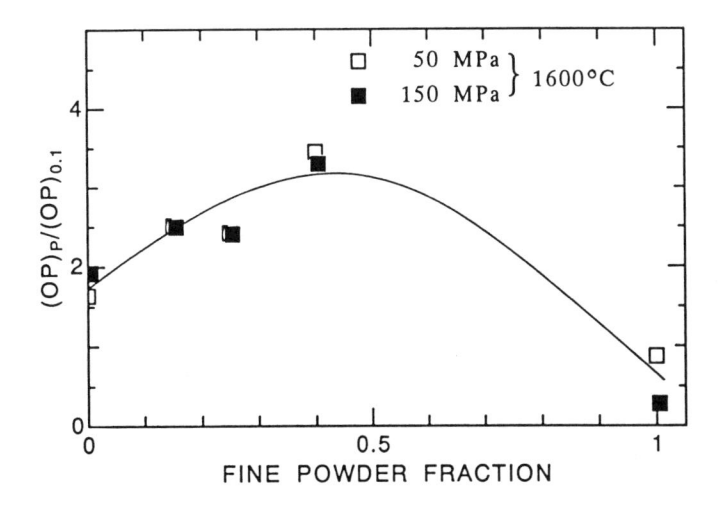

Figure 3 Ratio of open porosities versus fine powder fraction. $(OP)_P$ and $(OP)_{0.1}$ mean open porosity of samples sintered at pressure P and 0.1 MPA, respectively.

CONCLUSIONS

Porous alumina ceramic is obtained by capsule-free hot isostatic pressing of bimodally distributed alumina powders with large size difference (>10).

HIP sintering of bimodally distributed alumina with fine-grained powder fraction ≤0.4 at 1600°C produces a high amount of open porosity (31-35%) with a minimum of closed porosity (<0.5%). Since sintering occurs mainly with fine particles, increasing of bridges between coarse particles and consequently increasing of strength of porous ceramics are observed.

The proposed method is useful for production of porous alumina ceramics with controlled porosity.

REFERENCES

1. N. Kato, "Ceramics Filter for Microfiltration and Ultrafiltration", Seramikkusu, 23, 726(1988).
2. S. Okada, "Development and Applications of Porous Ceramic

Materials", edited by S. Hattori, publ. by CMC, Tokyo, 139 (1984).
3. K. Keizer and A.J. Burggraaf, "Porous Ceramic Materials in Membrane Application", Science of Ceramics, 14, 83(1988).
4. U. Hammon and M. Kotter, "Herstellung von Formkörpen mit definierten porenstructur", Chem.-Ing.-Tech., 56, 455-463(1984).
5. D.L.Trim and A.Stanislaus, "The Control of Porte Size in Alumina Catalyst Support", Appl. Catalysis, 21, 215-238(1986).
6. K. Ishizaki, A. Takata and S. Okada, "Mechanically Enchanced Open Porous Materials by HIP Process", J. Cer. Soc. Jpn., Int. Edition, 98, 15(1989).
7. A. Takata, K. Ishizaki and S. Okada, "Improvement if Porous Material Mechanical Property by Hot Isostatic Pressing", Mat. Res. Soc. Symp. Pro., 208, 135(1991).
8. M. Nanko, K. Ishizaki and T. Shioura, "Porous TiO_2 by HIP process", Proc. the 66th Annual Meeting Jpn. Ceram.Soc., Tokyo, May 22-24, p.585(1991).
9. J.P. Smith and G.L. Messing, "Sintering of Bimodally Distributed Alumina Powder", J. Amer. Cer. Soc., 67, 238(1984).
10. K. Futaki and Y. Tomita, "Manufacture of Porous Alumina Bodies", Jpn. Kokai Tokkyo Koho, JP 02,160,679 [90,160,679], Japan, Dec. 14, 1988.

SINTERING OF POROUS MATERIALS BY A CAPSULE FREE HIP PROCESS

Makoto Nanko, Kozo Ishizaki and Atsushi Takata[*]
Department of Materials Science and Engineering
School of Mechanical Engineering
Nagaoka University of Technology
Nagaoka, Niigata 940-21, Japan.
[*]Present Address;
Japan Grain Institute Co., Ltd.,
Kokubunjicho, Kagawa 769-01, Japan.

Porous materials produced by a capsule-free hot isostatic pressing (HIP) process pertain excellent properties such as high open porosity, high mechanical strength, high Young's modulus and narrow pore size distribution. In the capsule free HIP process to produce porous materials, high pressure gas delays shrinkage of open pores during sintering. By this HIP process, it is possible to sinter at higher temperatures than by conventional sintering, while keeping the same porosity for both cases. As a result, the HIPed porous materials have a well-grown bridging between particles, even though they have high open porosity. In this work, the sintering mechanism of the HIP process was evaluated experimentally by studying the sintering process under high gas pressure.

INTRODUCTION

Porous materials, especially those with open pores, are important for industrial applications, such as filters[1, 2], carriers for bioreactor or catalyst[3, 4] and gas bearings[5]. High open porosity and high mechanical strength are required for these applications. It is difficult, however, to obtain high strength porous materials with high open porosity, because these properties are contradictory.

Although there are different methods of manufacturing porous materials, the properties of porous materials produced by these methods are not satisfactory. For instance, in

---

To the extent authorized under the laws of the United States of America, all copyright interests in this publication are the property of The American Ceramic Society. Any duplication, reproduction, or republication of this publication or any part thereof, without the express written consent of The American Ceramic Society or fee paid to the Copyright Clearance Center, is prohibited.

sintering at lower temperature than a normal sintering, bridging parts between powder particles become weak[6]. Recently, a new method to obtain porous materials by capsule-free hot isostatic pressing (HIP) was reported[7, 8]. As a result, enhancement of mechanical properties and narrow pore size distribution with sufficient open porosity were achieved[7, 9].

In this work, density, porosity and internal structure were investigated as a function of sintering time. By the investigation, HIP sintering process formation of porous materials was evaluated.

EXPERIMENTAL PROCEDURE

Non-agglomerated TiO_2 powder (Nippon Soda Co. Ltd.) was used in the present work. The powder particles are spherical with size 1–10 μm (Fig. 1). The powder was uniaxially pressed at a pressure of 5 MPa to make pellets of 20 mm in diameter and about 3 mm thick, and then cold isostatically pressed (CIPed) at 60 MPa for 1 min.

The CIPed samples were sintered in flowing O_2 at normal pressure (NS) or HIP sintered (O_2 Professor HIP, System 3: Kobe Steel Co. Ltd.) under 100 MPa of total gas pressure in various conditions (Table 1). Oxygen partial pressure was 0.1 MPa in both sintering cases. Typical HIPing schedule is shown in Fig. 2. In all HIPing, gas pressure was controlled to reach to the holding pressure of 100 MPa before increasing the temperature to 500 °C. Otherwise the heating rate for all cases was 400 K/h.

Density and porosity were measured by water displacement, whose details were described elsewhere[8]. The internal structure of sintered samples was evaluated by observing their fractured surfaces.

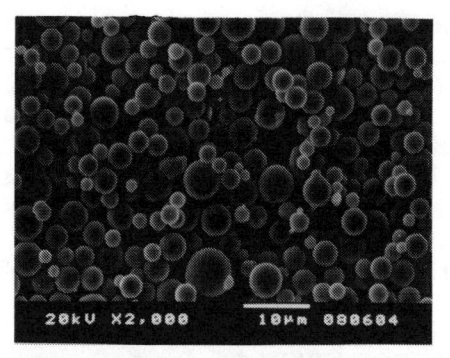

Figure 1 SEM photograph of the raw powder.

Table 1. Experimental Condition

	O2 HIP	NS
Sintering temp. / °C	1 200, 1 300	1 200, 1 300
Sintering time / min	60, 120, 300, 600	60, 120, 300, 600
Atmosphere	Ar + 0.1 vol% O2, 100 MPa (PO2: 0.1 MPa)	O2 flow (PO2: 0.1 MPa)
Quenching method	Adiabatic expansion	Gas flow

NS: normal sintering

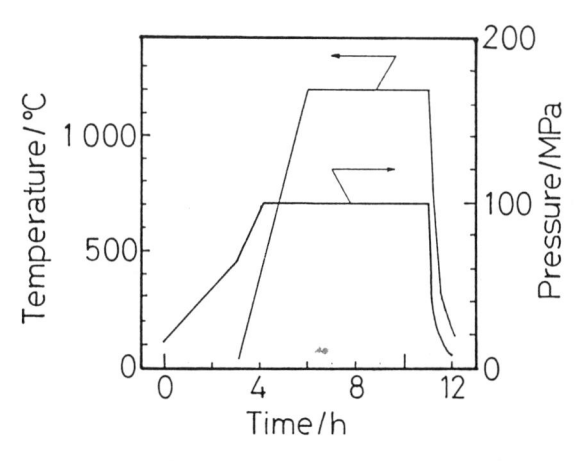

Figure 2 Typical HIP schedule.

RESULTS

Figure 3 shows the relationships between fraction of bulk density to the theoretical density, ρ, as well as open porosity, P_O, and sintering time, t. With increasing sintering time, the value of ρ increases, and open porosity decreases. The value of ρ of the HIPed samples is lower while the open porosity is higher compared with the normal sintered samples at the same temperature. This is in agreement with the previous reports[7, 9–12]. The value of ρ of CIPed bodies is about 0.5.

Porous Materials

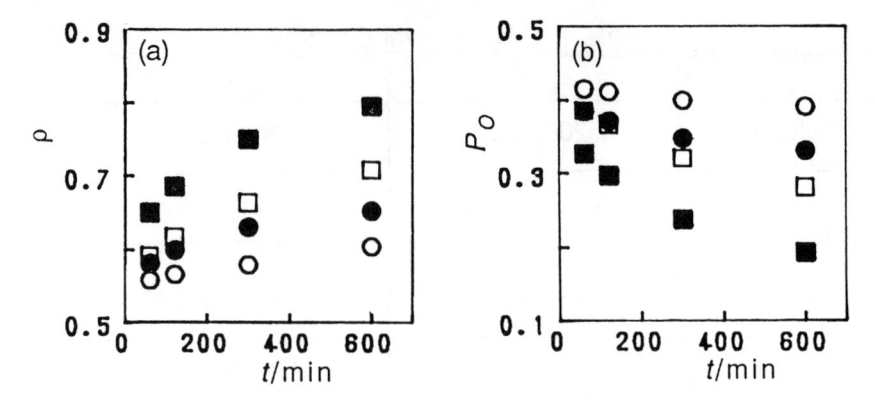

Figure 3 The relationships between (a) fraction of bulk density to the theoretical density, ρ, as well as (b) open porosity, P_O, and sintering time, t. The circle and square marks mean 1200°C and 1300°C of sintering temperature, respectively. The open and closed marks indicate HIPing and normal sintering, respectively.

Figures 4 and 5 show the fractured surfaces of the sintered samples with ρ of 0.60 and 0.65, respectively. The shape of particles changes drastically. Bridging between particles does not change significantly between those samples sintered by normal or HIP sintering. The shape of large particles (6-9 μm diameter in the raw powder) changed significantly.

DISCUSSION

Diffusion is one of the main driving forces of sintering processes, and is based mainly on high surface energy of the particles' connecting bridges[13]. The radius of bridge surfaces is negative and considerably smaller than the particle radius (assuming spherical particles). This radius difference gives difference in surface energy, i.e., high in bridge and low in particle surface[13]. Only surface and bulk diffusion mechanisms are considered in the present work. If surface diffusion is the only mass transport mechanism during the initial stage of sintering, no change in distance between particles occurs (i.e., no densification)[14]. In the intermediate condition of high surface diffusion rate and low in bulk, the densification rate, $d\rho/dt$, is small. If the configuration

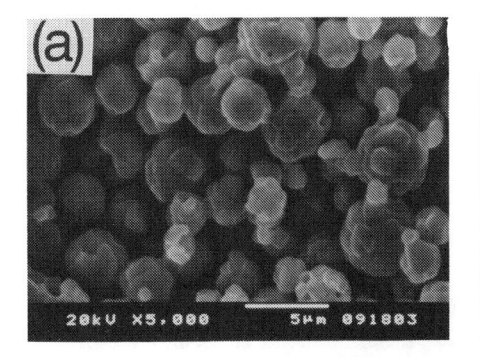

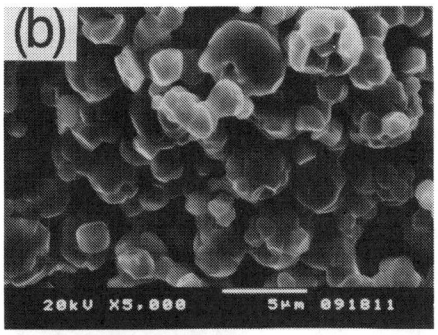

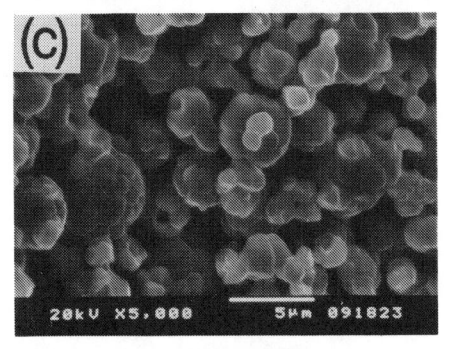

Figure 4 SEM photograph of the fractured surface of the sintered porous TiO$_2$ with ρ approximately equal to 0.6.
(a) Normally sintered at 1 200 °C for 60 min.
(b) HIPed at 1 300 °C for 60 min.
(c) HIPed at 1 200 °C for 600 min.

of the powder is the same, the value of dρ/dt depends on the ratio of the surface diffusion coefficient (D_s) to the bulk's (D_b), D_s/D_b. The dρ/dt decreases with increasing D_s/D_b.

In the second stage of sintering, surface diffusion does not produce shrinkage and tends to change the pore shape to an equilibrium shape. In this stage, the bridges grow well and the shape of the pores among particles becomes circular. But the influence of large D_s/D_b on the densification rate is similar to the one in the initial stage.

With increasing temperature, both D_s and D_b increase. The surface diffusion mechanism is predominant for sintering at lower temperatures, because the

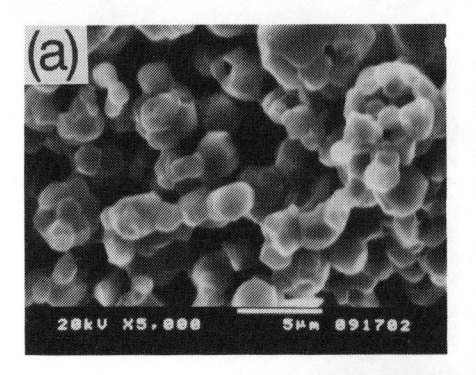

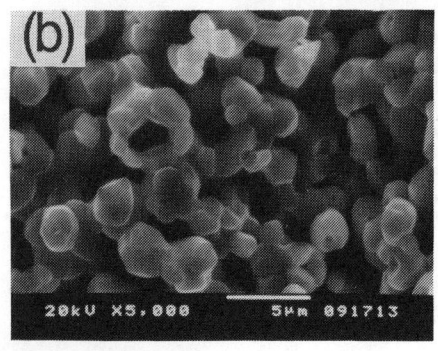

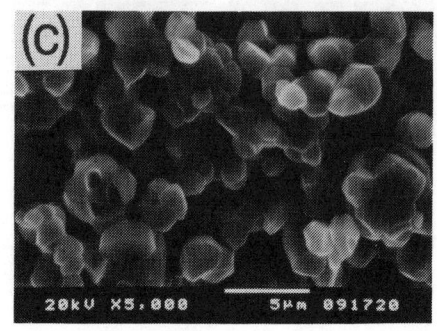

Figure 5 SEM photographs of the fractured surface of the sintered porous TiO_2 with ρ approximately equal to 0.65.
(a) Normally sintered at 1 300 °C for 60 min.
(b) HIPed at 1 300 °C for 300 min.
(c) Normally sintered at 1 200 °C for 600 min.

temperature dependence of D_b increment rate is larger than that of D_s [15, 16] and D_s at sintering temperatures is generally remarkably larger than D_b. This means that D_s/D_b is high at low temperatures and decreases with increasing temperature. In sintering at low temperatures, $d\rho/dt$ is considerably low because of not only low diffusion rate but also high D_s/D_b.

Figure 6 shows the relationships between ρ as well as P_O and t plotted on logarithmic coordinates. The $d\rho/dt$ at 1 200 °C was smaller than that at 1 300 °C. This is in good agreement with the aforementioned temperature dependance of $d\rho/dt$ which is low under high D_s/D_b at low temperature. The $d\rho/dt$ of HIPing is smaller than that of normal sintering. This behavior may be explained by increasing D_s/D_b. Assuming that D_b changes little by applying 100 MPa due to small molar volume change of solids by pressure, the increase of D_s/D_b is due to the increase of D_s. This assumption is

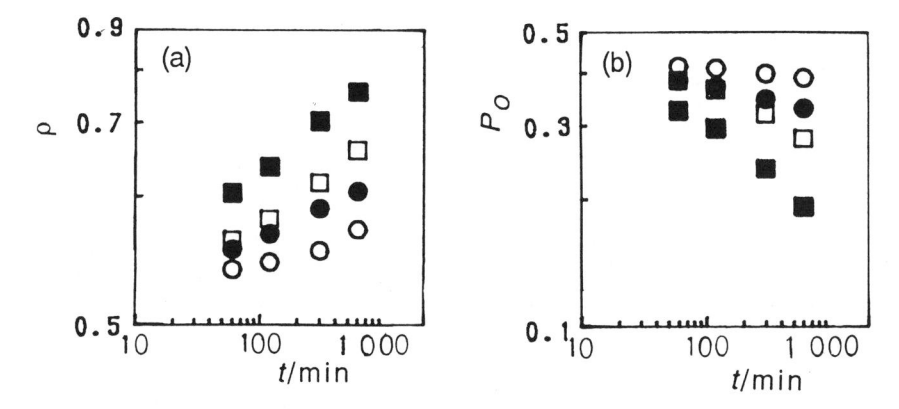

Figure 6 The relationships between (a)relative density, ρ, as well as (b)open porosity, P_O, and sintering time, t, in logarithmic scale. The circle and square marks mean sintering temperatures of 1 200 °C and 1 300 °C, respectively. The open and closed marks indicate HIPing and normal sintering, respectively.

also based on the fact that diffusion coefficients of metals decrease little by applying 100 MPa[17, 18].
The difference of particle shape between HIPed and normal sintered samples at same porosity as shown in Figs. 4 and 5 may be due to the increase of D_s by HIPing.

From the following observations, the increased surface diffusion can be also proved. In HIPed porous copper as shown in Fig. 7, formation of surface steps was observed clearly as HIPing pressure increased. The surface of porous copper approached to its equilibrium state faster by HIPing than by normal sintering. In general, the surface tends to become smoother in order to decrease surface energy for glasses or liquids. But metallic (crystalline) surfaces consisting of crystal planes with the crystallographic low energy are more stable than smooth curved surface.

On the other hand, amorphous material surfaces are smooth at their equilibrium state. In HIPed vitrified grinding wheels, the surface became smoother compared with normal sintering[7, 9]. The materials consist of alumina abrasive and additives, Y_2O_3 and AlN[7], or silicon carbide abrasive and frit, which is glassy binder consisting of

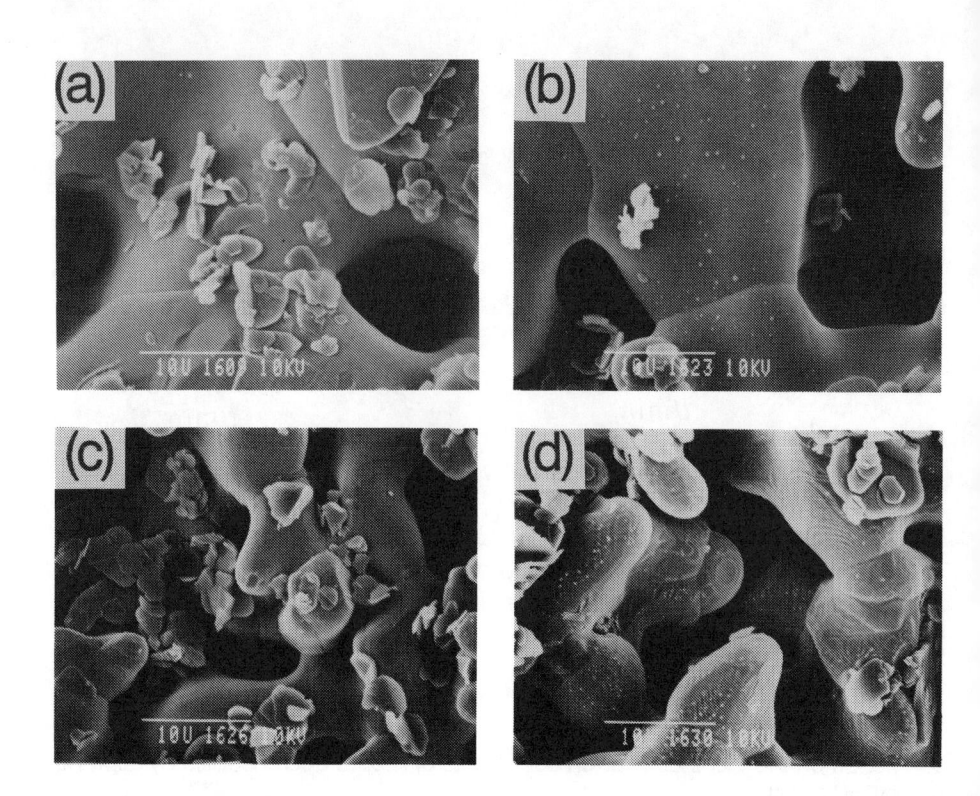

Figure 7 SEM photograph of the porous copper surfaces sintered at 850 °C for 1 h under various pressures; (a) under 0.1 MPa, (b) under 9 MPa, (c) under 20 MPa and (d) under 200 MPa.

silica, alumina and boron oxide as the main components[9]. The surfaces of these grinding wheels were covered by the glassified additives, and became smoother by HIPing than normal sintering. Thus, the surface equilibrium state is attained faster by HIPing than by normal sintering. Those phenomena also lead to conclude that HIPing enhances surface diffusion.

There have been no reports about the influence of high pressure gas on surface diffusivity. The present work proves the influence of surface diffusion by high pressure gas.

Another possibility for the low $d\rho/dt$ on HIPing is that discharging a high pressure gas from pores requires long time. The gas in pores must be discharged by shrinking the pores during sintering. Discharging gas in the pores under high pressure should require a longer time than that under ordinary pressure, because the higher pressure gas has a higher density, that is; a larger quantity of gas atoms or molecules than that of normal pressure in a given volume. This might be considered as an additional reason for higher porosity compared with normal sintering.

CONCLUSIONS

High pressure gas delays shrinkage of a compact or pre-sintered body, and is considered as a favorable factor to produce porous materials due to the following two reasons:
(1) During HIP sintering, the surface diffusion coefficient is higher than in normal sintering, and as a result the shrinkage is delayed.
(2) The shrinkage is delayed also due to the longer time required to discharge high pressure gas from pores.

REFERENCES

[1] K. K. Chan and A. M. Brownstein, "Ceramic Membranes – Growth Prospects and Opportunities", Ceram. Bull., **70**, 703–707 (1991).
[2] K. Keixer and A. J. Burggraaf, "Porous Ceramic Materials in Membrane Applications", Science of Ceramics, **14**, 83–93 (1988).
[3] Y. Onoda, T. Iwasaki, H. Hayashi and K. Torii, "Development of a New Composite Material Based on λ-MnO$_2$ and Porous Ceramics", J. Ceram. Soc. Jpn., **100**, 767–774 (1992).
[4] K. Iwasaki and N. Ueno, "Porous Alumina Ceramics for Immobilization of Soy Sauce Yeast Cells", J. Ceram. Soc. Jpn, **98**, 1186–1190 (1992).
[5] I. Kawashima, S. Togo, S. Sato and N. Tamada, "Study on Characteristics of Porous Ceramic Gas Bearings", J. Jpn. Soc. Prec. Eng., **56**, 153–1858 (1990).
[6] K. Koumoto, H. Shimizu, W. S. Seo and , C. H. Pai and H. Yanagida, "Thermal Shock Resistance of Porous SiC Ceramics", Br. Ceram. Trans. J., **90**, 32–33 (1991).
[7] K. Ishizaki, A. Takata and S. Okada, "Mechanically Enhanced Open Porous Materials by a HIP Process", J. Ceram. Soc. Jpn., **98**, 533–540 (1990).
[8] K. Ishizaki, T. Fujikawa, S. Okada and A. Takata, Patent Bending (Japan, U.S. and Germany).
[9] A. Takata, K. Ishizaki and S. Okada, "Improvement of a Porous Material Mechanical Property by Hot Isostatic Pressing", Mater. Res. Soc. Symp. Proc. Vol. 207: Mechanical Property of Porous and Cellular Materials, edited by L. J. Gibson, D. Green and K. Sieradzki, 135–140 (1991).
[10] A. Kuzjukevics and K. Ishizaki, "Porous Alumina by Hot Isostatic Pressing", Mater.

Res. Soc. Symp. Proc. Vol. 251: Pressure Effects on Materials Processing and Design, edited by K. Ishizaki, E. Hodge and M. Concannon, 121-125, 1992.

[11]A. Takata, K. Ishizaki, Y. Kondo and T. Shioura, "Influence of HIPping Pressure on the Strength and the Porosity of Porous Copper", Mater. Res. Soc. Symp. Proc. Vol. 251: Pressure Effects on Materials Processing and Design, edited by K. Ishizaki, E. Hodge and M. Concannon, 133-138 (1992).

[12]N. Saito, K. Ishizaki and A. Kuzjukevics, "The Production of Porous Si_3N_4 Ceramics with 100% β Phase", Mater. Res. Soc. Symp. Proc. Vol. 251: Pressure Effects on Materials Processing and Design, edited by K. Ishizaki, E. Hodge and M. Concannon, 149-152 (1992).

[13]G. C. Kuczynski, "Self-diffusion in Sintering of Metallic Particles", Trans. AIME, **185**, 169-178 (1949).

[14]W. D. Kingery and M. Berg, "Study of the Initial Stages of Sintering Solids by Viscous Flow, Evaporation-Condensation, and Self-Diffusion", J. Appl. Phys., **26**, 1205-1212 (1955).

[15]Y. Arai, "Material Chemistry in Powder", Baihu-kan, Tokyo, p150 (1987).

[16]W. D. Kingery, H. K. Bowen and D. R. Uhlmann, "Introduction to Ceramics, Second Ed.", John Wiley & Sons, NY, p250, (1976).

[17]Y. Minamino, T. Yamane, A. Shimomura, M. Shimada, M. Koiumi and N. Ogawa, "Interdiffusion in Al-Ag alloy under High Pressure", Light Metals, **34**, 174-181, (1984).

[18]Paul G. Shewmon, "Diffusion in Solids", McGrow-Hill Book Co., NY, pp81-83, (1963).

SINTERING BEHAVIOR OF TILE MATERIALS UNDER NORMAL AND HIGH GAS PRESSURE ATMOSPHERES

Toyokazu Kurushima
Central Research Laboratory, INAX Corporation, 36 Kume-aza-Yariba, Tokoname, Aichi 479, Japan

The sintering behavior of tile materials in the $CaO-Al_2O_3-SiO_2$ system was studied for different sintering gas pressures. Scrap tile material in powder form and the same material with 20 wt% glass powder, were cold isostatically pressed under 100 MPa. These green samples were sintered at temperatures between 900℃ and 1300℃ for 1 hour in air, or sintered at 1200℃ for 1 hour in Ar gas at 0.6, 7.5, 50 or 140 MPa. The scrap tile materials could be recycled using a normal sintering method. The recycled ceramics had a closed porosity of more than 20%, no open porosity, and a bending strength of more than 100 MPa. Ar gas at low pressure (0.6 and 7.5 MPa) accelerated the densification of the samples. However, Ar gas at high pressure (50 and 140 MPa) prevented the densification of the samples.

INTRODUCTION

The bending strength of ceramics with closed porosities is higher than the ones with the same amount of open porosities (1). Because the ceramics with a large amounts of closed porosities are light heat insulating materials with a high bending strength, these materials can be used as building materials.

Ishizaki et al. produced ceramics with high open porosities and high bending strengths by using a hot isostatic press (HIP) method (2-4). Kurushima, however, produced alumina and zirconia ceramics with no open porosity and a closed porosity of more than 10% by using a HIP method at higher temperature (5). The open and closed porosity vary with sintering temperature, sintering gas pressure, and chemical composition of the starting material.

Most scrap tile materials are not recycled. In the ideal case, these materials should be recycled with other ceramics. The recycled building material should be light and be a good heat insulation. It must also have an appropriate strength. A material with high open porosity also becomes dirty very early. Therefore, we hope to recycle the scrap tile material to a material having more closed pores, no open pores and a bending strength of more than 30 MPa.

In this study, the powdered scrap tile materials were sintered in air or HIPed. The porosity and the bending strength of the products were discussed in relation to the sintering temperature and gas pressure.

EXPERIMENTAL

Powder Preparation and Pretreatment for Sintering

The raw materials used were scrap tile material and glass powder. Table 1 shows the chemical composition of both. The tile material was crushed by using a hammer and ground into powder by an alumina ball mill for 24 hours. Both raw materials had an average particle size of about 10 μm.

Table 1. Chemical Composition of the Raw Materials

Composition	Tile / wt%	Glass / wt%
SiO_2	67.8	59.4
Al_2O_3	17.9	10.6
Fe_2O_3	0.5	0.1
CaO	8.0	9.8
MgO	0.3	18.6
K_2O	0.8	–
Na_2O	0.2	0.8
TiO_2	0.5	–
ZnO	1.5	0.1
ZrO_2	0.8	–
Ig-loss	0.3	0.4
Total	98.6	99.8

Two kinds of samples, consisting of the tile materials with 0 or 20 wt% glass powder, were prepared. The green body was pressed uniaxially in a steel mold under 20 MPa, and then cold isostatically pressed under 100 MPa.

Sintering and Assessment

The samples were sintered at temperatures between 900℃ and 1300℃ for 1 hour in air, or HIPed at 1200℃ for 1 hour without encapsulation in Ar gas at 0.6, 7.5, 50 or 140 MPa. The HIP chamber was pressurized to 0.6, 7.5, 50 or 50 MPa at room temperature. In the last case, the pressure in the chamber reached 140 MPa at 600℃, which was maintained up to 1200℃. All the other samples were heated to the sintering temperature under constant pressure. The samples sintered in air were cooled naturally to room temperature. The HIPed samples were cooled to room temperature in about 1 hour in the closed chamber. Therefore, the pressure decreased as the temperature decreased.

The density and porosity were calculated by the Archimedean method. The weight was measured under three different conditions, i.e., dry in air, water soaked in air and in water. The closed porosity was determined by : Closed Porosity = 1−[(Fraction to the Theoretical Density, ρ)+(Open Porosity)]. The three point bending strength of sintered specimens as measured by using a mechanical testing machine (Tensilon RTM-500, Orientec Co.).

RESULTS

Figure 1 shows the porosities as a function of the sintering temperatures (T). The closed porosity increased as the sintering temperature increased. The samples with 0 and 20 wt% glass powder had no open porosity when the surfaces of the samples were melted slightly at 1250℃ and 1200℃ respectively. There are no data above these temperatures because the samples were melted completely.

Figure 2 shows the porosities as a function of the sintering gas pressure (P). All the samples were sintered at 1200℃. The samples with 20 wt% glass powder hardly had any open pores, and the closed porosity is almost independent of the sintering gas pressure. The closed porosity of the samples without glass powder increased as the sintering gas pressure increased. But the samples without glass powder sintered under 50 MPa had 5% lower closed porosity than the ones sintered under 7.5 and 140 MPa.

Figure 3 shows the bending strength (σ) as a function of the sintering temperatures (T). Both samples with 0 and 20 wt% glass powder showed a similar behavior.

Figure 4 shows the bending strength (σ) as a function of the sintering gas pressure (P). All the samples were sintered at 1200℃. The bending strength increased as the sintering gas pressure increased. But both samples sintered under 50 MPa had a remarkable low bending strength.

DISCUSSION

Recycling of Scrap Tile Material

As shown in Fig. 1 and 3, the sample without glass powder sintered at 1250℃ had a closed porosity of more than 20%, no open porosity and a bending strength of more than 100 MPa. Therefore, the sample reached the requirements explained in the introduction and can be used as a building material.

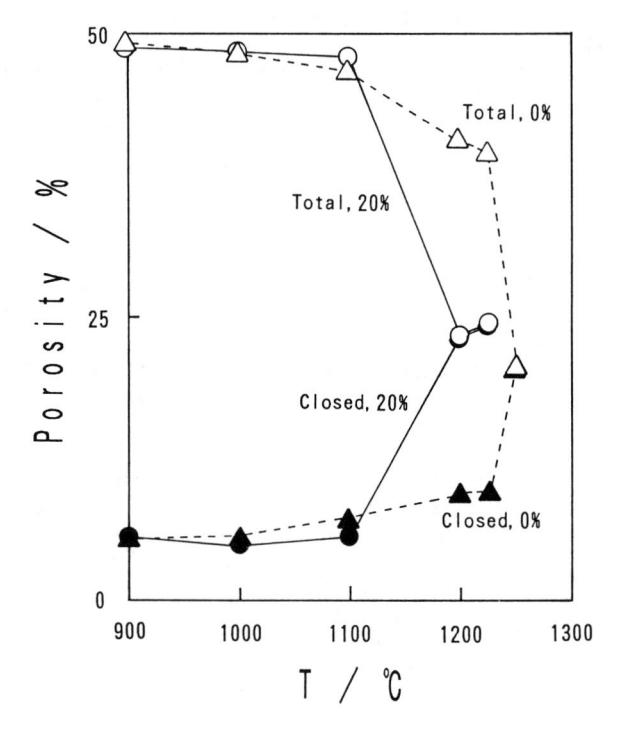

Figure 1 The porosities as a function of the sintering temperatures (T). The triangles and circles indicate the samples with 0 and 20 wt% glass powder respectively. The open and closed marks indicate the total and closed porosities respectively.

Effects of Glass Powder

When the samples had no open porosity, the sample with 20 wt% glass powder contained slightly more closed porosities than the one without glass powder, as shown in Fig. 1. The bending strength of both samples is almost the same as shown in Fig. 3. The glass powder lowered the melting point of the sample, and increased the closed porosities. Because the glass powder melts at a lower temperature than the powdered tile material, it connects the particles of the tile material. However, the melted glass may fill the closed pores.

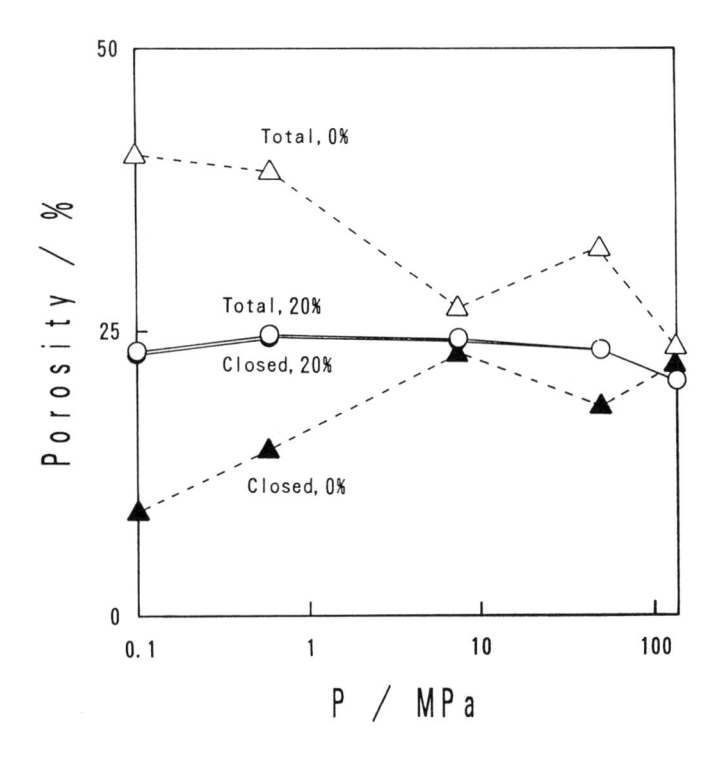

Figure 2 The porosities as a function of the sintering gas pressure (P). All the samples were sintered at 1200℃. The marks have the same significance as in Fig. 1.

Sintering Behavior under Normal Atmosphere

The samples without glass powder were hardly densified below 1225℃, and densified remarkably at 1250℃, as shown in Fig. 1. The same samples were melted completely at 1300℃. The samples with 20 wt% glass powder indicated the same tendency as the ones without glass powder at 50℃ lower temperature. Just before the melting of the samples, the samples had no open pores and achieved the maximum bending strength. The partially or

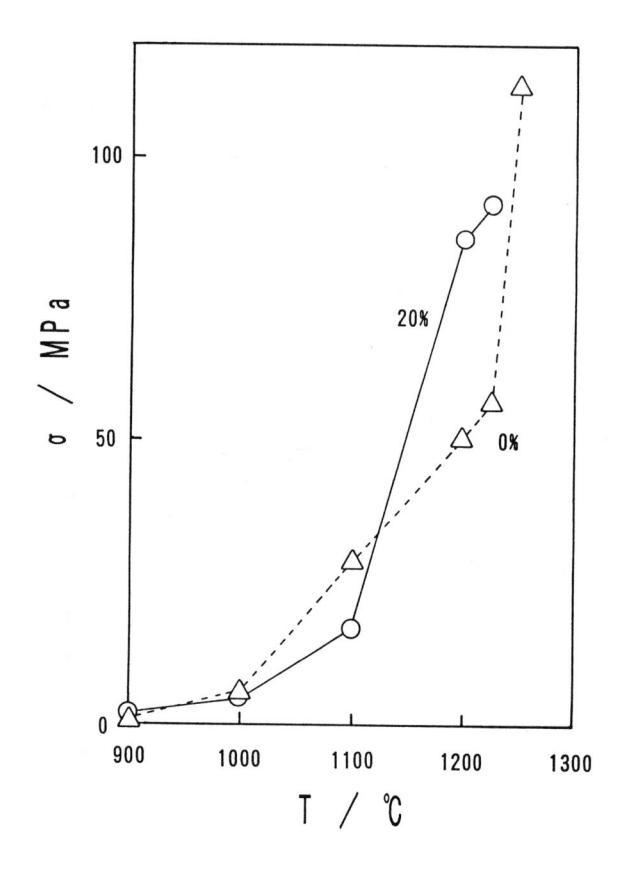

Figure 3 The bending strength (σ) as a function of the sintering temperatures (T). The triangles and circles indicate the samples with 0 and 20 wt% glass powder respectively.

completely melted samples had a lower bending strength than the ones just before melting, because bubbles were produced in the melted samples by expanding of air in the closed pores. In Fig. 3, the sample with 20 wt% glass powder sintered at 1225℃ was melted partially. If the sample was sintered at temperatures between 1200℃ and 1225℃, it might have a bending strength above 100 MPa. For the production of samples with high closed porosities and high bending strength, accurate temperature control during sintering is required.

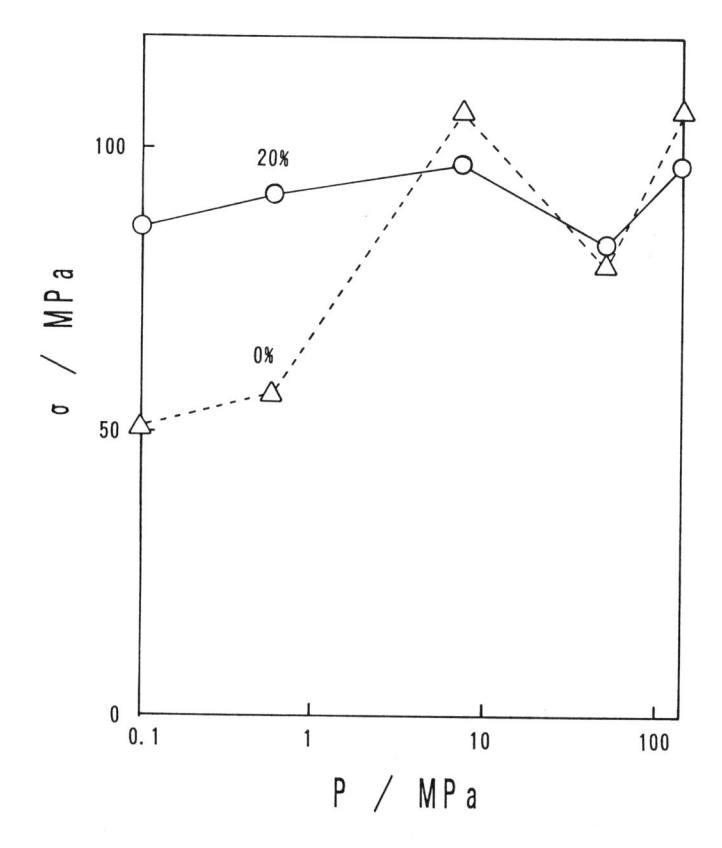

Figure 4 The bending strength (σ) as a function of the sintering gas pressure (P). All the samples were sintered at 1200℃. The σ of the sample with 20 wt% glass powder sintered under 140 MPa was measured by using the fractured small pieces. The marks have the same significance as in Fig. 3.

Porous Materials

Effects of Sintering Gas Pressure

Figure 5 shows the photographs of the samples with 0 and 20 wt% glass powder sintered at 1200℃. The sample with 20 wt% glass powder sintered under 7.5 MPa was melted partially and the one sintered under 140 MPa fractured. The sample without glass powder sintered under 50 MPa had a lower density than the ones sintered under 7.5 and 140 MPa. The experiments were carried out three times by using the same furnace. The results of ρ, melted behavior and fractured behavior of these experiments were similar.

Estimating from Fig. 2, 4 and 5, Ar gas at low pressure (0.6 and 7.5 MPa) accelerated the densification of the samples, which may be attributed to the surfaces of samples being activated by Ar gas molecules. Ar gas at high pressure (50 and 140 MPa) prevented the densification of the samples by packing a large amount of Ar gas in the pores. These possibilities were also reported briefly by Ishizaki et al. (2-4). Above 50 MPa, these effects seem to be more dominant than the accelerating effcts.

The closed pores of the sample with 20 wt% glass powder sintered under high gas pressure atmospheres contained the highly pressurized Ar gas. Because of this, the sample sintered under 140 MPa was fractured. The high gas pressure in the closed pores fractures the ceramics during cooling. Therefore, excess gas pressure should not be applied to the ceramics during sintering.

CONCLUSIONS

The tile materials were sintered at temperatures between 900℃ and 1300℃ in air, or at 1200℃ in Ar gas at 0.6, 7.5, 50 or 140 MPa. The closed porosities and bending strength were measured. From the results obtained, the following points were concluded:
(1) Ar gas at low pressure (0.6 and 7.5 MPa) accelerated the densification of the samples, which may be due to the surfaces of samples being activated by Ar gas molecules. Ar gas at high pressure (50 and 140 MPa) prevented the densification of the samples by packing a large amount of Ar gas in the pores.
(2) The powdered scrap tile material could be recycled using a normal sintering method to a material with a closed porosity of more than 20%, no open porosity and a bending strength of more than 100 MPa.
(3) The closed pores of the sample with 20 wt% glass powder sintered under high gas pressure (140 MPa) contained highly pressurized Ar gas. The high pressurized Ar gas in the closed pores fractured the ceramics during cooling.

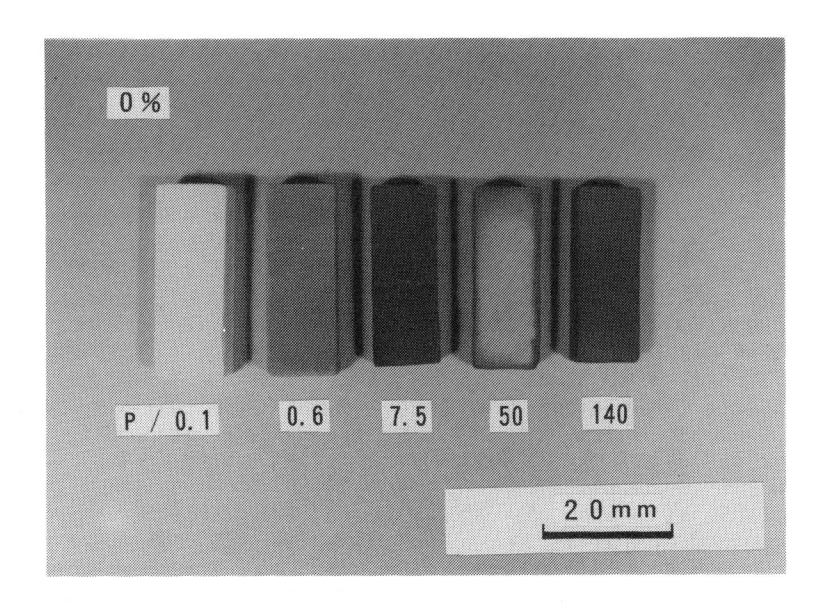

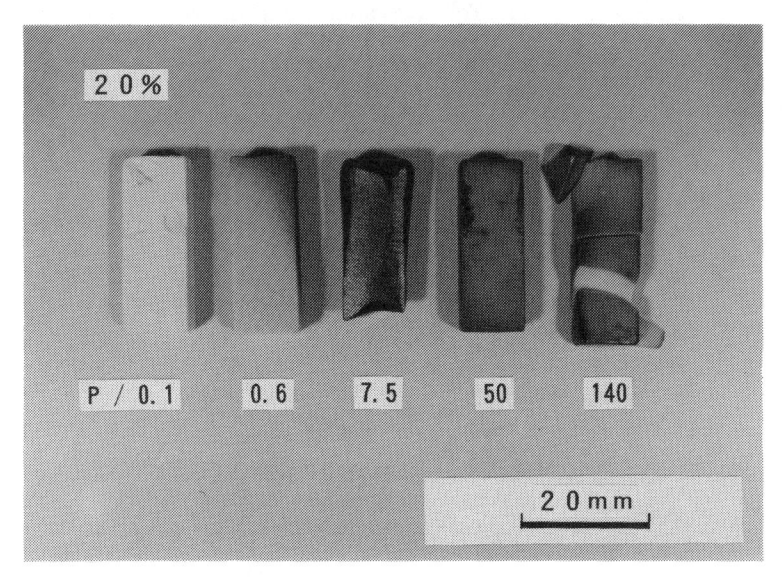

Figure 5 The photographs of the samples with 0 and 20 wt% glass powder sintered at 1200℃. The numbers under the samples indicate the sintering gas pressure in MPa.

(4) The samples with 20 wt% glass powder were melted at 50℃ lower temperature and had higher closed porosities than samples without glass powder.

REFERENCES

(1) P.Boch, J.C.Glandus, J.Jarrige, J.P.Lecompte and J.Mexmain, Sintering, Oxidation and Mechanical Properties of Hot Pressed Aluminium Nitride, Ceram. Int., 8, 34-40 (1982).
(2) K.Ishizaki, A.Takata and S.Okada, Mechanically Enhanced Open Porous Materials by a HIP Process, J. Jpn. Ceram. Soc., 98, 533-40 (1990).
(3) A.Takata, K.Ishizaki and S.Okada, Improvement of a Porous Material Mechanical Property by Hot Isostatic Process, in Mechanical Properties of Porous and Cellular Materials, Mater. Res. Soc. Proc., 207, Pittsburgh, PA (1991) pp.135-40.
(4) K.Ishizaki, Stronger Open Porous Ceramics and Metals Produced by HIP, Mater. & Proc. Rep., 5, 4-5 (1991).
(5) T.Kurushima, Closed Porous Alumina and Zirconia Ceramics Sintered under High Gas Pressure Atmospheres, Abst. General Mtg. Jpn. Ceram. Soc. Tokai Branch, submitted (1992).

FABRICATION AND PROPERTIES OF MULLITE CERAMICS WITH NEEDLE-LIKE CRYSTALS

Hiroaki KATSUKI, Hiroyoshi TAKAGI and Ohsaku MATSUDA*
Saga Ceramics Research Laboratory, Arita-machi, Saga 844 JAPAN
*Saga University, 1, Honjyo-machi Saga-shi, Saga 840 JAPAN

Needle-like mullite crystals ($2.97Al_2O_3 \cdot 2SiO_2$), of 3-5 μm in length and 0.5-1.0 μm in diameter, were synthesized by acid leaching out the coexisting silica glass from New Zealand kaolin fired at 1650°C. Porous mullite ceramics composed of needle crystals with porosity of 23-57% were prepared by pressing or slip casting. Pore characteristics (porosity, pore size and pore volume), three-point bending strength, thermal shock resistance, and thermal conductivity were examined on the porous mullite ceramics sintered at 1550-1700°C in air. With an increase in sintering temperature, the porosity of ceramics decreased, which brought the increase of the bending strength. However, the pore size distribution was almost independent of the sintering temperature. The average pore size of mullite ceramics sintered at 1500-1700°C were in the range of 0.46-0.61 μm. In the case of slip casting, the sintering was slightly promoted due to an increase in orientation of needle crystals. The calculated porosity of samples showed a good agreement with that measured by a mercury porosimeter, so it could be concluded that most of the pores in the sample were of open pore structure, and needle crystals couldn't be fully sintered below 1650°C. Thermal conductivity at room temperature decreased linearly with increasing porosity.

INTRODUCTION

Porous ceramics are attractive materials for use in various industrial fields as reactive filters, membranes for separation and gas sensors(1-4).

Mullite is increasingly becoming one of very useful materials for high-temperature engineering applications due to its chemical stability, low-thermal expansion coefficient, high bending strength, and good creep resistance over 1300°C(5-8). Therefore, porous ceramics based on mullite will be expected to have many applications, such as gas filters, catalyst supports, and refractories. Needle- and pillar-like mullite have been traditionally derived from the decomposition of silicate minerals such as sericite, kaolin, silimanite and kyanite, resulting in a product which contains silica glass and some oxide impurities(9-10).

In previous research(11-12), we revealed that fine needle-like mullite crystals with sizes of 10-20 μm in length and 0.5-1.5 μm in diameter were obtained by leaching out the coexisting silica glass from New Zealand kaolin fired at 1550-1650°C. These crystals, with the growth direction of [001], were of stoichiometric composition. Their morphologies did not change after reheating up to 1650°C. Generally, sinterability of fibers or large particles tends to be lower than that of fine particles. So, ceramics composed of only fibrous (needle or whisker) crystals will result in the formation of open pore structure by sintering.

In this study, porous ceramics based on needle-like mullite prepared from New Zealand kaolin were fabricated, and some properties such as pore characteristics, thermal expansion, thermal conductivity, bending strength, and thermal shock resistance were investigated.

EXPERIMENTAL

Preparation of Needle-like Mullite and Fabrication of Porous Mullite Body

Needle-like mullite crystals were prepared by the same procedure as reported previously(11-12). New Zealand kaolin powder fired at 1650°C for 2 hr was sealed and left for 48 hr in 18% HF solution to leach out the coexisting silica glass completely. Mullite sediments were collected and washed fully in the pure water. The morphology of mullite crystals was observed by a TEM and SEM and their chemical composition was examined by a fluorescent X-ray analyzer.

To fabricate porous bodies, needle crystals were formed by the following methods; (1) pressed to disks in a uniaxial pressure device at 30 MPa, (2) slip casted into plates using a gypsum mold. If impurity oxides that easily form a glassy phase are contained in the samples, the porous properties may be reduced at high temperature. So no inorganic agents were added to source materials in this study.

Evaluation of Properties of Porous Body

The bulk density of porous bodies was measured by Archimede's method at 25°C. Porous properties, such as the porosity, the average pore size and pore distribution, were evaluated by a mercury porosimeter. Sample size for three point bending strength was 3x5x25 mm. Thermal shock resistance was evaluated by measuring the residual bending strength of specimens water-quenched and air-cooled from various temperatures. The thermal expansion coefficient was measured with a dilatometer and thermal conductivity was measured in air or vacuum from 25 to 700°C by a laser flash method. Structure of sintered bodies was observed by SEM and the degree of orientation of needle crystals in sintered bodies was evaluated by comparing the X-ray diffraction intensity of (210) and (110) planes with data of a JCPDS card.

RESULTS AND DISCUSSION

Sintering of Needle Crystals and Porous Properties of Fired Body

Figure 1 shows the morphology of needle crystals used in this study. Fired kaolin was crushed into particles of 2-3 mm in size so that obtained crystals were smaller, with an average

Figure 1 The morphology of needle-like mullite crystals from New Zealand kaolin fired at 1650°C.

length of 3-5 µm, thickness of 0.5-1 µm, and aspect ratio of 3-5. The chemical composition of mullite crystals was 2.97Al$_2$O$_3$·2SiO$_2$ and the total content of Na$_2$O, K$_2$O, CaO, and MgO was 0.1 wt% or less. Furthermore, the specific surface area of crystals was 4.1 m^2/g by a BET method and the morphology was stable up to 1700°C(12).

Table 1 shows the effect of sintering temperatures on pore characteristics and bending strength of sintered bodies after pressing. Samples were too brittle to measure some properties below 1500°C. Samples sintered at 1550-1650°C had a porosity as high as 48-57%, while advanced densification reduced the porosity to 26% at 1700°C. But, the average pore size was almost constant (0.46-0.58 µm) at 1550-1700°C. The bending strength increased with temperature to reach 72 MPa at 1700°C.

Table 1 Some Properties of Porous Mullite Ceramics Composed of Needle Crystals After Pressing.

Properties	Firing Temperature (°C)			
	1550	1600	1650	1700
Bulk Density (g/cm^3)	1.37	1.52	1.65	2.32
Relative Density (%)	43.6	48.3	52.4	73.5
Porosity (%)	57.4	52.5	48.1	26.1
Pore Volume (cc/g)	0.43	0.35	0.29	0.11
Average Pore Size (µm)	0.58	0.53	0.54	0.46
Bending Strength (MPa)	7	18	52	72

Figure 2 shows the fractured surface of samples sintered at 1650 and 1700°C. Pores submicron in size were observed between entangled crystals. The result of SEM observation almost agreed with pore size measured by mercury porosimetry. Whereas, sintering proceeded as shown in Fig. 2(b) but crystals maintained needle shape.

Figure 3 gives the pore size distribution of samples sintered at 1550 and 1700°C. Both distribution curves were sharp and showed that the effect of sintering temperature on pore structure was relatively low. The main reason is probably that no substantial change occured in the morphology of needle crystals even after sintering at 1700°C. Moreover, the calculated porosity by Archimede's method of samples showed

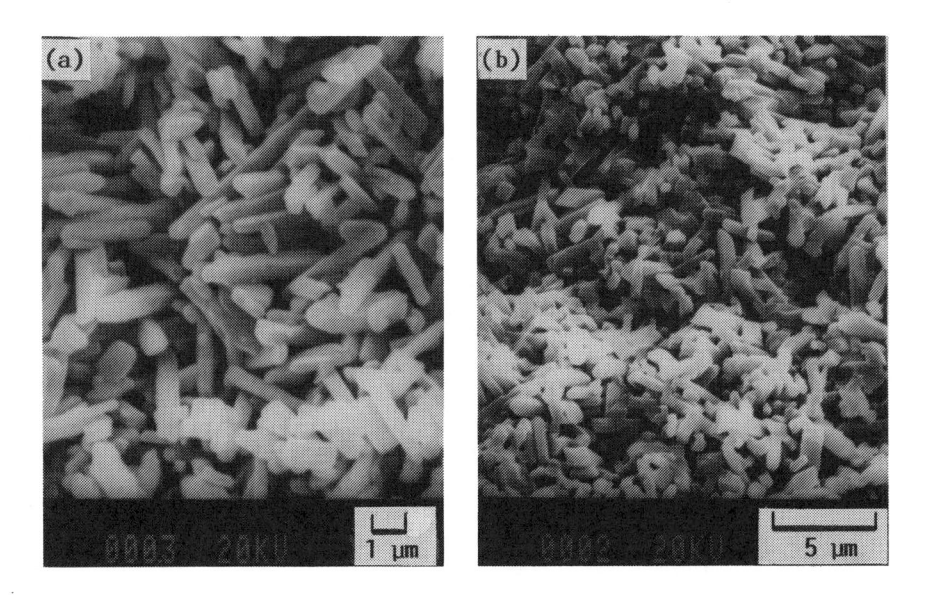

Figure 2 Fractured surface of porous mullite ceramics with needle-like mullite sintered at (a) 1650 and (b) 1700°C.

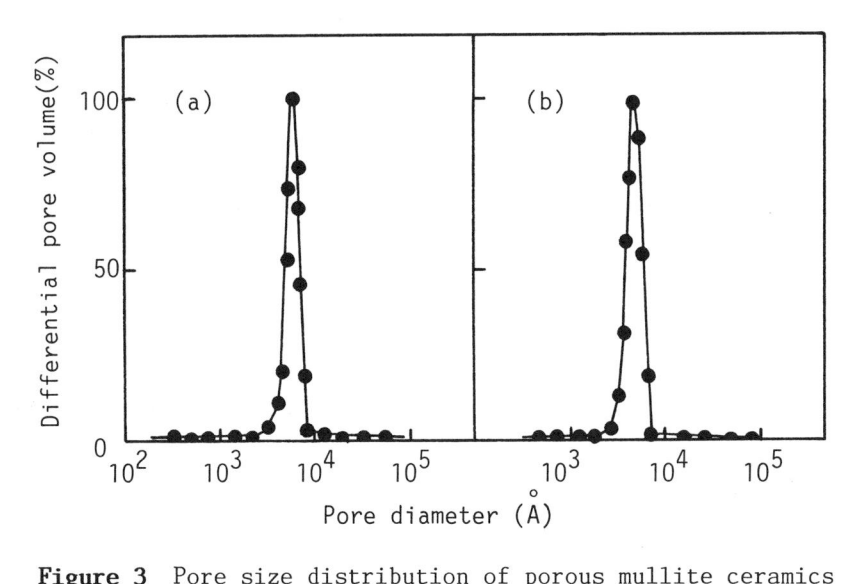

Figure 3 Pore size distribution of porous mullite ceramics sintered at (a) 1550°C and (b) 1700°C. Green bodies were formed by pressing.

Porous Materials 141

a good agreement with that measured by mercury porosimetry, so it could be concluded that most pores in the sample were of open pore structure, and densely entangled needle crystals couldn't be fully sintered below 1650°C.

The orientation of needle- or fibrous-like crystals in green bodies usually tends to be promoted by slip casting or extrusion methods; therefore, the effect of slip casting on pore characteristics was examined. Table 2 shows some properties of sintered bodies at 1550-1700°C after slip casting. Compared with the results of a pressing (Table 1), the sintering was promoted and the porosity decreased by 4-9% at the same temperature. Consequently, the promotion of the sintering increased the bending strength, which was about 70 MPa at 1650°C. From SEM observations of fractured surface of sintered bodies at 1650°C after slip casting, needle crystals sedimented parallel to the surface of the gypsum mold, and a contact of lateral plane of each crystal could be confirmed as shown in Fig. 4. Furthermore, the ratio of relative XRD intensity ratio,

$$\frac{I(110)}{I(210)\ sintered} \Big/ \frac{I(110)}{I(210)\ JCPDS} = 0.99 \text{ by pressing and}$$

1.48 by slip casting, respectively, revealed that the degree of orientation of needle crystals was promoted by slip casting more than by pressing.

Table 2. Some Properties of Porous Mullite Ceramics Composed of Needle Crystals After Slip Casting.

Properties	Firing Temperature (°C)			
	1550	1600	1650	1700
Bulk Density (g/cm^3)	1.57	1.75	1.92	2.44
Relative Density (%)	49.8	55.5	61.0	77.4
Porosity (%)	50.6	44.2	39.6	22.4
Pore Volume (cc/g)	0.32	0.28	0.22	0.09
Average Pore Size (μm)	0.60	0.61	0.59	0.58
Bending Strength (MPa)	19	31	68	102

Figure 4 Fractured surface of porous mullite ceramics with needle crystals sintered at 1650°C (by slip casting).

Thermal Properties of Porous Mullite Body

Thermal expansion coefficient of porous mullite bodies with needle crystals sintered at 1550–1700°C by pressing and slip casting were 4.8–5.0 x10^{-6}/°C up to 1400°C. There was no difference between the two forming methods. Since mullite has relatively lower thermal conductivity, porous mullite ceramics having high porosity are expected to be used as adiabatic materials at high temperature. Figure 5 shows the thermal conductivity at room temperature of bodies sintered at 1550–1700°C as a function of porosity. The specific heat in vacuum was 0.196–0.201 J/°C·g, and no difference was encountered by temperature. On the other hand, thermal conductivity decreased linearly with an increase in porosity, and reached 0.63 W/m·K for a body of 57.4% porosity sintered at 1550°C. The value is about 1/10 that of commercial dense mullite ceramics, and such a porous texture which has needle crystals leads to lower thermal conductivity as in the case of other porous ceramics(13). In addition, thermal conductivity up to 750°C was examined on samples having 52% porosity, but it decreased slightly from 1.1 to 0.8 W/m·k at 100–750°C.

Thermal Shock Resistance of Porous Mullite Body

Figure 6 shows the residual bending strength of samples after water-quenching from various temperatures, which have 48 and

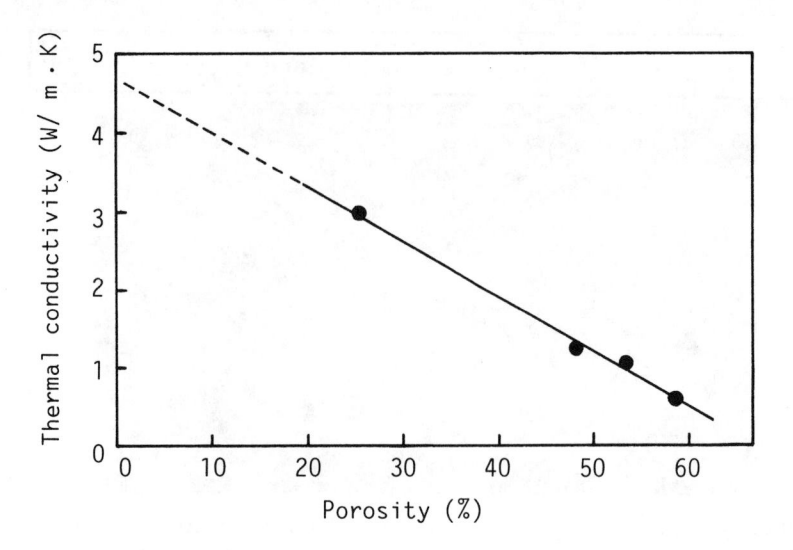

Figure 5 Thermal conductivity of porous mullite ceramics with needle crystals at 25°C. Green bodies were formed by pressing.

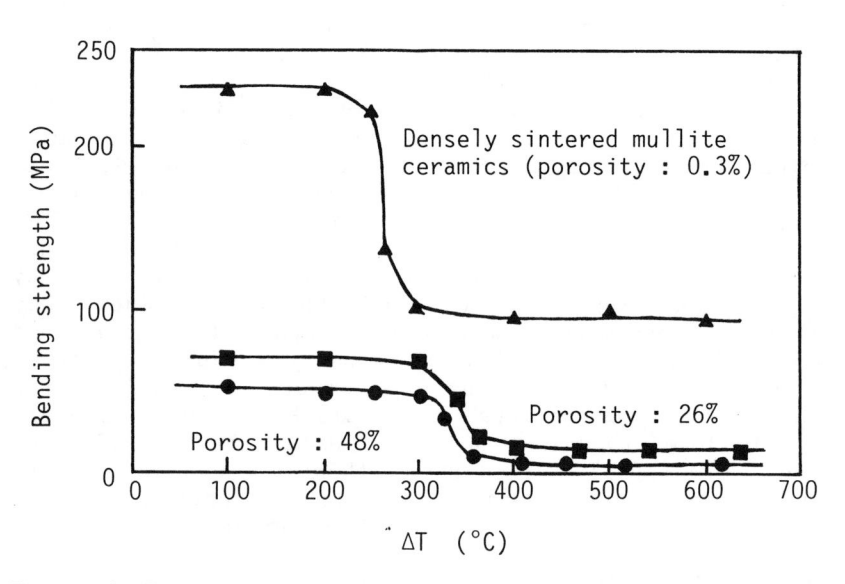

Figure 6 Thermal shock resistance test of porous mullite and densely sintered mullite derived from commercial fine powder sintered at 1650°C.

26% porosity. In the case of air-cooled samples from 1100°C to room temperature, the residual bending strength didn't decrease, and porous mullite bodies had excellent thermal shock resistance. But they could not withstand thermal shock (ΔT) exceeding 350°C by water-quenching. Compared with dense mullite ceramics prepared from commercial mullite powder, the critical temperature difference (ΔTc) of porous mullite bodies increased by about 100°C. It is expected that resistance to crack propagation can be improved by introducing many initial cracks or more generally by introducing microstructural heterogeneities of any form, which serve as stress concentrators and heat radiation in porous materials. So, an increase of the critical temperature difference in Fig.6 may be due to the existence of small open pores in mullite ceramics.

CONCLUSIONS

Porous mullite ceramics composed of needle crystals were prepared and the various properties were examined and the following results were obtained.
(1) The porosity of mullite ceramics formed by pressing with densely-entangled needle crystals, decreased with increasing sintering temperatures and yet samples sintered at 1650°C had a porosity as high as 48.1%. At 1700°C, sintering proceeded and the porosity decreased to 26.1%. In the case of slip casting, needle crystals tended to orient to the surface of the gypsum mold, the sintering at 1550-1700°C was promoted and the porosity decreased.
(2) The pore size distribution of samples sintered at 1550-1700°C after both compacting methods was sharp. Most of the pores in the samples were of open structure and their average pore size was 0.46-0.61 μm.
(3) The average thermal expansion coefficient of porous bodies from 25 to 1400°C was $4.8-5.0 \times 10^{-6}$/°C. The thermal conductivity of porous bodies at room temperature decreased linearly with an increase in porosity, and 0.63 W/m·k was obtained in the case of samples sintered at 1550°C with a porosity of 57.4%. The thermal conductivity from 25 to 750°C was almost independent of the measured temperature.
(4) The bending strength of porous mullite bodies increased with sintering temperature to 72 MPa at 1700°C. The critical temperature difference after the thermal shock test by water-quenching of mullite ceramics (48% porosity) was about 320-340°C.

REFERENCES

(1) H.Hosono, Y.Sakai, M.Fasano, and Y.Abe, "Preparation of Monolithic Porous Titania-Silica Ceramics", J.Am.Ceram.Soc., 73, 2536-38(1990)
(2) Y.C.Yen, T.Y.Tseng, and D.A.Chang, "Electrical Properties of Porous Titania Ceramics Humidity Sensors", J.Am.Ceram.Soc., 72, 1472-75(1989)
(3) L.C.Klein and N.Giszpenc, "Sol-Gel Processing for Gas Separation Membranes", Am.Ceram.Soc.Bull., 69, 1821-30(1990).
(4) J.F.Zievers, P.Eggerated and E.C.Zievers, "Porous Ceramics for Gas Filtration", ibid., 70, 108-11(1991)
(5) A.P.Hynes and R.H.Doremus, "High-Temperature Compressive Creep of Polycrystalline Mullite", J.Am.Ceram.Soc., 74, 2469-75(1991)
(6) M.G.M.U.Ismail, Z.Nakai and S.Somiya, "Microstructure and Mechanical Properties of Mullite Prepared by the Sol-Gel Method", ibid., 70, C-7(1987)
(7) I.A.Aksay, D.M.Dabbs and M.Sarikaya, "Mullite for Structural, Electronic, and Optical Applications", ibid., 74, 2343-58(1991)
(8) H.Ohnishi, T.Kawanami, A.Nakahira and K.Niihara, "Microstructure and Mechanical Properties of Mullite Ceramics", Yogyo Kyokaishi, 98, 541-47(1990)
(9) M.Slaughter and W.D.Keller, "High Temperature Phases from Impure Kaolin Clays", Am.Ceram.Soc.Bull., 38, 703-707(1959)
(10) J.A.Pask and A.P.Tomsia, "Formation of Mullite from Sol-Gel Mixtures and Kaolinite", ibid. 74, 2367-73(1991)
(11) H.Katsuki, S.Furuta, H.Ichinose and H.Nakao, "Preparation and Some Properties of Porous Ceramics Sheet Composed of Needle-like Mullite", Seramikkusu-Ronbunshi, 88, 1081-86(1988)
(12) H.Katsuki, A.Kawahara, H.Ichinose, S.Furuta and S.Yoshida, "High-Temperature Properties of Needle-like Mullite Obtained from New Zealand Kaolin", ibid., 89, 1521-24(1989)
(13) T.Yamaguchi and H.Yanagida (ed.), "Seramikkusu to Netsu", Ceramics Science Series 6, Gihodo Suppan(1987)p.87

SINTERING BEHAVIOR AND THERMAL SHOCK RESISTANCE OF POROUS
ZrO_2 CERAMICS WITH ZrO_2 FIBER, Al_2O_3 FIBER OR SiC WHISKER

Kazuhiko HATTORI* and Toyokazu KURUSHIMA*, **
*Central Research Laboratory, INAX Corporation,
36 Kume-aza-Yariba, Tokoname, Aichi 479, Japan
**Materials Science Course, Graduate School, Nagaoka
University of Technology, Nagaoka, Niigata 940-21, Japan

Three kinds of ZrO_2 ceramics which contain 0, 4, 8, or 12 vol% ZrO_2 fibers, Al_2O_3 fibers or SiC whiskers, were sintered at 1400, 1500 and 1600 °C for two hours. All samples were porous with ceramic fibers or whiskers dispersed. The sintering behavior, bending strength and strength ratio before and after thermal shock tests of the samples with ZrO_2 fibers were similar to the ones with Al_2O_3 fibers, but they were different from the ones with SiC whiskers. The thermal shock resistance was improved by dispersing ceramic fibers or whiskers in the ZrO_2 matrix. The samples sintered at successively higher temperature had rather low thermal shock endurance. To obtain a high thermal shock endurance, the uniform dispersion of fibers or whiskers, sintering temperature and porosity control are required.

INTRODUCTION

Thermal shock resistance of ceramics is an important property for high temperature applications. Zirconia ceramics (ZrO_2) have a very high bending strength but do not have a high thermal shock endurance (1).

The dence ZrO_2 and Al_2O_3 ceramics, which were hot pressed for reinforced with SiC whiskers were reported up to the present (2,3).

Porous ZrO_2 reinforced with ZrO_2 fibers has not yet been discussed for high temperature applications. The authors thought that porous ZrO_2 with ceramic fibers should have a high thermal shock resistance. In this study, porous ZrO_2 with ZrO_2 fibers, Al_2O_3 fibers or SiC whiskers was discussed, with emphasis on its high thermal shock resistance.

EXPERIMENTAL PROCEDURE

ZrO_2 powder containing 2.7 wt% Y_2O_3, ZrO_2 fibers containing 8 wt% Y_2O_3, Al_2O_3 fibers, and SiC whiskers were used as starting materials. These materials were mixed in the desired volume ratio by using an ultrasonic generator. Table 1 shows the composition of the mixtures.

The mixtures were pressed uniaxially in a steel mold under 20 MPa to form plates, and were cold isostatically pressed under 100 MPa. The samples were normally sintered at 1400, 1500 and 1600°C for 2 hours. The sintered samples were cut into bars of 5 x 5 x 30 mm.

Table 1. The composition of the raw materials

Sample No.	Composition / vol%			
	ZrO_2	ZrO_2-Fiber	Al_2O_3-Fiber	SiC-Whisker
1	100	0	0	0
2	96	4	0	0
3	92	8	0	0
4	88	12	0	0
5	96	0	4	0
6	92	0	8	0
7	88	0	12	0
8	96	0	0	4
9	92	0	0	8
10	88	0	0	12

Several properties of sintered samples were analyzed. The density were calculated by the Archimedean method after weighting in air and water. The bending strength of sintered samples was measured by a three point bending test with a cross head speed of 1 mm/min and a span length of 25 mm. The thermal shock resistance of each sample was evaluated by the bending strength change before and after water quenching from

625°C to 25°C (ΔT=600 K).

RESULTS and DISCUSSION

Sintering Behavior

Figure 1 shows the bulk density (ρ) as a function of sintering temperature (T). Figures a, b and c show the samples with ZrO_2 fibers, Al_2O_3 fibers and SiC whiskers, respectively. The samples with ZrO_2 fibers had higher density than the other samples with Al_2O_3 fibers or SiC whiskers. This is due to the fact that ZrO_2 fibers have a lower sintering temperature than Al_2O_3 fibers and SiC whiskers. The samples with 12 vol% Al_2O_3 fibers had a much lower density than the other samples with Al_2O_3 fibers, and had a different sintering behavior than the ones with 4 and 8 vol% Al_2O_3 fibers. The density of the samples with SiC whiskers was lower than the other samples, and decreased as the sintering temperature increased. SiC whiskers oxidized easily in air. The samples with fibers or whiskers had always a lower density than the samples without fibers or whiskers. Porous ZrO_2 was obtained by dispering ceramics fibers or whiskers, even if it was sintered above 1400°C.

Bending Strength

Figure 2 shows the bending strength ($\sigma 1$) before the thermal shock test (ΔT=600 K) as a function of sintering temperature (T). The samples with ZrO_2 fibers had a higher bending strength as the sintering temperature increased and fiber content decreased. The bending strength behavior of the samples with Al_2O_3 fibers was similar to the one of the samples with ZrO_2 fibers. The samples with SiC whiskers had nearly equal bending strength regardless of the sintering temperature. The samples with SiC whiskers were not densified because SiC whiskers have difficulty sintering below 1600°C, and they oxidized above 1500°C.

Thermal Shock Endurance

Figure 3 shows the bending strength ($\sigma 2$) after thermal shock test

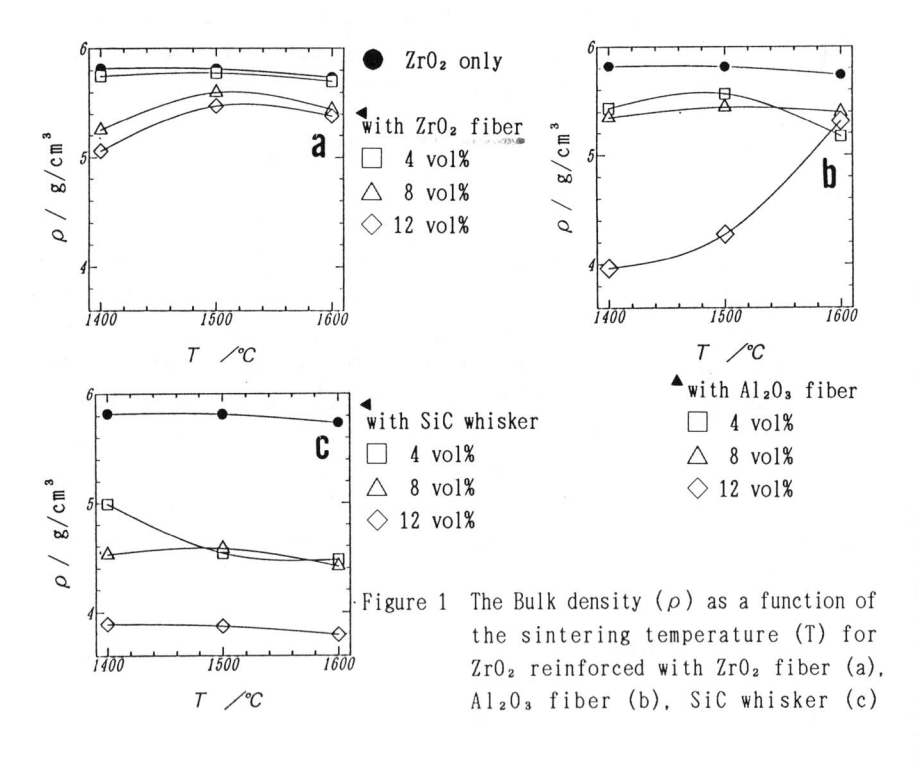

($\Delta T=600$ K) as a function of sintering temperature (T). Many samples had the maximum values for the sintering temperature of 1500°C. The samples were fully densified and could not adsorb thermal shock at 1600°C. Kurushima et al. previously reported that the sample sintered at high temperature had a low thermal shock endurance because the fiber disappeared in to the matrix. In this study, this same effect might have occurred too. The samples with 12 vol% ZrO_2 fiber, 8 and 12 vol% Al_2O_3 fiber had a higher bending strength as the sintering temperature increased. Those samples were not densified at 1500°C because the fiber contents were too high.

The maximum strength will appear at a sintering temperature higher than 1500°C. The sample with 8 and 12 vol% SiC whisker had a lower bending strength as the sintering temperature increased, since the oxidized phase increased as the sintering temperature increased. The bending strength of the samples with ZrO_2 fiber was more than 150 MPa even after the thermal shock test.

Figure 1 The Bulk density (ρ) as a function of the sintering temperature (T) for ZrO_2 reinforced with ZrO_2 fiber (a), Al_2O_3 fiber (b), SiC whisker (c)

Figure 4 shows the strength ratio between $\sigma 1$ and $\sigma 2$ as a function of sintering temperature (T). The strength ratio $\sigma 1/\sigma 2 =1$ indicates that there is no decrease in bending strength by the water quenching. All the samples with fibers or whiskers had a higher $\sigma 1/\sigma 2$ than the samples without fibers or whiskers. The thermal shock resistance was improved by dispersing fibers or whiskers. The strength ratio before and after the thermal shock test of the samples with ZrO_2 fibers were similar to the ones with Al_2O_3 fibers, but were different from the ones with SiC whisker. The samples with ZrO_2 and Al_2O_3 fibers had a maximum strength ratio at a temperture of 1500 C. This is an appropriate sintering temperature for the samples. The samples sintered at excessively higher temperature had low thermal shock resistance. Diffusion of the fibers in the matrix (4), as described before and the porosity seem to be related to the thermal shock resistance. For a high thermal shock rsistance, control of sintering temperature and porosity are required.

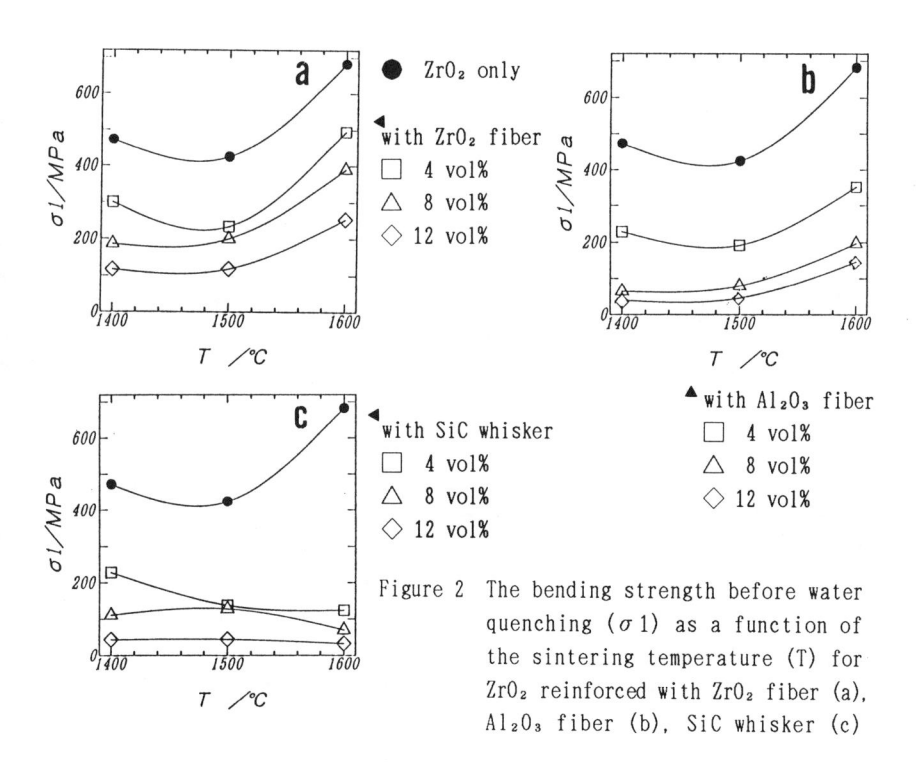

Figure 2 The bending strength before water quenching ($\sigma 1$) as a function of the sintering temperature (T) for ZrO_2 reinforced with ZrO_2 fiber (a), Al_2O_3 fiber (b), SiC whisker (c)

Porous Materials   151

Figure 3 The bending strength after water quenching ($\sigma 2$) as a function of the sintering temperature (T) for ZrO$_2$ reinforced with ZrO$_2$ fiber (a), Al$_2$O$_3$ fiber (b), SiC whisker (c)

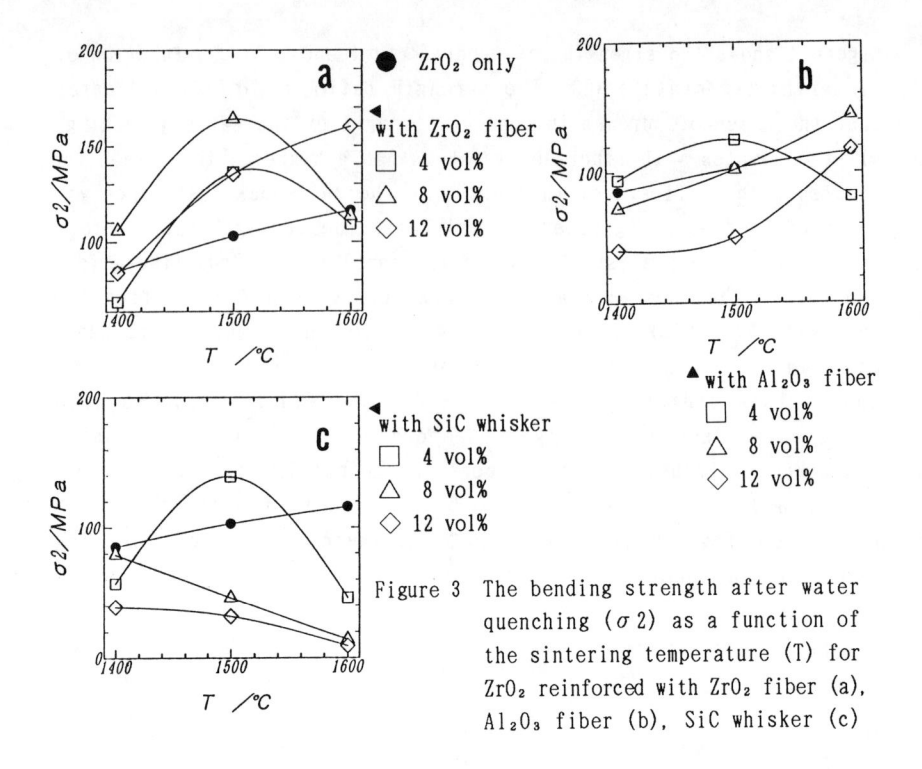

CONCLUSION

ZrO$_2$ ceramics with 0, 4, 8 and 12 vol% of ZrO$_2$ fibers, Al$_2$O$_3$ fibers or SiC whiskers were sintered at 1400, 1500 and 1600°C, and their density, bending strength and thermal shock resistance were examined. From the results obtained, the following points were concluded:
(1) Porous ZrO$_2$ ceramics were obtained by dispersing ceramic fibers or whiskers even if they were sintered above 1400°C.
(2) The thermal shock resistance was significantly improved very much by dispersing ceramic fibers or whiskers, and the bending strength after the thermal shock testwas more than 150 MPa.
(3) There is an appropriate sintering temperature for ZrO$_2$ ceramices with fibers or whiskers. For high thermal shock resistance, control of sintering temperature and porosity are also required.
(4) The sintering behavior, bending strength and strength ratio before and after the thermal shock test of the samples with ZrO$_2$ fibers were

similar to the ones with Al₂O₃ fibers, but were different from the ones with SiC whiskers.

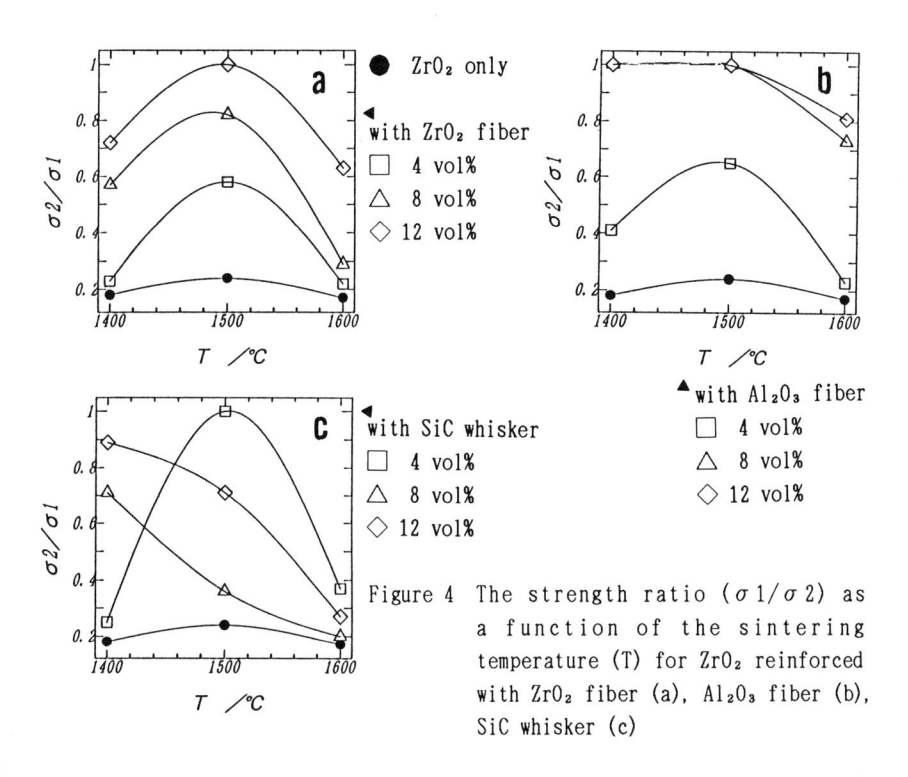

Figure 4 The strength ratio ($\sigma 1/\sigma 2$) as a function of the sintering temperature (T) for ZrO₂ reinforced with ZrO₂ fiber (a), Al₂O₃ fiber (b), SiC whisker (c)

REFERENCES

(1) Reckziegel , Properties and applications of advanced Zirconia ceramics. , Ceram. Forum. Int. , 54, 378-385 , 1986
(2) T.N.Tiegs, P.F.Becher , Thermal shock behavior of an Alumina-SiC composite. , J. Am. Ceram. Soc., 70, 109-111 , 1987
(3) N.Claussen, K.L.Weisskopf, M.Ruhle , Tetragonal Zirconia polycrystals reinforced with SiC whiskers. , J. Am. Ceram. Soc. , 69, 282-292 , 1986
(4) T.Kurushima, K.Ishizaki , Influence of sintering behavior on the thermal shock endurance and the mechanical properties of fiber dispersed ZrO₂ ceramics. , J. Ceram. Soc. Jpn. , submitted , 1992

Porous Materials

MICROPOROUS METAL INTERCALATED CLAY NANOCOMPOSITES

Sridhar Komarneni, Materials Research Laboratory and Department of Agronomy, The Pennsylvania State University, University Park, PA 16802, U.S.A.

Swelling-type of layer silicates, such as montmorillonite, have been intercalated with Cu and Ni metal (zero-valent) clusters of 4-5Å by *in-situ* reduction of Cu^{2+} and Ni^{2+} ions using the polyol process which involves the reduction of transition metal ions, such as Cu^{2+} and Ni^{2+} by ethylene glycol at about 175°C. The metal-cluster intercalates prop the silicate layers apart, just as ceramic oxides do in pillared clays, leading to microporous materials. Although the metal intercalated clay nanocomposites are not as stable as pillared clays, intercalated metal clusters of these dimensions are expected to behave very differently from the bulk metal and may prove to be versatile catalysts. The novel metal intercalated clays have been characterized by powder x-ray diffraction and transmission electron microscopy. The micropore structure of the metal intercalated clay nanocomposites was probed by measuring water adsorption isotherms at 25°C and nitrogen surface area by BET. The surface areas estimated from BET monolayer capacity of water and nitrogen are 187 m^2g^{-1} and 165 m^2g^{-1}, respectively, for the Ni metal intercalated clay. Although these surface areas are smaller than those of pillared clays (200-400 m^2g^{-1}), further optimization in processing may lead to higher surface areas. These novel microporous metal intercalated clays are expected to find unique applications in catalysis, adsorption, molecular selectivity, etc.

INTRODUCTION

Clay minerals are, in general, aluminosilicates and magnesium aluminosilicates and have two-dimensional layer structures (1,2). Among these, the smectite group of clay minerals such as montmorillonite, beidellite, hectorite, saponite and nontronite are extremely interesting because of their cation exchange property in the interlayers and their swelling-shrinking of the interlayers. These properties have been found to be highly amenable in the design and synthesis of pillared clays which are

---

To the extent authorized under the laws of the United States of America, all copyright interests in this publication are the property of The American Ceramic Society. Any duplication, reproduction, or republication of this publication or any part thereof, without the express written consent of The American Ceramic Society or fee paid to the Copyright Clearance Center, is prohibited.

microporous materials with wide-ranging applications (3-13). The preparation of pillared clays involves cation exchange with hydroxy polymeric cations followed by dehydration and dehydroxylation of the polymeric species to prop the layers apart with ceramic oxide pillars in the interlayers (10). The pillared clays are, in fact, nanocomposites, i.e., two or more Gibbsian solid phases mixed on a nanometer scale.

While the pillaring of swelling clays with numerous ceramic oxides has been remarkably successful, attempts to intercalate or pillar the swelling clays with zero-valent metals, by H_2 reduction, for example, have not been fruitful (14-16), because of the migration of metal particles to the external surfaces. However, recently we have been able to overcome this difficulty, and have successfully synthesized Cu metal intercalated clays by *in-situ* reduction of Cu^{2+} ions on the interlayer exchange sites by ethylene glycol at temperatures below 200°C (17, 18) using the polyol process (19). The metal clusters are apparently 4-5Å in size (17, 18) and metal clusters of these dimensions supported on microporous substrates are expected to behave very differently from bulk metals and may exhibit unique catalytic and adsorption properties, molecular selectivity and polyfunctional activity. Thus, the objective of the present study was to intercalate zero-valent metals of Cu and Ni in a montmorillonite clay by a modified ethylene glycol reduction process and thus obtain metal intercalated clay nanocomposites which are microporous.

EXPERIMENTAL

Montmorillonite Clay Host

A natural Na montmorillonite from Wyoming, USA, was used in this study. The clay (SWy-1) was supplied by the Source Clay Minerals Repository, Department of Geology, University of Missouri, Columbia, Missouri, 65201, U.S.A. The chemical composition of this sample was reported (20) to be 62.9% SiO_2, 19.6% Al_2O_3, 3.35% Fe_2O_3, 0.32% FeO, 3.05% MgO, 1.68% CaO, 1.53% Na_2O, 0.53% K_2O, 0.09% TiO_2, 0.049% P_2O_5, 0.05% S, 0.006% MnO, 0.111% F and 1.33% CO_2. The losses on heating below 550°C and 550-1000°C are 1.59 and 4.47%, respectively. The cation exchange capacity of this clay was reported to be 76.4 meq/100g while the N_2 surface area was found to be 31.82 ± 0.22 m^2/g (20).

Cu^{2+} and Ni^{2+} Exchange

The montmorillonite sample was ion exchanged with Cu^{2+} or Ni^{2+} by repeated treatment (3x) with an excess of 0.5M $Cu(NO_3)_2$ or $Ni(NO_3)_2$ at room temperature followed by washing with deionized water to remove excess Cu or Ni salts in the samples. These samples were then dried in an oven at about 60°C before use in the reduction experiments.

Reduction of Exchanged Metal Ions in Ethylene Glycol

About 0.5g of the above exchanged samples were mixed with 25 ml of ethylene glycol and treated for 2 hours at about 175°C using a modified (proprietory) polyol process. The reduction, however, can also be carried out by refluxion in ethylene glycol as has been reported previously (17-19). After the reduction treatment, the solid was separated from ethylene glycol by centrifugation. The sample was then washed with methanol repeatedly to remove all the ethylene glycol and subsequently stored in a vacuum desiccator.

Thermal Treatments

The reduced samples were heated under different conditions to determine the thermal stability of the intercalated metals. The heat treatments are (a) 200°C for 12 hours under evacuation (b) 400°C in H_2 for 4 hours and (c) 400°C in air for 4 hours.

Powder X-ray Diffraction

Slide mounts of the Cu^{2+} and Ni^{2+} exchanged samples, and their reduced and heated counterparts were prepared. Powder x-ray diffraction (XRD) patterns of the samples were recorded with a Scintag diffractometer using Ni-filtered CuKα radiation.

Water-sorption Isotherms and B.E.T. Nitrogen Surface Area

A computer-interfaced water sorption apparatus (21) was used for measuring water sorption isotherms at 25°C by a volumetric method. Samples were evacuated at 200°C for five hours before sorption measurements. The B.E.T. nitrogen surface areas were measured by a multi-point method using Autosorb-1, Quanta chrome apparatus. Samples were evacuated at 200°C for two hours before the nitrogen surface area measurements.

Transmission Electron Microscopy (TEM)

Powder samples dispersed in isopropyl alcohol were deposited on carbon-coated TEM grids and dried before TEM analysis using a Philips EM 420 microscope with an accelerating voltage of 120 kV. The smallest condenser aperture (50µm) was used to protect the sample from electron damage. Furthermore, the electron beam was spread during focussing to minimize sample damage. Selected area and/or convergent-beam electron diffraction patterns were obtained to identify the metal particles.

RESULTS AND DISCUSSION

Powder XRD

Powder XRD patterns of Cu^{2+}-exchanged sample, reduced sample and reduced and heated samples are compared in Figure 1. The Cu^{2+}-exchanged sample (Fig. 1a) showed a typical d-spacing of 12.4Å representing Cu^{2+} ions with a one-layer of water molecules between the layers (17). The reduced and degassed, at 200°C, sample (Fig. 1b) showed a d-spacing of about 13.39Å whereas the Cu^{2+}-exchanged sample when degassed at 200°C showed (figure not shown) a d-spacing of 9.6Å. This result shows that the reduction treatment has led to a stabilization of the interlayer spacing presumably due to the presence of Cu° metal particles of approximately 4Å formed by the reduction treatment. The size of the Cu° particles can be obtained by subtracting the thickness of the basic clay layer (9.6Å) from the 'd' spacing (13.39Å) of reduced i.e. metal intercalated sample. When the reduced sample was heated in H_2 at 400°C, the 'd' spacing remained the same (Fig. 1c) without any migration of the metal particles to the external surfaces of the clay. Treatment of the reduced sample in air at 400°C did cause some reduction in 'd' spacing (12.9Å) but did not lead to complete oxidation of the metal particles. This result shows that the Cu metal particles appear to be quite stable in the interlayers. There are only trace amounts of CuO as indicated by peaks at 2.56 and 2.32Å in the diffraction patterns and this may have formed by oxidation of the Cu metal particles present on the external surfaces. The fact that the intensity of these peaks did not significantly increase by treatment in air at 400°C suggests that the metal particles in the interlayers are stabilized. If CuO had formed in the interlayers by oxidation, it would have been expelled because CuO and other transition metal oxides are unstable in the interlayers (14-16). These results obtained with the modified polyol process clearly support our earlier findings with a different montmorillonite and its intercalation with Cu° metal particles.

Figure 2 shows powder XRD patterns of Ni^{2+}-exchanged sample, reduced sample and reduced and heated samples. The Ni^{2+}-exchanged sample (Figure 2a) had a somewhat broad 'd' spacing of 13.3Å which indicates the presence of about 1 layer of water molecules in the interlayers of montmorillonite along with the Ni^{2+} ions.

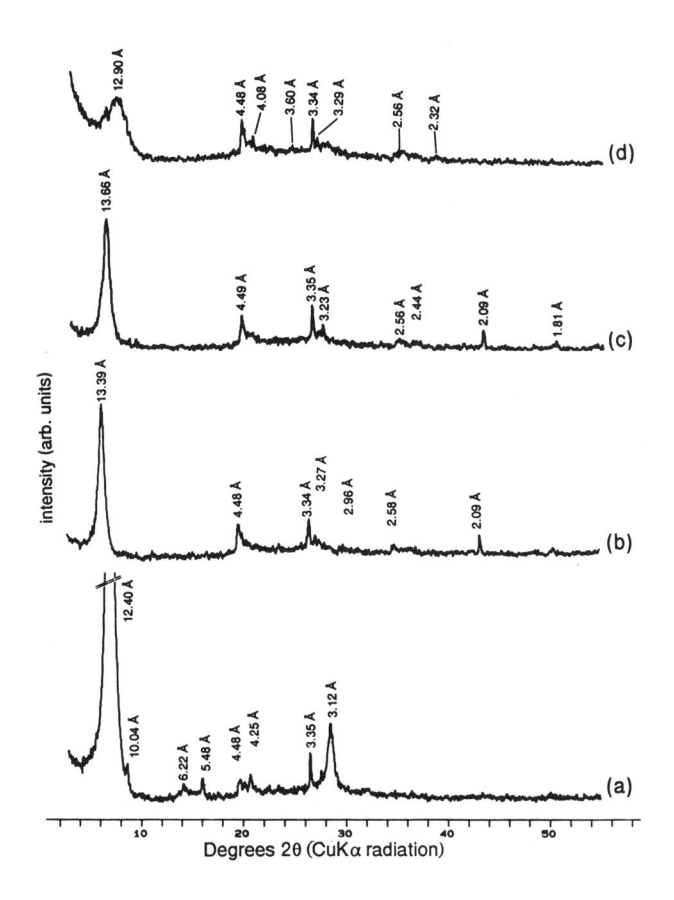

Figure 1 X-ray powder diffraction patterns of variously treated montmorillonites: (a) Cu^{2+}-exchanged (b) Cu°-intercalated by reducing in ethylene glycol followed by evacuation at 200°C (c) Cu°-intercalated and heated in hydrogen at 400°C for 4 hours and (d) Cu°-intercalated and heated in air at 400°C for 4 hours. The montmorillonite has quartz and feldspar impurities as indicated by the peaks at about 3.34 and 3.23Å, respectively.

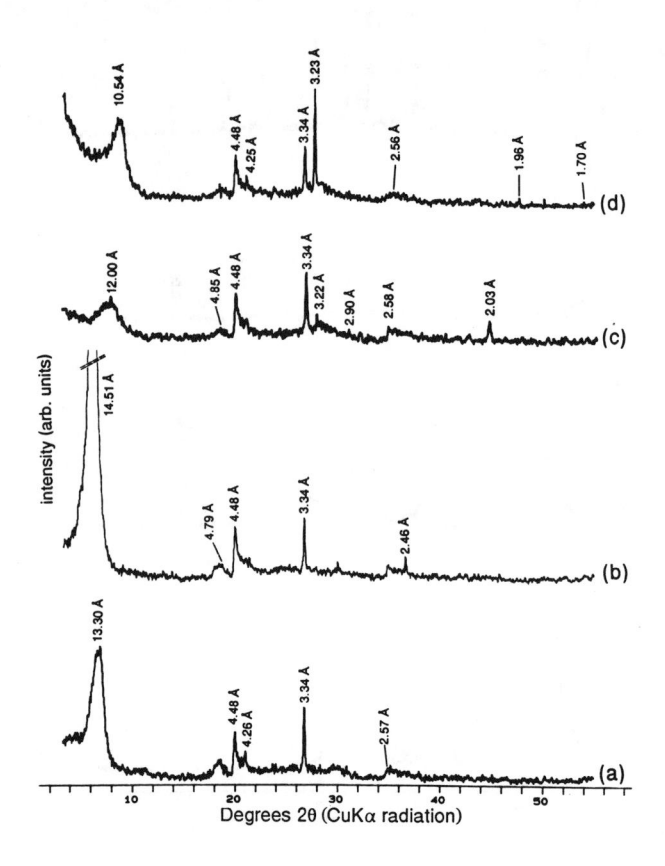

Figure 2 X-ray powder diffraction patterns of variously treated montmorillonites: (a) Ni^{2+}-exchanged (b) Ni°-intercalated by reducing in ethylene glycol followed by evacuation at 200°C (c) Ni°-intercalated and heated in hydrogen at 400°C for 4 hours and (d) Ni°-intercalated and heated in air at 400°C for 4 hours. Note quartz and feldspar impurities at 3.34 and 3.23Å, respectively, in the montmorillonite sample.

When the above sample was reduced in ethylene glycol and degassed at 200°C, it showed a 'd' spacing of 14.51Å. Subtraction of a layer thickness of 9.6Å from this 'd' spacing gives about 5Å for the thickness of metal particles present in the interlayers, the 9.6Å being the layer thickness of a Ni^{2+} exchanged sample after degassing at 200°C. Heating the sample in H_2 at 400°C led to a broad 'd' spacing of 12Å which indicates some removal of Ni metal particles from the interlayers. This inference is supported by the fact that Ni metal peak can be detected at 2.03Å in the XRD pattern of the sample heated in H_2 at 400°C (Figure 2c) but not in the sample which was reduced and evacuated at 200°C (Figure 2b). When the sample was heated in air at 400°C, it showed a 'd' spacing of 10.54Å indicating the collapse of the layers because of oxidation of Ni to NiO. The fact that the basal spacing did not collapse to 9.6Å shows the resistance to oxidation of the Ni metal because of its presence in the interlayers.

Neither Ni° nor NiO could be detected in the XRD pattern of the sample which was heated in air at 400°C (Figure 2d). The stabilization of the interlayers in samples heated at 400°C in H_2 or air (Figure 2c,d) could not have been caused by the presence of NiO because NiO is unstable in the interlayers (13-16) and would have been expelled from the interlayers. The XRD results clearly show that both Cu° and Ni° intercalated clay nanocomposites can be synthesized using the low temperature polyol process.

Water Sorption Isotherms and Nitrogen Surface Areas

In order to determine the microporous nature of the metal intercalated clay nanocomposites water sorption isotherms and water and nitrogen surface areas were determined. Figure 3 shows the water adsorption isotherms of Cu^{2+} exchanged and 200°C evacuated sample and Cu^{2+} exchanged, reduced and 200°C evacuated sample. The latter sample showed an initial enhancement in water sorption (Figure 3a) compared to the former (Figure 3b). The adsorption of water is also higher in the reduced sample up to P/Po = 0.9 (Figure 3a) because of the the exchanged and 200°C evacuated sample (Table 1) because of the presence of micropores in the former. In addition to the micropores, there are mesopores and micropores available in the interlayers. The microporous nature of the Cu° intercalated sample can also be inferred from the high H_2O and N_2 surface areas (Table 1) compared to the Cu^{2+} exchanged and 200°C evacuated sample. Water adsorption isotherms of Ni^{2+} exchanged and 200°C evacuated sample and Ni^{2+} exchanged, reduced and 200°C evacuated sample are compared in Figure 4. The reduced sample with Ni° metal clusters in the interlayers clearly shows a higher adsorption at all water vapor pressures compared to the Ni^{2+} exchanged and 200°C evacuated sample. The former also shows a much higher adsorption capacity than the exchanged and 200°C evacuated sample (Table 1) because of the presence of micropores in the former. In addition to the micropores, there are mesopores and micropores in all the samples as can be deduced from the large condensation regions which is typical of montmorillonites. The water adsorption isotherms of metal intercalated clays are of Type II and the surface areas reported here for these are comparable to those for alkylammonium intercalated montmorillonites (22) having about 4.5Å interlayer free distances.

The results clearly show that the metal intercalated clay nanocomposites are microporous. Further control of the microporous nature may be possible by optimizing the parameters such as Cu^{2+} loading, layer charge, time and temperature of treatment, type of clay etc.

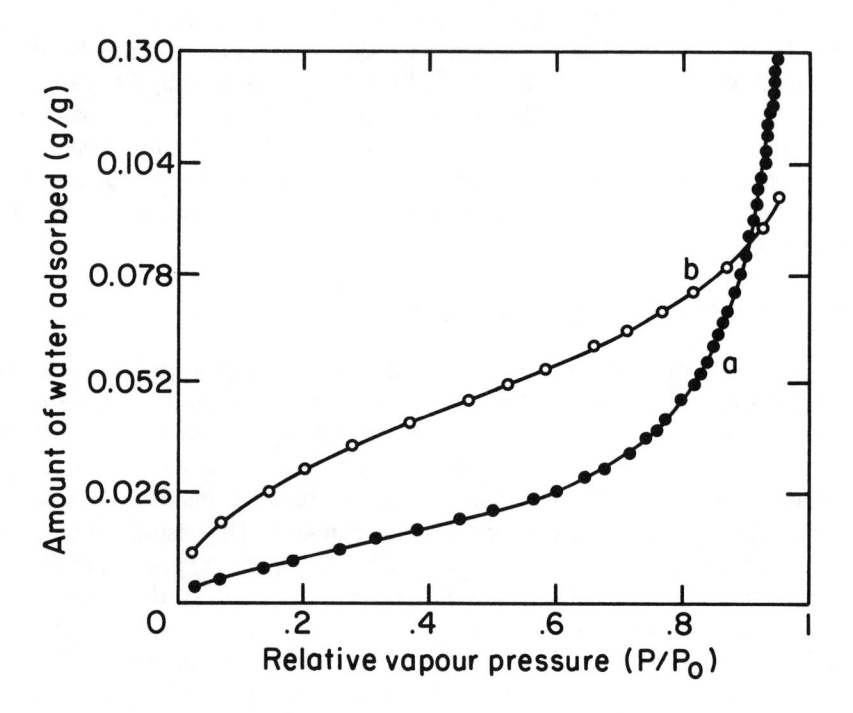

Figure 3 Water adsorption isotherms of differently treated Cu montmorillonites: (a) Cu^{2+} exchanged and degassed at 200°C and (b) same as above but reduced in ethylene glycol and degassed at 200°C.

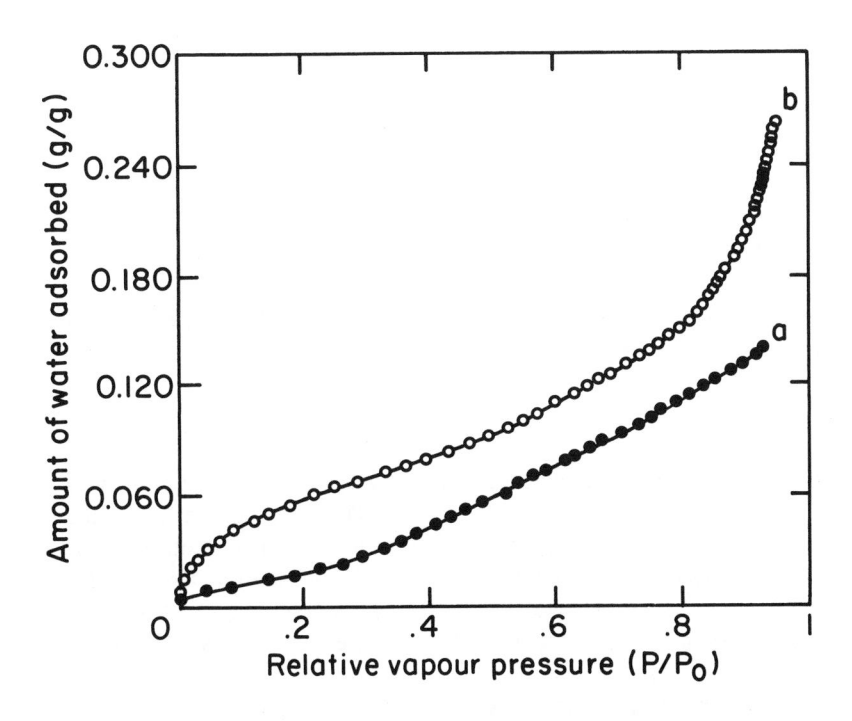

Figure 4 Water adsorption isotherms of differently treated Ni montmorillonites: (a) Ni^{2+} exchanged and degassed at 200°C and (b) same as above but reduced in ethylene glycol and degassed at 200°C.

Table 1 BET nitrogen and water surface areas of different samples

Sample	Surface area, m²/g	
	N_2	H_2O
Cu^{2+} exchanged	37	48
Cu^{2+} exchanged and reduced	64	110
Ni^{2+} exchanged	34	115
Ni^{2+} exchanged and reduced	165	187

Transmission Electron Microscopy

The Cu and Ni metal intercalated clay nanocomposites were examined by TEM to detect the metal particles. Figure 5 shows the TEM photographs of Cu intercalated clay nanocomposite before and after its destruction in the electron beam. No Cu particles can be seen on the clay surface before the destruction of clay in the electron beam (Figure 5A). However, note the coarsening of the structure of montmorillonite after the destruction of clay in the electron beam (Figure 5B).

Although the metal particles formed by the aggregation of the metal clusters expelled from the interlayers are not clearly discerned (arrows), electron diffraction showed a weak and diffuse Cu (111) reflection confirming the presence of Cu°. The transmission electron micrographs of Ni° intercalated clay nanocomposites before and after destruction in the electron beam are shown in Figure 6. A few Ni° particles of 50-200 nm were seen (Figure 6A) before destruction (see arrow). After the destruction of the clay in the electron beam numerous Ni° particles of 5-20 nm appeared besides the initially present large particles (Figure 6B). Note the rounding (Figure 6B, arrows) of the Ni° after exposure to the electron beam. It is clear from these micrographs that the destruction of the Ni-clay nanocomposite in the electron beam led to the aggregation and growth of the Ni° clusters from the interlayers and their expulsion to the surfaces of the destroyed clay.

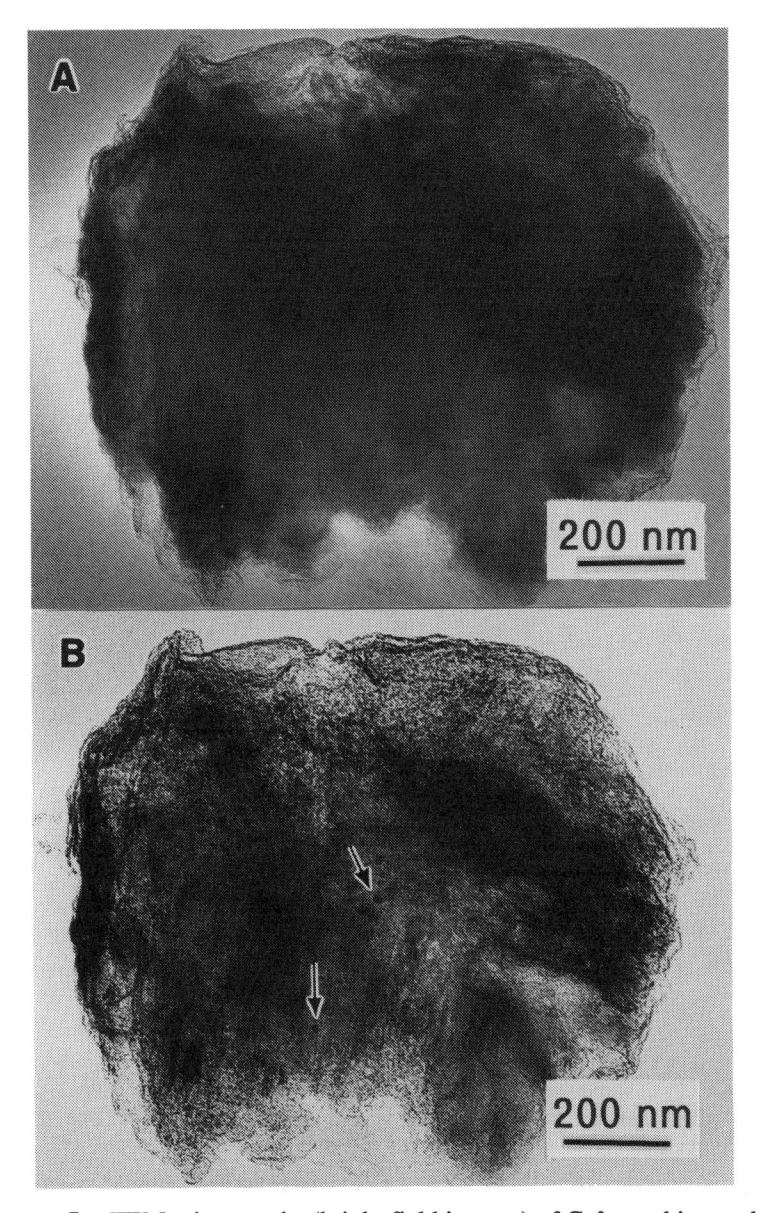

Figure 5. TEM micrographs (bright field images) of Cu° metal intercalated clay nanocomposite: (A) after 10-20 seconds exposure to the electron beam, and (B) after about 10 minutes. Note the appearance of about 20-30 nm Cu° particles (arrows) in B.

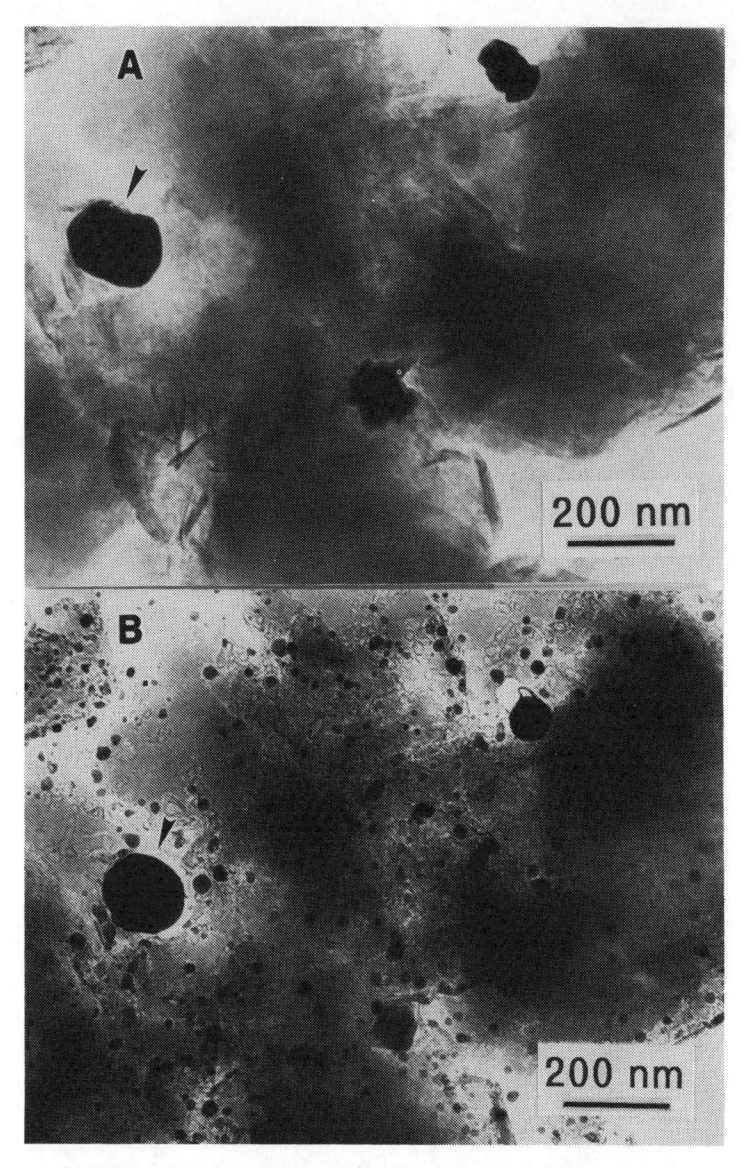

Figure 6 TEM micrographs (bright field images) of Ni° metal intercalated clay nanocomposite: (A) after 10-20 seconds exposure to the electron beam, and (B) after about 10 minutes. Note the appearance of about 5-20 nm Ni° crystallites in B. Note also the rounding of the initial Ni° particles (arrows in A and B).

CONCLUSIONS

Montmorillonite clay has been successfully intercalated with about 4-5Å Cu° and Ni° metal clusters using the polyol reduction process. Water adsorption isotherms and BET N_2 and water surface areas confirm the microporous nature of the Cu° and Ni° intercalated clay nanocomposites.

ACKNOWLEDGEMENTS

This research was supported by the Division of Materials Science, Office of Basic Energy Sciences, U. S. Department of Energy under grant No. DE-FG02-85ER45204. Experimental assistance by Drs. M. Z. B. Hussein and E. Breval and Ms. C. L. Liu, is gratefully appreciated.

REFERENCES

1. R. E. Grim, Clay Mineralogy, McGraw-Hill, New York, 1968.
2. G. W. Brindley and G. Brown, Crystal Structures of Clay Minerals and Their X-Ray Identification, Mineralogical Society of Great Britain, 1980.
3. T. J. Pinnavaia, "Intercalated Clay Catalysts" Science 220, 365-371 (1983).
4. S. Yamanaka and G. W. Brindley, "High Surface Area Solids Obtained by Reaction of Montmorillonite with Zirconyl Chloride" Clays Clay Miner. 27 [2] 119-124 (1979).
5. T. J. Pinnavaia, M. S. Tzou and S. D. Landau, "New Chromia Pillared Clay Catalysts", J. Am. Chem. Soc. 107, 4783-4785 (1985).
6. S. Yamanaka, T. Doi, S. Sako and M. Hattori, "High Surface Area Solid Obtained by Intercalation of Iron Oxide Pillars in Montmorillonite", Mat. Res. Bull. 19 [2] 161-168 (1984).
7. S. Yamanaka, G. Yamashita and M. Hattori, "Reaction of Hydroxy-Bismuth Polycations with Montmorillonites", Clays Clay Miner. 28 [4] 281-284 (1980).
8. J. Sterte and J. Shabtai, "Cross-Linked Smectites V. Synthesis and Properties of Hydroxy-Silicoaluminum Montmorillonites and Fluorohectorites", Clays Clay Miner., 35 [6] 429-439 (1987).
9. J. Sterte, "Synthesis and Properties of Titanium Oxide Cross-Linked Montmorillonite", Clays Clay Miner., 34 [6] 658-664 (1986).
10. S. Yamanaka, "Design and Synthesis of Functional Layered Nanocomposites", Am. Ceram. Soc. Bull., 70 [6] 1056-1058 (1991).
11. P. B. Malla, S. Yamanaka, and S. Komarneni, "Unusual Water Vapor Adsorption Behavior of Montmorillonite Pillared with Ceramic Oxide", Solid State Ionics, 32/33, 354-362 (1989).
12. S. Yamanaka, P. B. Malla, and S. Komarneni, "Water Adsorption Properties of Alumina Pillared Clay", J. Colloid Interface Sci. 134 [1] 51-58 (1990).
13. S. Yamanaka, T. Nishihara, and M. Hattori, "Preparation and Properties of Titania Pillared Clay", Mat. Chem. Phys. 17, [1] 87-101 (1987).

14. M. Patel, "Reduction of Interlayer Ni^{2+} and Cu^{2+} in Montmorillonite with Hydrogen", <u>Clays Clay Miner.</u>, 30 [5] 397-399 (1982).
15. K. Ohtsuka, M. Suda, M. Ono and M. Takahashi, "Preparation of Nickel (II)-Hydroxide-(Sodium Fluoride Tetrasilicic Mica) Intercalation Complexes and Formation of Ultra Fine Nickel Particles by H$_2$ Reduction", <u>Bull. Chem. Soc. Jpn.</u>, 60 [3] 871-876 (1987).
16. K. Ohtsuka, J. Koga, M. Suda, M. Ono and M. Takahashi, "Preparation and Properties of Cobalt (II) Hydroxide-(Sodium Fluoride Tetrasilicic Mica) Intercalation Complexes and of Highly Dispersed Cobalt on Mica", <u>Bull. Chem. Soc. Jpn.</u>, 60 [8] 2843-2847 (1987).
17. P. B. Malla, P. Ravindranathan, S. Komarneni and R. Roy, "Intercalation of Copper Metal Clusters in Montmorillonite", <u>Nature</u> 351, 555-557 (1991).
18. P. B. Malla, P. Ravindranathan, S. Komarneni, E. Breval and R. Roy, "Reduction of Copper Acetate Hydroxide Hydrate Interlayers in Montmorillonite by a Polyol Process, A New Approach in the Preparation of Metal Supported Catalysts", <u>J. Mat. Chem.</u>, 2 [5] 559-565 (1992).
19. F. Fievet, J. P. Lagier, B. Blin, B. Beaudoin and M. Figlarz, "Homogeneous and Heterogeneous Nucleations in the Polyol Process for the Preparation of Micron and Submicron Size Metal Particles", <u>Solid State Ionics</u>, 32/33 [1] 198-205 (1989).
20. H. Van Olphen and J. J. Fripiat, "Data Handbook for Clay Materials and Other Non-metallic Minerals", Pergamon, Oxford, 1979.
21. S. Yamanaka and S. Komarneni, "Apparatus for Measuring Liquid Vapor Adsorption and Desorption Characteristics of a Sample", United States Patent 5,058,442, October 22, 1991.
22. R. M. Barrer, "Zeolite and Clay Minerals as Sorbents and Molecular Sieves", Academic Press, London, 1978.

MICRO- AND MESOPOROUS MATERIALS DERIVED FROM TWO-DIMENSIONAL SILICATE LAYERS OF CLAY

Shoji Yamanaka and Koichi Takahama[*]
Department of Applied Chemistry, Faculty of Engineering, Hiroshima University, Higashi-Hiroshima 724, Japan
[*] Central Research Laboratory, Matsushita Electric Works Ltd., 1048 Kadoma, Osaka 571, Japan

Two-dimensional silicate layers of montmorillonite clay were pillared with SiO_2-TiO_2 mixed oxide sols. The small and positively charged sol particles were packed between the silicate layers, expanding the basal spacing up to a value larger than 4 nm. Micropores were formed in the interstices between the packed sol particles and the silicate layers. The sol particles were rearranged by incorporation of organic cations; mesopores were formed by burning off the organic part. Supercritical drying of the sol pillared clay resulted in the preservation of a cardhouse type unique porous structure.

INTRODUCTION

Aluminosilicate zeolites are typical microporous crystals which are constructed of TO_4 tetrahedral units as primary building blocks, where T is either Si or Al. These units are connected through their corners to form a variety of secondary building units such as single or double 4-, 5-, and 6-membered rings, which in turn connect to form the extended frameworks of zeolites having cavities and channels within the structures.[1] Although owing to the recent developments of characterization methods, some of the building units have been found in aluminosilicate solutions,[2] little is known about the mechanisms or the routes by which these building units are linked to generate the structures.

If it is very difficult to design the synthetic route to obtain a desired structure, a possible choice that we can make next is to use a larger building unit, from which we can anticipate the final structure easily.[3] The two-dimensional silicate layers of expandable clays can be versatile building units for such a purpose; the silicate layers have a thickness as thin as a molecular level (0.96 nm), and have a sufficient stiffness for a building unit.

There are two types of combination for the silicate layers to be assembled with other building units as schematically shown in Fig. 1;
  (i) Two dimension + one (or zero) dimension (2D+1D or 2D+0D),
  (ii) Two dimension + two dimension (2D+2D).

The first combination is obtained from the so-called pillaring process, where the two-dimensional silicate layers are kept apart by one- (or zero-) dimensional ceramic oxide particles and micropores are formed between the silicate layers. The pore dimensions can be roughly estimated from the size of pillars introduced. Pillared clays with a variety of ceramic oxide particles have been synthesized and their adsorption and catalytic properties have been extensively studied.[4-12] In the second combination, the silicate layers are stacked with a different type of layer structured compound such as metal hydroxide by a face-to-face association,[13] which is called intergrowth or interstratified structure. If the layers are linked by a face-to-edge association, it results in a card-house structure, and much larger pores would be formed between the layers.

This paper briefly reviews our recent studies on the pillaring of montmorillonite clay with SiO_2-TiO_2 sol particles, and some attempts to modify the structure to derive micro- to mesoporous materials.

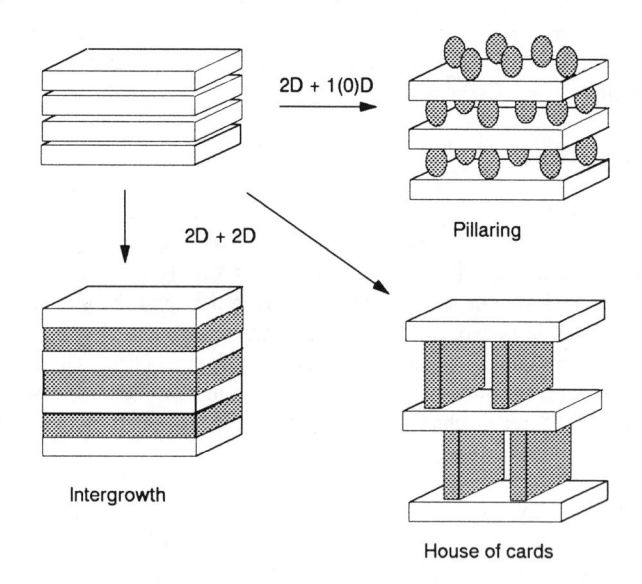

Figure 1 Types of association of two-dimensional silicate layers with other building units.

PILLARING WITH SOL PARTICLES

The clay used in this study is mainly Na-montmorillonite supplied from Kunimine Ind. with the name Kunipia F; the cation exchange capacity (CEC) is 100 meq/100 g. A silica sol solution was prepared by mixing silicon tetraethoxide $Si(OC_2H_5)_4$, 2 M HCl and ethanol in a ratio of 41.6 g/10 ml/12 ml at room temperature according to Sakka et al..[14] Titanium tetraisopropoxide $Ti(OC_3H_7)_4$ was hydrolyzed by adding it to 1 M HCl solution in such a way that the molar ratio of $HCl/Ti(OC_3H_7)_4$ is about 4. The resulting slurry was peptized to a clear sol solution by continuous stirring for 3 h at room temperature.[15]

The preparation of the pillared clays is very simple. The silica and the titania sol solutions thus prepared were mixed in various ratios, and stirred for 30 min at room temperature, which were then added to the clay suspension.[16] The mixed ratios will be hereafter designated as 30/3/1 for example, where the first number and the second number refer to the molar ratios of silica and titania to the CEC equivalent (the third number) of the clay, respectively. The reaction temperature as well as the mixing ratio were varied. After stirring for different reaction times, the products were separated by centrifugation, washed with water several times, and dried in a stream of dry air at room temperature.

STRUCTURE AND ADSORPTION PROPERTIES

The silica sol particles obtained by the hydrolysis of silicon tetraethoxide are negatively charged, which can be converted into positively charged sols by the modification with small amount of titanium ions. The modified sol particles can be ion-exchanged with the interlayer cations of montmorillonite. The spacing increased rapidly soon after the clay was mixed with the sol solutions and then attained constant values at ~4 nm after about 6 h.[16] As shown in Fig. 2, the amount of SiO_2-TiO_2 sol 30-50 times the equivalent of the CEC of the clay is required to obtain the fully expanded pillared clays. Although the spacing of the sol pillared clay is remarkably larger than those ever reported for the clays pillared with other ceramic oxides, the nitrogen adsorption isotherm (Fig. 3) is fitted on the BET plot for the limited number of adsorption layers, suggesting that the pore sizes are much smaller than the pillar heights. The large basal spacing and a high surface area of about 400 m^2/g were maintained at least up to 500°C. Adsorption-desorption isotherms for various kinds of solvent vapors were also measured at 25°C on the 30/3/1 sample. The adsorption isotherms for large molecules such as xylenes and mesitylene fitted on the Langmuir linear plot, whereas those for smaller molecules such as water and methanol fitted rather on the BET plot. This finding also supports that the pore sizes are on the order of the molecular dimensions examined. A structural model for the SiO_2-TiO_2 sol pillared clays is proposed in Fig. 4, where the average size of the SiO_2-TiO_2 sol particles must be much smaller

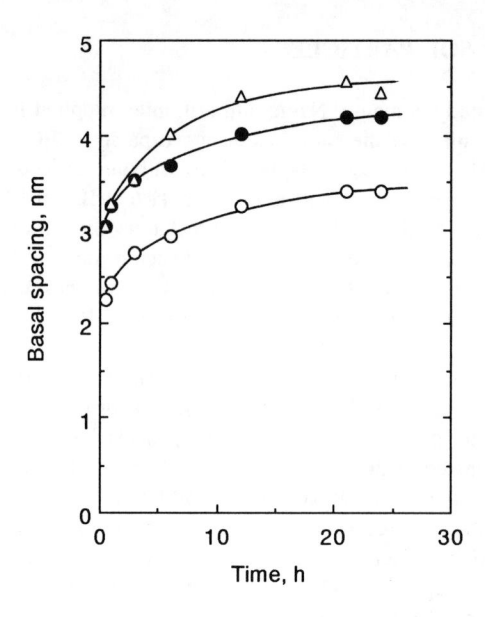

Figure 2 Basal spacing of the products as a function of reaction time at 50°C. Three different mixtures were used: ○, 10/1/1; ●, 30/3/1; △, 100/10/1.[16]

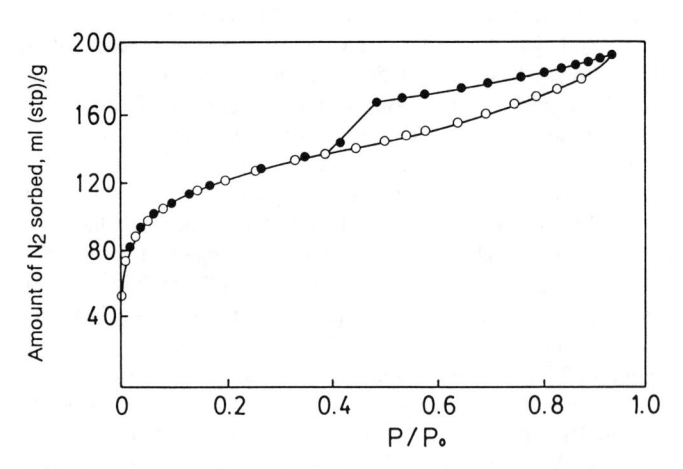

Figure 3 Adsorption–desorption isotherm of nitrogen for a typical reaction product (50/5/1) calcined at 500°C; ○, adsorption and ●, desorption.[6]

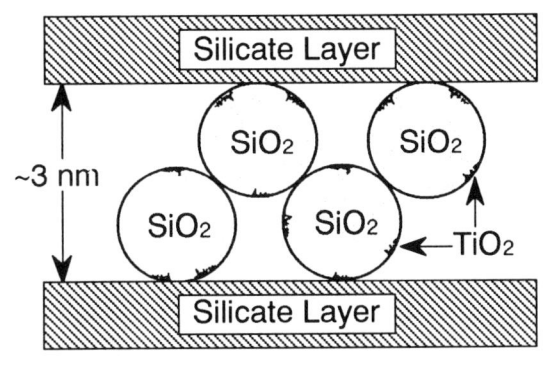

Figure 4 Schematic structural model of SiO_2-TiO_2 pillared clays.[16)]

than the interlayer spaces and such small sol particles are packed uniformly so as to form small pores in between the sols and silicate layers. The SiO_2 sols are modified by the TiO_2 on the surface and positively charged.

INTERLAYER REARRANGEMENT

Although the SiO_2-TiO_2 pillared clays have a very large basal spacing, the interlayers are occupied by the silica sol particles, and thus only the interstices between the pillars and silicate layers are used as pores. An attempt was made to replace a part of the sol particles in the interlayer space with large organic cations which could be subsequently removed by burning.[17)] The SiO_2-TiO_2 pillared clay (4/0.4/1) was treated with octadecyltrimethylammonium (OTMA) chloride as an organic cation agent by the following two methods:

Process I: OTMA in various ratios was added to a 0.8 wt% montmorillonite suspension and stirred; SiO_2-TiO_2 sols were then added.
Process II: SiO_2-TiO_2 sols were added to a 0.8 wt% montmorillonite suspension and stirred; OTMA was then added.

Each mixture was stirred for 1.5 h at room temperature, and centrifuged and washed several times.

The order of the addition was found to be important; if OTMA was added first before the ion-exchange with sol solutions (Process I), the OTMA ions interacted with silicate layers so strongly that the montmorillonite did not swell for further ion exchange with the sol particles. On the other hand, if the sol pillared clays were first prepared, and then OTMA was added (Process II), the sol particles could be partly replaced with OTMA with the rearrangement of the interlayer sol particles. The basal spacing increased from 4 to 4.4–4.9 nm by incorporation of OTMA ions,

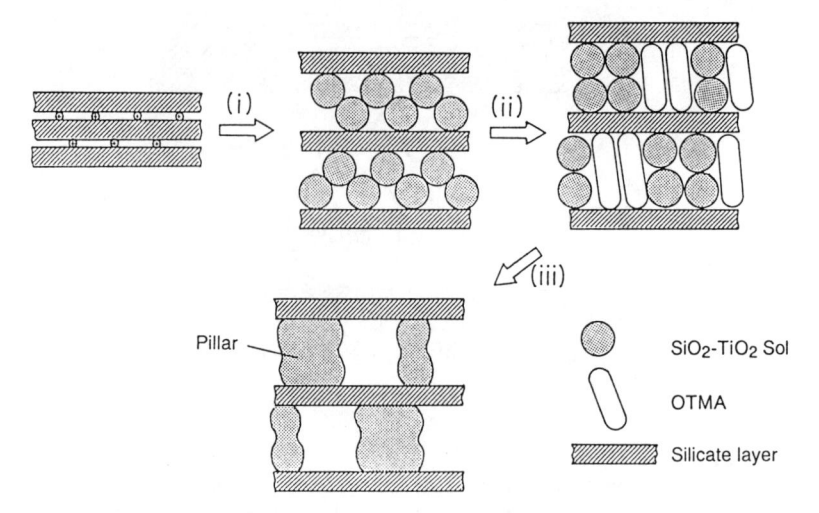

Figure 5 Schematic structural model for the formation of mesopores in the interlayer spaces of montmorillonite by the Process II: (i) ion-exchange with sol particles, (ii) uptake of OTMA along with the rearrangement of the interlayer sol particles, (iii) removal of OTMA by calcination and formation of mesopores.[17]

and the enlarged spacing was maintained even after calcination at 500°C. The most porous material obtained by this method had a pore volume as large as 0.8 ml/g and a BET surface area larger than 600 m^2/g. The pore size distribution was determined on the basis of the adsorption isotherm of nitrogen, which indicated that the increase in the pore volume came mainly from mesopores ranging from 2 to 3 nm in diameter. Figure 5 shows a schematic structural model for the formation of mesopores in the interlayer spaces of montmorillonite by Process II.

CARD-HOUSE STRUCTURE

The ion exchange of the interlayer cations of montmorillonite with precursory cations for pillars is usually carried out in aqueous media, and the samples separated after the ion exchange are extremely swelled with water. The swelled sample experiences remarkable shrinkage during drying, as in the case of the drying of aqueous gels formed by solution-sol-gel route.[18] The shrinkage of the gel is caused by the large liquid-vapor interfacial (capillary) forces, which act to disrupt the delicate microporous gel structure during an evaporative procedure.

Recently a new drying technique, supercritical drying, has been developed in which the solvent in the gel pores is substituted by a fluid with a relatively low critical temperature, and the fluid is extracted supercritically avoiding the formation of liquid-vapor interfaces.[19,20] The supercritical drying technique was applied to the drying of SiO_2-TiO_2 sol pillared clays, and the resulting pore structure was investigated in comparison with that of the sample prepared by conventional air-drying procedures.[21]

The SiO_2-TiO_2 pillared clay (4/0.4/1) was rinsed with ethanol three times by centrifugation in order to replace the water in the product with the ethanol, which was then extracted with supercritical fluid of CO_2 under a pressure of 120 atm at 40°C. The CO_2 was vented for 5 h. For comparison, a separate sample of SiO_2-TiO_2 pillared clay was similarly prepared and air-dried at 60°C. The samples thus prepared were calcined at 500°C for 3 h in air.

The BET surface areas and micropore volumes of the two samples were measured by nitrogen adsorption at the liquid nitrogen temperature. Pore-size distributions in a range larger than about 10 nm in diameter were analyzed by mercury intrusion measurements. The total pore volumes measured by the nitrogen adsorption study and the mercury porosimetry are summarized in Table 1. It is clearly shown that the air-dried sample has few macropores, while the supercritically dried (SCD) sample has an extremely large pore volume of 9.82 ml/g, the most part of which comes from the pores ranging from 5-0.5 μm in diameter. It is interesting to note that the total mercury intrusion volume for the air-dried sample is only 1% of that for the SCD sample.

Figure 6 shows SEM photographs of the two dried samples. It is shown that the SCD sample has a well developed card-house structure, and macropores are constructed by thin silicate layer walls. From the photograph, it is also seen that most of the pores are approximately 3-0.3 μm in diameter, in accordance with the data given by the mercury porosimetry. In contrast to the SCD sample, in the air-dried sample the silicate layers are stacked in a face-to-face fashion, and only a small number of large pores are observed.

Table 1. BET Surface Areas and the Pore Volumes Measured by Nitrogen Adsorption Analysis and Mercury Porosimetry.

Drying procedure	BET surface area, $m^2 g^{-1}$	Pore volumes	
		N_2 adsorption [1] $ml\ g^{-1}$	Hg porosimetry $ml\ g^{-1}$
Supercritical drying	459	1.13 (0.09)	9.82
Air drying	325	0.25 (0.19)	0.098

1) Micropore volumes are shown in parentheses.

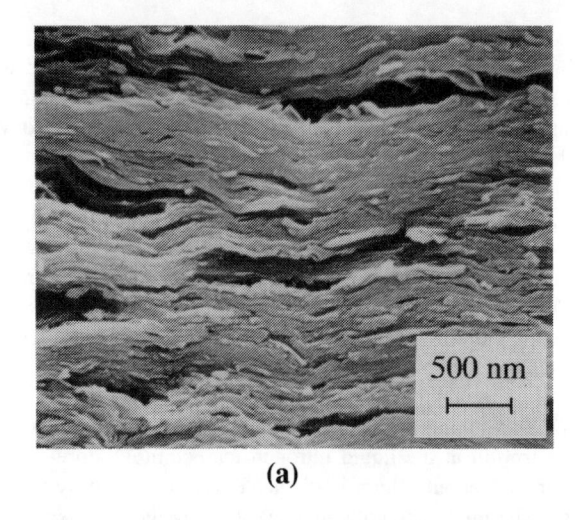

Figure 6 SEM photographs of the two pillared clays. (a) Air-dried, (b) supercritically dried.[21]

The XRD study revealed that the pillared structure with a basal spacing of 3-4 nm was preserved on supercritical drying. The packing of the small sol particles in the interlayer space would not be much influenced by the drying method. Evidently,

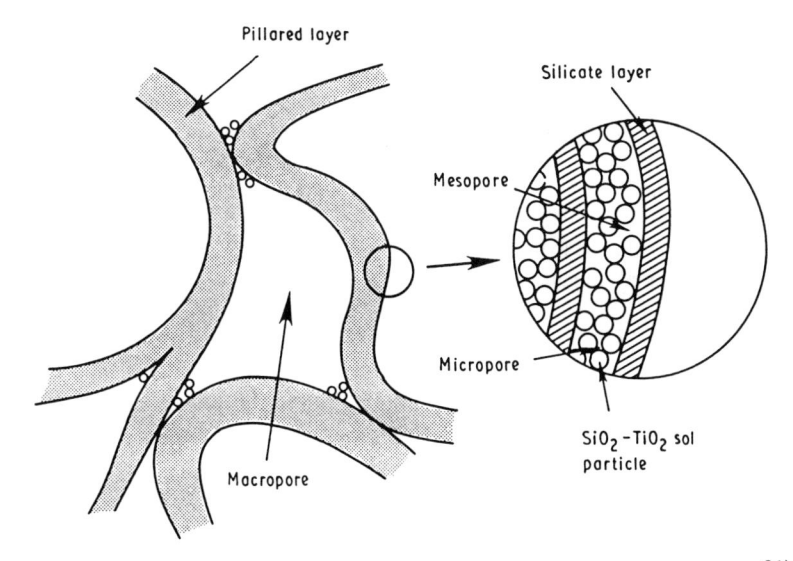

Figure 7 Schematic structural model of the pillared clay supercritically dried.[21]

the macropores which were measured by mercury intrusion had been filled with water before drying. The macropores would disappear in a conventional evaporative drying procedure, because the strong interfacial forces between the swelled silicate layers lead to collapse of the pores on evaporation of the water. In the supercritical drying procedure, however, the water occupying the macropores is replaced with alcohol and then with a supercritical CO_2 fluid. In this case, the formation of liquid-vapor interfaces is avoided and the delicate card-house structure constructed by the pillared silicate layer can be preserved during drying. The pore-volume data given in Table 1 show that the macropores shrink even to 1% of the original volume on air drying.

A schematic structural model showing the dual-pore structure of the SCD sample is shown in Fig. 7. The micro- and mesopores are located in between the loosely packed sol particles in the interlayer space of the pillared layers. The macropores are included in the highly porous card-house structure developed by the connection of the pillared layers.

CONCLUSION

The two-dimensional silicate layers of montmorillonite are useful building units for the preparation of micro- and mesoporous materials in combination with SiO_2-

TiO$_2$ sol particles. The positively charged sol particles are packed between the silicate layers by the ion exchange with the interlayer cations of montmorillonite. Micropores are formed in the interstices between the packed sol particles and the interlayer surface of the silicate layers. The sol particles can be rearranged by incorporation of organic cations such as octadecyltrimethylammonium; mesopores are formed by burning off the organic part. The supercritical drying of the swelled pillared clays can preserve the delicate card-house type porous structure; a dual-pore structure results with micropores between the silicate layers and macropores between the pillared silicate layers.

ACKNOWLEDGMENT

This study was partly defrayed by Grant-in-Aid for Scientific Research on Priority Area of the Ministry of Education, Science and Culture.

REFERENCES

1. R. Szostak, *Molecular Sieves, Principles of Synthesis and Identification*, Van Nostrand Reinhold, New York (1989).
2. A. V. McCormick, A. T. Bell and C. J. Radke, "Application of ^{29}Si and ^{27}Al NMR to determine the distribution of anions in sodium silicate and sodium aluminum silicate solutions," in *Proc. 7th Intern. Zeolite Conf.*, Y. Murakami, A. Iijima, and J. W. Ward eds., Kodansha/Elsevier, Tokyo (1986) p.247-254.
3. S. Yamanaka, "Design and synthesis of functional layered nanocomposites," *Ceramic Bull.*, **70**, 1056-1058 (1991).
4. R. Burch, ed., "Pillared Clays," *Catalysis Today*, **2**, 1-185 (1988).
5. B. Delmon and P. Grange, "New materials in catalysis," *Erdol Erdgas Kohle*, **107**, 376-382 (1991).
6. EM. Farfan-Torres and P. Grange, "Pillared Clays," *J. Chim. Phys.*, **87**, 1547-1560 (1990).
7. F. Figueras, "Pillared clays as catalysts," *Catal. Rev. Sci. Eng.*, **30**, 457-499 (1988).
8. T. J. Pinnavaia, "Intercalated clay catalysts," *Science*, **220**, 365-371 (1983).
9. S. Yamanaka and M. Hattori, "Clays pillared with ceramic oxides" in *Chemistry of Microporous Crystals*, T. Inui, S. Namba and T. Tatsumi (eds.), Kodansha/Elsevier, Tokyo, 1991, p.89-96 (1991).
10. S. Yamanaka and M. Hattori, "Preparation and adsorption properties of pillared clays," *Hyomen*, **27**, 290-301 (1989) (Japanese).
11. R. A. Schoonheydt, "Clays: From two to three dimensions," *Stud. Surf. Sci. Catal.*, **58**, 201-239 (1991).
12. M. L. Occelli and H. E. Robson eds., *Expanded Clays and Other Microporous Solids*, Van Nostrand Reinhold, New York (1992).
13. S. Yamanaka and G. W. Brindley, "Hydroxy-nickel interlayering in montmorillonite by titration method," *Clays Clay Miner.*, **26**, 21-24 (1978).
14. S. Sakka, K. Kamiya, K. Makita, and Y. Yamamoto, "Formation of sheets and coating films from alkoxide solutions," *J. Non-Cryst. Solids*, **63**, 223-235 (1984).
15. S. Yamanaka, T. Nishihara, M. Hattori, and Y. Suzuki, "Preparation and properties of

titania pillared clay," *Mat. Chem. Phys.*, **17**, 87–101 (1987).
16. S. Yamanaka, Y. Inoue, M. Hattori, F. Okumura, and M. Yoshikawa, "Preparation and properties of the clays pillared with SiO_2-TiO_2 sol particles," *Bull. Chem. Soc. Jpn.*, **65**, 2494–2500 (1992).
17. K. Takahama, M. Yokoyama, S. Hirao, S. Yamanaka, and M. Hattori, "Preparation and properties of pillared clays with controlled pore structures (part 1) – Effects of organic cation additives–," *J. Ceramic Soc. Jpn. Int. Ed.*, **99**, 14–18 (1991).
18. G. W. Scherer, "Theory of drying gels" in *Ultrastructure Processing of Advanced Ceramics*, J. D.Mackenzie and D. R. Ulrich (eds.) Wiley, New York (1988) p.295–302.
19. D. W. Matsonard and R. D. Smith, "Supercritical fluid technologies for ceramic processing applications," *J. Amer. Ceramic Soc.*, **72**, 871–881 (1989).
20. P. H. Tewari, A. J. Hunt, and K. D. Lofftus, "Ambient–temperature supercritical drying of transparent silica aerogels," *Mater. Lett.*, **3**, 363–367 (1985).
21. K. Takahama, M. Yokoyama, S. Hirao, S. Yamanaka, and M. Hattori, "Supercritical drying of SiO_2-TiO_2 sol–pillared clays," *J. Mat. Sci.*, **27**, 1297–1301 (1992).

PROPERTIES AND APPLICATIONS OF FUNCTIONAL POROUS GLASS-CERAMICS COMPOSED OF A TITANIUM PHOSPHATE CRYSTAL SKELETON

Hideo HOSONO and Yoshihiro ABE
Department of Materials Science and Engineering, Nagoya Institute of Technology, Gokiso-cho, Showa-ku, Nagoya 466, JAPAN

Bulk microporous glass-ceramics with a skeleton of two types of functional titanium phosphate crystals were prepared by a controlled crystallization of glasses in the pseudo-binary $RTi_2(PO_4)_3$ ($R=Li$, Na, $1/2Ca$) - $Ca_3(PO_4)_2$ system and subsequent chemical treatments without accompanying serious cracking, deformation and shrinkage. One is a NASICON-type crystal with a 3-dimensional network structure comprised of TiO_6 and PO_4. The other is a zirconium phosphate-type crystal $Ti(HPO_4)_2 \cdot 2H_2O$ with a 2-dimensional layered structure. The mean pore diameter and porosity are 30-60 nm (controllable to ~150 nm) and 40-60 vol %, respectively. It was found that porous $CaTi_4(PO_4)_6$ and $LiTi_2(PO_4)_3$ glass-ceramics have excellent properties as a supporting material for immobilization of enzymes or a humidity sensor and an ion exchanger, respectively.

INTRODUCTION

Since thermal, chemical, and mechanical properties of ceramics are much better than those of metals and organic polymers, porous ceramics are being used in various engineering fields such as catalytic supports, membranes, and gas-sensors. Novel uses, such as immobilized enzymes and packing materials in chromatography, are also anticipated in biotechnology. Porous glasses and glass-ceramics (ceramics prepared by controlled crystallization of glass utilizing spinodal-type phase separation (both of separated phases are continuous) in glass have great advantages over conventional porous ceramics prepared by sintering of powdery materials because they have well controlled pores and may be obtained in a desired shape and dimension. Research and development on porous glasses and glass-ceramics, however, have mainly been focused on silica-based systems following the invention of porous silica glass(Vycor)[1] in 1934. If porous glass-ceramics with a skeleton of crystals with active functions can be fabricated, they will have the combined advantages of porous bodies (pores, large surface area), glass-ceramics (ease of fabrication) and crystal skeletons (functionality).

In 1989 the authors reported[2] that porous glass-ceramics with a skeleton of $CaTi_4(PO_4)_6$ crystal could be prepared by controlled crystallization of glasses in the

CaO-TiO$_2$-P$_2$O$_5$ system and subsequent acid leaching of the resulting dense glass-ceramics composed of CaTi$_4$(PO$_4$)$_6$ and β-Ca$_3$(PO$_4$)$_2$ phases (which interlock with each other). The porosity is ~ 50 vol %, and the median pore diameter and specific surface area are approximately 40 nm and 50 m^2, respectively. This is the first porous glass-ceramic based on phosphates, to the authors' knowledge. Since then, the authors have prepared various porous glass-ceramics with a skeleton of functional titanium phosphate crystals and have examined applications, so far. In this paper, the preparation, modification , and some applications of these novel porous glass-ceramics are described.

FUNCTIONAL CRYSTALS IN TiO$_2$-P$_2$O$_5$ SYSTEM

Two types of high performance crystals are known in the TiO$_2$-P$_2$O$_5$ system. One is NASICON-type crystal such as RTi$_2$(PO$_4$)$_3$, i.e. PO$_4$ tetrahedra share their corners with TiO$_6$ octahedra to form a 3-dimensional network. When Li$^+$ ions (4 possible sites per formula unit) are located in two different sites of the conduction channels, LiTi$_2$(PO$_4$)$_3$ and their solid solutions have attracted a considerable attention as a chemically stable fast Li-conductor for electrode materials of a secondary battery. The other is a zirconium phosphate-type crystal, Ti(HPO$_4$)$_2$・nH$_2$O. This type of crystal has a 2-dimensional layered structure, i.e. Ti^{4+} ions lie very nearly in a plane and are bridged by phosphate groups, which are situated alternatively above and below the metal planes. Three of oxygen atoms of each PO$_4$ group are bonded to three different Ti^{4+}. The fourth oxygen atom of the PO$_4$ group has an exchangeable proton. The presence of such a proton and layer structure are the structural origin of excellent ion-exchange properties and intercalatability of polar organic molecules.

OUTLINE OF PREPARATION PROCEDURES

Although neither composition of NASICON-type crystal nor zirconium phosphate-type crystal can be vitrified by conventional quenching technique, It was found that there is a glass formation range around 1:1 in the pseudo-binary system of RTi(PO$_4$)$_3$ (R=Li, Na, 1/2Ca) and Ca$_3$(PO$_4$)$_2$ (neither of which forms a stable glass). A mixture of starting materials such as metal carbonates, TiO$_2$ (anatase), and H$_3$PO$_4$(85 %) liquid were placed with water in a silica glass beaker and stirred well to make a slurry. The slurries were dried at about 200 °C overnight. The resulting dried products were melted in Pt crucibles at 1300 °C for 1 h in air. Corrosion of Pt crucibles with the melt was not perceptible. The melts were poured into a carbon mold and subsequently annealed. The resulting glasses were transparent but deep violet in color. This color is due to absorptions of Ti^{3+} (~0.1 % of total Ti) in the glasses. The glasses were cut into rectangular plates (~1.5 x 1.5 x 0.5 cm) and finely polished.

Glass plates were crystallized by a two-step heat-treatment. First, the specimens were heated to a temperature of T$_g$ + 20 °C (T$_g$ is determined by DTA employing a heating rate of 10 K/min) for 20 h to promote nucleation. Subsequently, the specimens were heated to a temperature of T$_c$ - 40 °C (T$_c$ is the peak temperature of crystallization on the DTA curve) for 12 h to crystallize. The resulting dense glass-ceramics were immersed in a 1N-HCl aqueous solution and kept for 24 h.

Crystalline phases were identified by powder X-ray (Cu Kα) diffraction(XRD). Surface area and the pore-size distribution were measured by mercury porosimetry.

It was assumed in calculation that pores are open and cylindrical capillaries in shape.

PHYSICAL PROPERTIES

Figure 1 shows some examples of powder XRD pattern of glass-ceramics with acid leaching. The following are evident from the figure: (i) The dense glass-ceramics are composed of RTi$_2$(PO$_4$)$_3$ (R=Li, 1/2 Ca) and β-Ca$_3$(PO$_4$)$_2$. (ii) The acid dissolves β-Ca$_3$(PO$_4$)$_2$ selectively. (iii) LiTi$_2$(PO$_4$)$_3$ (abbreviated as LTP, hereafter) are converted into Ti(HPO$_4$)$_2$·2H$_2$O (TP) at around 100 °C. It is worth noting that LTP with a 3D-network structure is converted into TP with a 2D-layered structure under such a mild condition (this reaction is specific to LTP, NTP and CTP are stable under the same condition).[3] Figure 2 shows the pore size distribution and microstructure of porous glass-ceramics [4] with a skeleton of Li$_{1+x}$Ti$_{2-x}$Al$_x$(PO$_4$)$_3$ (x=0.4), which is a chemically stable fast Li-conductor reported in 1989.[5] As a first approximation, the distribution is log Gaussian in shape and the median pore diameter is approximately 0.2 μm (this diameter is the maximum among our porous glass-ceramics. The diameter of the other porous glass-ceramics is in the range 20-60 nm). Figure 3 is the photograph of the specimen. Although a weight loss of ~ 50 % was observed, no changes in the shape and dimension of the sample were seen in the course of crystallization and leaching. Such a feature is common to all the porous glass-ceramics described here.

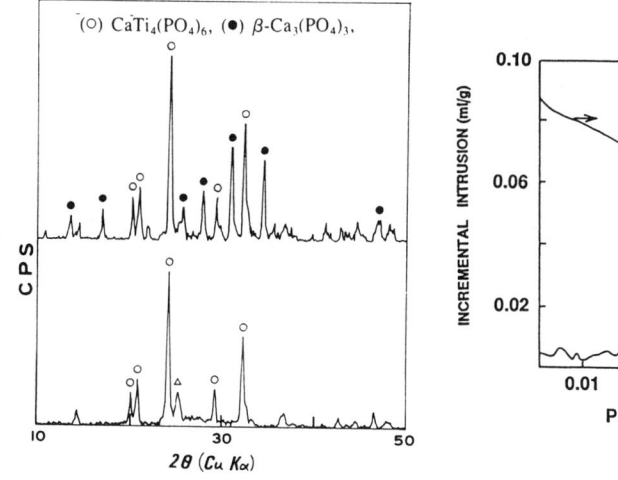

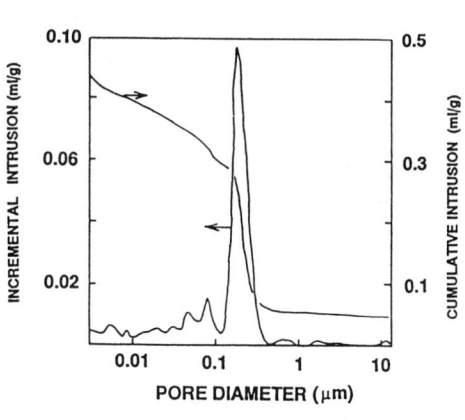

Figure 1 Powder XRD pattern of glass-ceramics in the CaTi$_4$(PO$_4$)$_6$ - Ca$_3$(PO$_4$)$_2$ system. leaching temp. = 100 °C, (a) : before leaching, (b): after leaching.

Figure 2 Pore-size distribution of porous glass-ceramics with a skeleton of Li$_{1+x}$Al$_x$Ti$_{2-x}$(PO$_4$)$_3$ (x=0.4).

Figure 3 Photograph of specimens and SEM photograph of specimen (3).
(1) as-prepared glass, (2) dense glass-ceramics, (3) porous glass-ceramics.

MODIFICATION

Pore Size Control

Tailoring the pore dimension is demanded for various applications. Chemical etching is not applicable for the present porous glass-ceramics because $RTi_2(PO_4)_3$ is stable against both acidic and alkaline solutions. A method for pore-size control was found by using a post crystallization heat treatment and subsequent acid leaching.[6] Here, the case of CTP is explained as an example. The dense glass-ceramics before leaching are composed of CTP and β-$Ca_3(PO_4)_2$, which are closely interlocked with each other. When these glass-ceramics are reheated at temperatures (T_{pc}) higher than the temperature for crystallization of the starting glass, the following reaction occurs at the boundaries between CTP and $Ca_3(PO_4)_2$ phases :

$$CaTi_4(PO_4)_6 + 5Ca_3(PO_4)_2 \longrightarrow 8Ca_2P_2O_7 + 4TiO_2. \quad (1)$$

Since $Ca_2P_2O_7$ phases are easily dissolved in acidic solutions as $Ca_3(PO_4)_2$, control of pore-size will be possible by this method. Figure 4 shows the variation in median pore diameter and pore volume of the resulting porous glass-ceramics as a function of T_{pc}. The mean pore diameter increases from 40 nm to 150 nm with increasing T_{pc} in the range of 900-960 °C.

Intercalation of Organic Diamine Molecules into Skeleton Structure

When the porous TP glass-ceramics were immersed in a benzene solution (~ 10 wt. %) of hexamethylene diamine (HMDA, $H_2N\text{-}C_6H_{12}\text{-}NH_2$) and the solution was ultrasonically stirred, a characteristic peak at $2\theta = 8.7°$ was shifted to $4.2°$ (Fig. 5). This peak shift corresponds to an increase in the interlayer spacing of 0.83 nm, which results from intercalation of HMDA into the layers. A model structure of TP intercalated with HMDA is shown in Fig. 6. This result indicates that the skeleton structure itself may be modified by such a soft-chemical technique as intercalation.

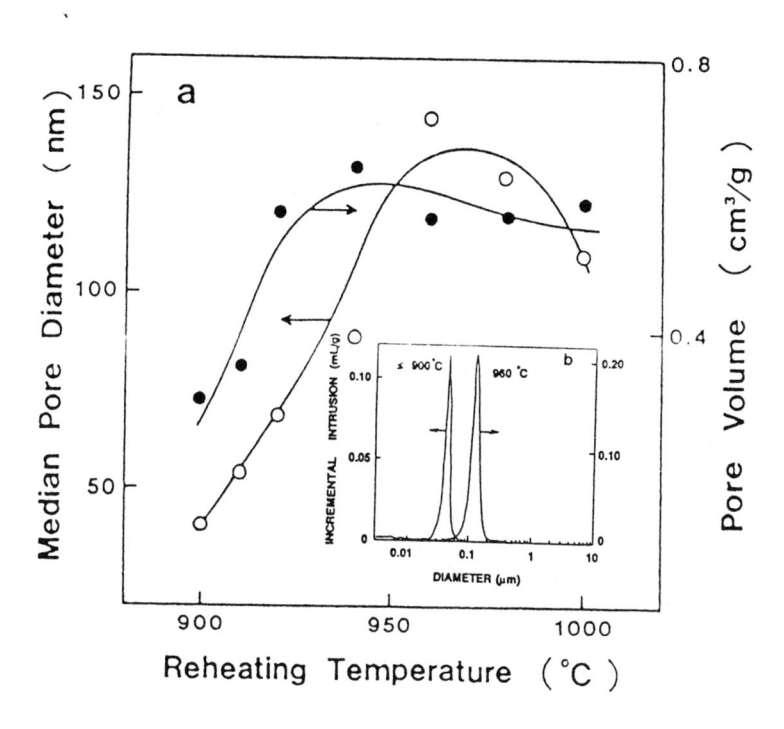

Figure 4 Mean pore diameter and pore volume as a function of post-crystallization temperature(T_{pc}). The inset is the examples of pore-size distribution.

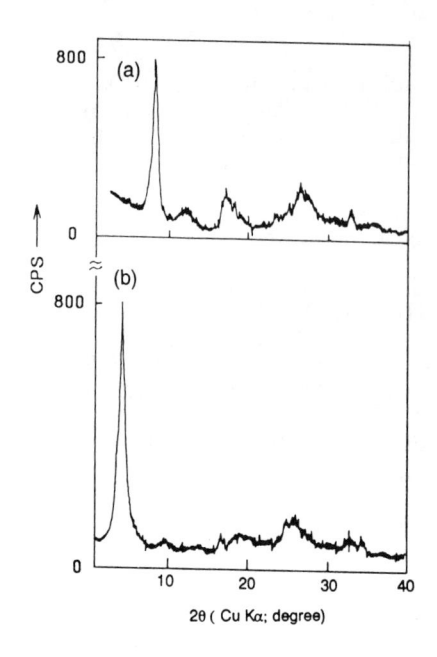

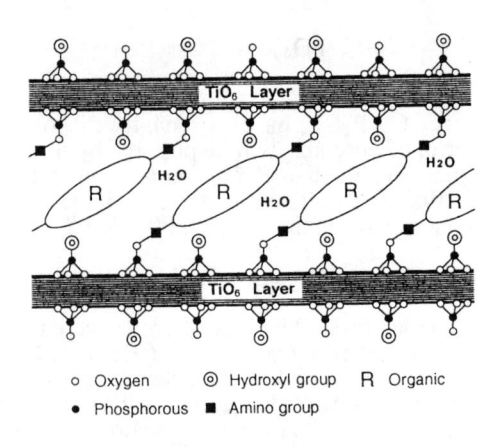

Figure 5 Changes in powder XRD pattern of porous $Ti(HPO_4)_2 \cdot 2H_2O$(TP) glass-ceramics (a) before and (b) after diamine treatment.

Figure 6 Schematic structural model of TP intercalated with hexamethylene diamine.

APPLICATION

The authors have been examining various applications utilizing characteristics of each porous glass-ceramic. Table 1 summarizes characteristics of porous glass-ceramics prepared in this laboratory together with their applications. Some examples of application are described, hereafter.

Immobilized Enzymes

Enzymes are precious biological catalysts with selectivity specific to substrate and very high efficiency. The application of enzymes is limited by their scarcity and cost. Therefore, immobilization is indispensable for their practical uses. The optimum pore diameter for immobilizing an enzyme is approximately twice the major axis of the unit cell of the enzyme. In most cases, the optimum pore diameter corresponds to 30-60 nm. The mean pore diameter of the porous $CaTi_4(PO_4)_6$ glass-ceramics is approximately 40 nm, which is in an appropriate range.

Figures 7 shows changes in activities of three kinds of enzymes with different

Table 1. Porous Glass-Ceramics Prepared and Their Applications

TYPE	Skeleton	Median pore diameter(nm)	Surface area (m²/g)	Porosity (vol%)	Characteristics	Applications
I NASICON-TYPE (3-dimensional)	$CaTi_4(PO_4)_6$ [1]	40-160 (controllable) [6]	50-90	50	alkali durability weak affinity to water low thermal expansion	immobilized enzyme[9] humidity sensor[10]
	$NaTi_2(PO_4)_3$ [2]	70	45	50	zero thermal expansion	
	$Li_{1+x}Ti_{2-x}Al_x(PO_4)_3$ [3]	200	50	55	fast Li-conductor[7] ion-exchanger[8]	intercalation electrode
II Zirconium Phosphate-(2-dimensional) TYPE	$Ti(HPO_4)_2 \cdot 2H_2O$ $+Ti(OH)PO_4$ [4]	15-20	120	45	good ion-exchanger affinity to protains intercalatable with polar organic molecules[4]	cation exchanger protein purification[11]
III	TiO_2 (<1μm dia) $+ a\text{-}SiO_2$ (30%)[5]	1200	40	65	large pore diameter stable to 1200°C[12]	O_2 gas sensor[12]

[1] J.Amer.Ceram.Soc.7 2,1587(1989), [2] J.Electrochem.Soc.1 3 7,3149(1990), [3] J.Amer.Ceram.Soc.in print, [4] J.Non-Cryst.Sol.1 3 9,86(1992), [5] J.Amer.Ceram.Soc.7 3,2536(1990), [6] J.Non-Cryst.Sol.1 3 9,90(1992), [7] Solid State Ionics,4 4,293(1991), [8] J.Electrochem.Soc.to be published, [9] J.Ferment.Bioeng.7 2,384(1991), [10] Phosphorous Res. Bull.1,351(1991), [11] To be submitted, [12] Jap.Pat.Application #4-26573.

optimal pH immobilized on porous CaTi$_4$(PO$_4$)$_6$ glass-ceramics (CTP) and porous silica glass (CPG), both of which have the similar pore size (~40 nm) and surface area (~80 m^2/g). Enzymes were cross-linked to aminosilanizated CTP and CPG carriers using glutaraldehyde (this method is frequently used for CPG). Detailed coupling procedures were described in reference 7. Hydrolysis reaction of proteins (saccharose, ONPG, or milk casein) was used to measure the activity of each enzyme. It is evident from figure 7 that although the initial activities of CPG derivatives were about 4 times higher than those of CTP, they lost their activities rapidly, i.e. the activities of CTP derivatives of enzymes with optimal pH > 7 are higher than those of CPG after 5 days of operation. Figure 8 shows chemical

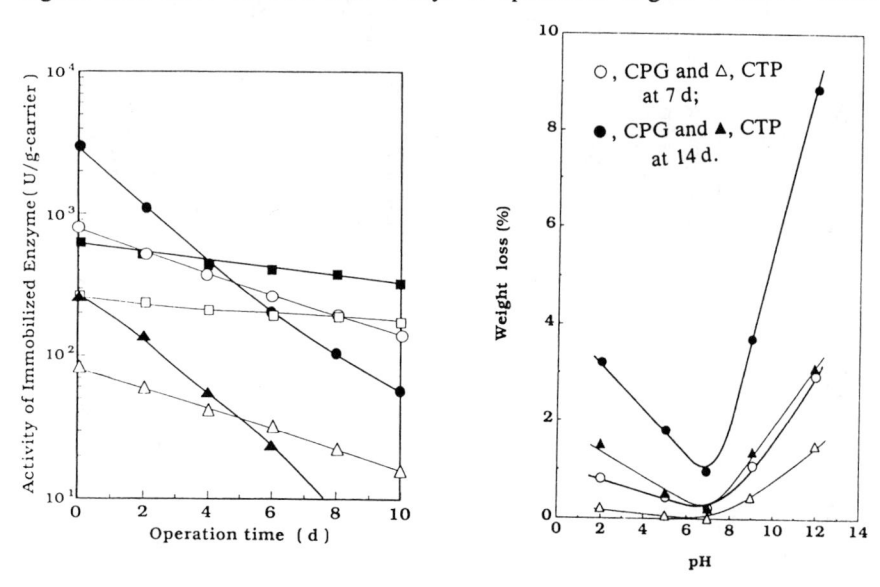

Figure 7 Time courses of activities of immobilized enzymes on porous CTP glass-ceramics and CPG glass. Invertase (optimal pH=4.6) on CTP (□) and CPG (■), β-gallactosidase(pH=7) on CTP (○) and CPG (●), alkalophilic proteinase on CTP (△) and CPG (▲). The temperature of reactor is 37 °C.

Figure 8 Chemical durability of CTP and CPG in various pH solutions at 37 °C. Weight loss was measured after 7 day or 14 day soaking.

durability of CTP and CPG as a function of pH. The durability of CTP is better than CPG, especially in the alkaline region. The excellent stability of CTP derivatives in pH >7 is believed to originate from better chemical durability of CTP in alkaline solutions. If a different coupling method, which is appropriate to the present CTP is chosen, results are expected to be improved.

Humidity Sensor

Humidity sensing properties were examined by measuring electrical resistance as a function of relative humidity (R.H.). Rectangular specimens (4 x 3 x 1 mm) sputtered with RuO$_2$ (Au and Pd`/ Ag were inadequate to have a good Ohmic contact) were used for the measurement. Electrical resistance was measured at 20 °C by biasing AC (300 Hz) voltage of 1 V. R.H. was changed in the range 20 - 90 %. Each datum was monitored after keeping for 3 min at a temperature. Figure 9 shows changes in electrical resistance with R.H. The resistance decreased almost linearly on a logarithmic scale with increasing R.H. The changes were reversible and no significant hysteresis were observed. The resistance was changed over 2 orders of magnitude for change in the R.H. of 70 %. These results are favorable for applications to humidity sensors.

Ion Exchanger

There is an increasing demand for cation exchangers in various engineering fields such as biotechnology and environmental engineering to separate or remove a specific ion. Inorganic ion exchangers have several advantages over organic exchangers because of their excellent resistance to heat, chemicals, and radiation. Although a variety of inorganic ion exchangers have been discovered so far, there are no materials which can be fabricated via glass, to the authors knowledge. If an excellent ion exchanger material can be synthesized via the glass process, materials can be fabricated in a desired shape and dimension unlike conventional ceramic processing. Porous bodies of ion exchangers are very favorable for application because ion exchange occurs at the surface and near surface region of materials. Therefore, if porous materials composed of continuous pores and ion-exchange crystals are obtained via glass, one may prepare porous materials with a skeleton of ion-exchange crystals in a desired shape appropriate to each application.

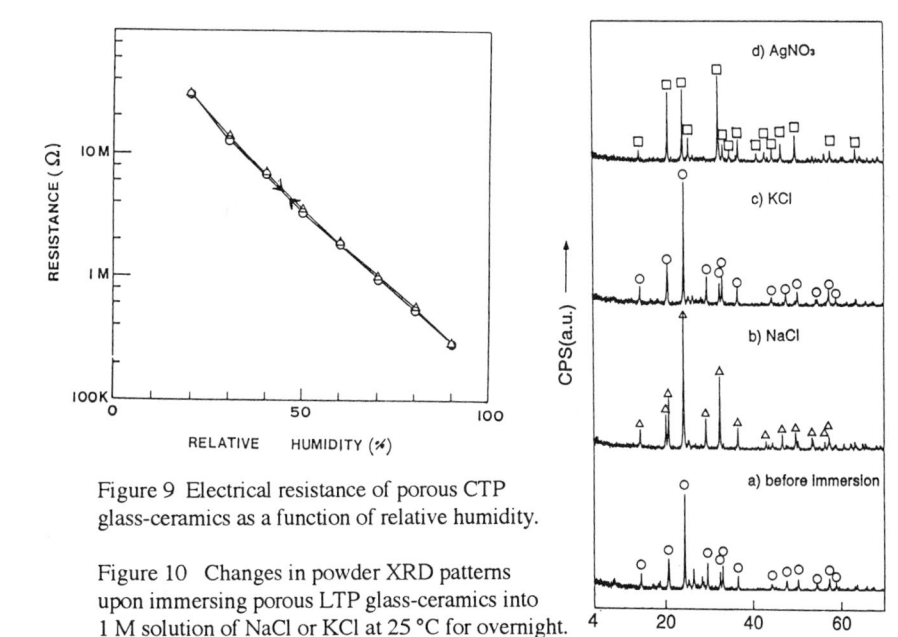

Figure 9 Electrical resistance of porous CTP glass-ceramics as a function of relative humidity.

Figure 10 Changes in powder XRD patterns upon immersing porous LTP glass-ceramics into 1 M solution of NaCl or KCl at 25 °C for overnight.

Since ion conductivities of grain bulk of LiTi$_2$(PO$_4$)$_3$ (LTP) ceramics are very high

($\sigma_{300\ K}$ = 1 x 10^{-4} S•cm^{-1}) [8], there is a possibility that these materials have excellent ion exchange properties. Recently, It has been found that LTP ceramics are easily exchanged with Na$^+$ and Ag$^+$ and porous glass-ceramics with a skeleton of LTP crystal show excellent ion exchanging properties.[9] Figure 10 shows changes in powder XRD diffraction pattern of the specimen after porous LTP glass-ceramics were immersed in 1M aqueous solution of NaCl, KCl, or AgNO$_3$ at 25 °C for overnight. It is evident that CTP was not converted into KTi$_2$(PO$_4$)$_3$ but into NaTi$_2$(PO$_4$)$_3$. Li ions in the materials exchange with monovalent ions with ionic radius which are smaller than ~130 pm and have high affinity to Ag$^+$ ions.

CONCLUDING REMARKS

Novel porous glass-ceramics with a skeleton of two-types of high performance titanium phosphate crystals, RTP and TP have been prepared, and their applications examined. The next goal is to fabricate integrated porous glass-ceramics with different functional crystals. If integrated porous RTP / TP ceramics can be prepared from a glass plate, multi-functional porous glass-ceramics can be obtained because these two types of crystals have different properties originating from different dimensionality. Recently, it has been found that the conversion rate of LTP to TP in acidic solutions strongly depends on temperature, i.e. the conversion proceeds at temperature > 50°C but does not at < 50°C. When one side of a porous LTP glass-ceramic plate is immersed in a hot acid solution, integrated porous glass-ceramics composed of LTP and TP can be fabricated. Details will be published elsewhere.

REFERENCES

[1] H.P.Hood and M.E.Nordberg, "Treated Borosilicate Glass, " U.S.Pat. No. 2 106 744, 1934.
[2] H.Hosono, Z.Zhang, and Y.Abe,"Porous Glass-Ceramics in the CaO-TiO$_2$-P$_2$O$_5$ System," J.Am.Ceram.Soc., 7 2[9]1587-90(1989).
[3] H.Hosono and Y.Abe, "Porous Glass-Ceramics with Skeleton of Two-dimensional Layered Crystal Ti(HPO$_4$)$_2$•2H$_2$O," J.Non-Cryst.Sol. 139, 86-89(1992).
[4] H.Hosono and Y.Abe, "Porous Glass-Ceramics with a Skeleton of the Fast-Lithium-Conducting Crystal Li$_{1+x}$Ti$_{2-x}$Al$_x$(PO$_4$)$_3$," J.Amer.Ceram.Soc. in press.
[5] H.Aono, E.Sugimoto, Y.Sadaoka, N.Imanaka, and G.Adachi, "Ionic Conductivity of the Lithium Titanium Phosphate System," J.Electrochem.Soc. 136, 590-591(1989).
[6] H.Hosono, Y.Sakai, and Y.Abe, "Pore Size Control in Porous Glass-Ceramics with Skeleton of NASICON-type Crystal CaTi$_4$(PO$_4$)$_6$," J.Non-Cryst.Sol.,139, 90-92(1992).
[7] T.Suzuki, M.Toriyama, H.Hosono, and Y.Abe, "Application of a Microporous Glass-Ceramics with a Skeleton of CaTi$_4$(PO$_4$)$_6$ to Carriers for Immobilization of Enzymes," J.Ferment. Bioeng., 7 2[5] 384-391(1991).
[8] H.Aono, E.Sugimoto, Y.Sadaoka, N.Imanaka, and G.Adachi, "Ionic Conductivity of Solid Electrolytes Based on Lithium Titanium Phosphate," J.Electrochem.Soc., 137, 1023-27 (1990).
[9] H.Hosono, K.Imai, and Y.Abe, "Cation Exchange Properties and Synthesis of Its Microporous Materials via Glass," J.Electrochem.Soc., in press.

PREPARATION OF POROUS GLASSES FOR PHOTONIC MATERIALS

Makoto Umino, Tetsuji Yano, Atsuo Yasumori, Shuichi Shibata and Masayuki Yamane
Department of Inorganic Materials, Tokyo Institute of Technology,
2-12-1 Ookayama, Meguro-ku, Tokyo 152, Japan

INTRODUCTION

Porous glasses prepared by phase separation and subsequent acid leaching of separated glasses have been applied in various fields of engineering as filters, supports for catalysts or enzymes, adsorbents for chemicals, etc. [1-3]. For these applications, it is desirable that the porous glasses have a high mechanical strength, a high chemical durability, a high thermal stability, in addition to the specific pore structure desired for the purpose.

Besides applications in these fields, porous glasses have recently been used as starting materials to make high-performance glasses for optical and photonic devices. These include Gradient-Index (GRIN) lenses obtained by a molecular stuffing technique [4], quantum dots materials containing nanometer-scale semiconductor particles [5], optically-integrated circuits fabricated by a photo-patterning technique [6]. In these applications, a porous glass is impregnated with a solution containing a metal salt or photosensitive organometallic compound. Then, the glass is subjected to various heat-treatments to precipitate microcrystallites of the dopant on the micropore walls, followed by drying and sintering into a fully dense glass.

The desirable properties of the porous glasses for these applications are, therefore, entirely different from those for filters, supports, etc.. It is desirable for the pores to collapse at low temperatures rather than to remain stable up to high temperatures, in order to embed the doped microcrystallites before they decompose. The pores of the glasses for photo-patterning must be small enough to minimize the scattering loss of the UV light irradiated for photolitic decomposition of impregnated organometallic compounds. To hinder the growth of incorporated particles before densification, a small pore size is also important for the fabrication of quantum dots materials.

Among the porous glasses developed to date, the high-silica porous glass of the Vycor-type (HSG) with pores 2-50 nm in diameter which is appropriate for the desired applications. The pores of HSG, however, did not collapse up to temperatures above 900 °C [7], which is too high. Controlled-pore glass (CPG), a product of Electro-Nucleonic, Inc., has pores of about 10 nm in diameter and is densifiable at temperatures near 800°C. However, CPG is supplied in particulate form. Porous glasses based on the Shirasu composition (SPG) are densifiable at lower temperatures than HSG, considering from its mother composition [8]. However, the pore sizes of SPG ranging from 30 nm to several microns are generally too large to be applied to optical materials.

Nakashima and Kuroki [9] investigated the effects of CaO and Al_2O_3 additions on the phase separation behavior of Shirasu-type glasses and obtained porous glasses with pores of about 10 nm in radius in the pulverized sample. However, no reference was made to the integrity of their porous glass. From the crudely estimated porosity of 50%, the fabrication of porous plates with such a pore structure will be difficult, because the skeleton of such a glass is generally thin and weak. Since integrity is essential for glasses used in optical and photonic devices, some modifications must be made on this porous glass in order to obtain a material suitable for application.

In the present study, preparation of a new porous glass was accomplished using the Shirasu-type glass reported by Nakashima and Kuroki for developing a glass with pore structure similar to HSG but densifiable at lower temperatures.

EXPERIMENTAL

Preparation of Mother Glass, Heat-Treatment for Phase Separation, and Acid Leaching

Chemical composition of the mother glass was determined and shown in Table 1 along with that of the reference glass, SPG. Taken into consideration were the following:
(1) The densification of a porous glass is accomplished by viscous sintering. Therefore, the skeletal glass should have a much lower viscosity than the HSG

glass in order to densify at lower temperatures.

(2) The Heat-treatment for phase separation must be stopped before the pores begin to coalesce in order to retain the small pores. In such a case, the porosity of the glass should not be too large in order to maintain the integrity of the glass during acid leaching.

(3) The main components of the dissolved phase of the

Table 1. Chemical composition of the mother glass. (in mol%)

	Present Study	Shirasu-Type Glass [8]
SiO$_2$	64.0	51.99
Al$_2$O$_3$	6.2	2.81
B$_2$O$_3$	9.5	21.81
Na$_2$O	4.1	5.46
K$_2$O	1.7	0.67
CaO	14.2	16.84
MgO	0.3	0.19
Fe$_2$O$_3$	-	0.24

Shirasu-type glass are B$_2$O$_3$ and CaO [9]. Higher SiO$_2$ content and lower B$_2$O$_3$ and CaO contents were used to increase the volume fraction of the skeletal glass part and to suppress the rapid development of phase separation. This is in order to obtain the formation of large pores.

A batch composition of reagent grade SiO$_2$, H$_3$BO$_3$, Al$_2$O$_3$, CaCO$_3$, MgO, Na$_2$CO$_3$ and KNO$_3$ was melted in a platinum crucible at 1400°C and kept for 2 hrs. The melt was quenched and pulverized to about 0.5 mm or smaller. Then the pulverized glass was remelted at 1550°C and kept for 2 hrs in order to improve homogeneity.

Heat-treatments for phase separation were made at various temperatures in the range of 20-40°C above T$_g$ i.e., at 670, 680 and 690°C for 6-96 hrs. The phase separated glasses were sliced into 0.3-0.4 mm thick samples and polished until they had lustrous surfaces. Then, the glasses were subjected to leaching in 0.1N or 1N H$_2$SO$_4$ aqueous solution at room temperature for 2 days. Leaching was repeated again with a fresh solution, followed by drying at 60°C.

Characterization

Specific surface area, S$_g$, pore size distribution, and pore volume were measured by a nitrogen adsorption technique. The composition of the skeletal glass was estimated from porosity and the composition of the dissolved phase in the H$_2$SO$_4$ solution, which was analyzed by inductively coupled plasma emission spectroscopy (ICP).

Densification

Changes in pore structure at elevated temperatures were monitored after isothermal heat-treatment of the glasses at 700, 750 and 800°C. S_g and the pore size distribution were measured with a Quantachrome AUTOSORB1 on samples collected during heat-treatment at intervals of 2-3 hrs. Ultraviolet-visible (UV-Vis) transmission spectra of the porous glass and fully-densified glass were measured from 200 to 800 nm with JASCO UVIDEC- 610C Spectrophotometer.

RESULTS

Pore Structure

The dependence of the integrity of the porous glass after the leaching with 0.1N H_2SO_4 solution on heat-treatment temperature and time for phase separation is depicted in Fig. 1. All the glasses soaked at 670°C did not maintain integrity during the leaching process. Heat-treatment at 680°C for longer than 40 hrs was necessary to develop a skeletal structure of glass which maintained its integrity, while 10 hrs was enough to maintain integrity for the sample heat-treated at 690°C. These relations were also found by the leaching with 1N H_2SO_4 solution.

The relationships between pore structure and conditions of heat-treatment are shown in Figs. 2 and 3. The average

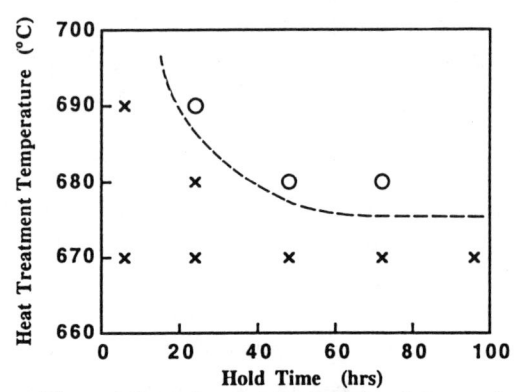

Figure 1 Dependence of the integrity of porous glass on heat-treatment temperature and time.
○: maintain integrity.
×: disintegrated.

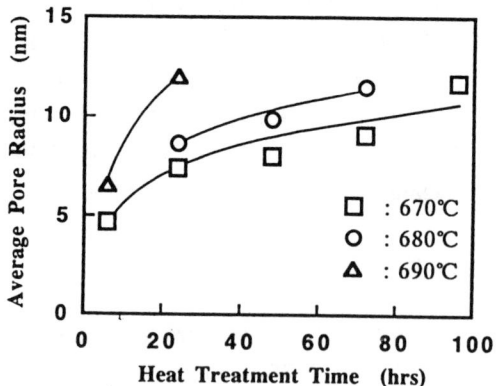

Figure 2 Changes in average pore radius of porous glass with the temperature and time of heat-treatment.

pore radius, R_{ave}, increased from 5 to 9-10 nm on the samples obtained by leaching with 0.1N H_2SO_4 solution, and S_g decreased from 50 to 30-40 m^2/g when the skeleton of the glass was well-developed. The porosity of the glass with a well-developed skeleton was estimated to be 40%.

A typical pore size distribution is shown in Fig. 4 for samples heat-treated at 680°C for 48 hrs and leached with 0.1N H_2SO_4 solution. In principle, the distribution is unimodal, although it looks bimodal with one mode at 10-20 nm and the other mode below 3 nm in radius. This effect is due to residual silica gel which precipitated during leaching and remained in the larger pores. The pore size distribution is more clear in the glass leached with 1N H_2SO_4 solution than that leached with 0.1N H_2SO_4 solution.

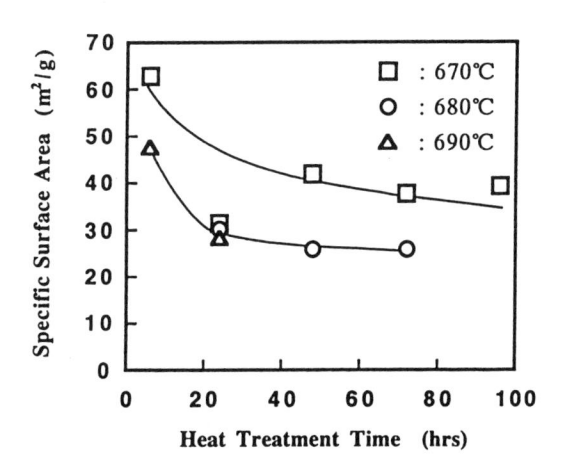

Figure 3 Changes in specific surface area of porous glass with heat-treatment temperature and time.

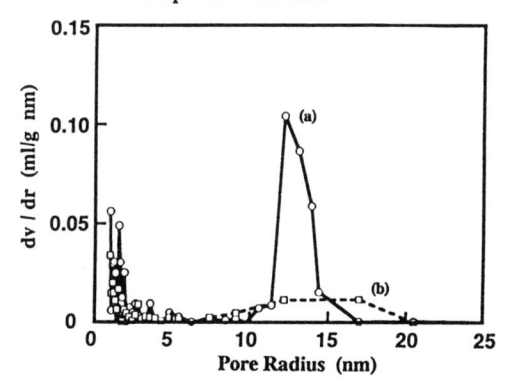

Figure 4 Change in pore size distribution of porous glass with the condition of leaching;
(a) : 96 hrs in 1N-H_2SO_4.
(b) : 48 hrs in 0.1N-H_2SO_4.

Table 2 shows the estimated composition of the skeletal glass along with those of SPG and the composition of the leached phase. The SiO_2 content in the new porous glass is comparable to or slightly higher than that of SPG. The glass contains an unexpectedly large amount of CaO, despite that it was obtained from a mother glass of smaller CaO content than SPG.

Table 2. Estimated compositions of skeletal glasses and leached phases. (in mol%)

	Present Study skeleton	Present Study Leached phase	Shirasu-Type Glass [9] skeleton	Shirasu-Type Glass [9] Leached phase
SiO_2	75~84	28~34	75.24	30.05
Al_2O_3	5	7~9	4.31	1.40
B_2O_3	5	20~26	13.52	29.46
Na_2O	4~8	3	5.10	5.78
K_2O	2~5	1	0.93	0.42
CaO	3~8	29~32	0.13	32.39
MgO	-	2	0.64	0.17
Fe_2O_3	-	-	0.14	0.33

Change in Pore Structure with Heat-Treatment

Changes in S_g and bulk density with isothermal heat-treatment are shown in Fig. 5 and 6. The glass phase was separated at 680°C-48 hrs and then leached with 1N H_2SO_4 solution. The value of S_g quickly decreased by heating at 800°C, and the glass was completely densified within 6 hrs. Also, the collapse of pores proceeded at an appreciable rate at 750°C, with complete densification occurring within 25 hrs. During this time, the bulk density increased from 1.42 g/cm³ to that of the skeletal glass at 2.40 g/cm³. The porous glass soaked at 700°C, however, remained uncollapsed, although its S_g decreased slightly in the early stage of heat-treatment. Densification at 750°C is lower by about 150°C compared to the same HSG with similar pore structure. This is low enough to fabricate quantum dots materials containing some semiconductor particles such as CdS before their sublimation becomes intensive. But the temperature is still too high to embed PbS without decomposing into PbO or $PbSO_4$.

Figure 7 shows the UV-Vis spectra for the porous glass measured in both air and water and for the completely densified glass. The scattering loss observed on the spectrum for the porous glass measured in air was reduced from that measured in water to an acceptable level for effective irradiation of UV light for photo-patterning. The fully dense glass had good transparency over the wavelength region of interest, showing that the glass can be used in optical devices.

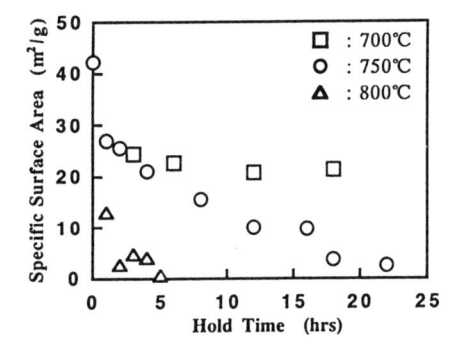

Figure 5 Changes in specific surface area of porous glass with isothermal heat-treatment.

Figure 6 Change in bulk density of porous glass with isothermal heat-treatment at 750°C.

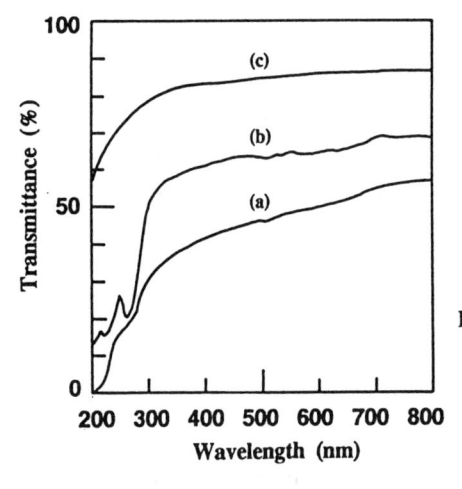

Figure 7 UV-vis transmission spectra of the porous glass and densified glass;
(a) : Porous glass in air.
(b) : Porous glass in water.
(c) : Densified glass.

DISCUSSION

The remarkable reduction in densification temperature and good transparency of the new porous glasses show that the composition of the mother glass was properly selected. Although it is still desirable to modify the glass composition further to reduce densification temperature, it is also of significance to discuss the sintering behavior of the new porous glass with respect to pore structure and viscosity of the skeletal glass at this stage.

According to Scherer [10], the sintering behavior of low density, open-pore materials which are densified by viscous flow and driven by surface energy, can be described by using a model consisting of a cubic array formed by intersecting cylinders. In such a model, the pore diameter, D, is given by Eq.(1):

$$D = 2(L-2R)/\sqrt{\pi} \tag{1}$$

where L is the length of the cubic cell, and R is the cylinder radius. The volume of the solid phase in the cubic cell, V_s, and the bulk density, ρ, are given by Eqs. (2) and (3):

$$V_s = 3\pi R^2 [L - (8\sqrt{2}/3\pi) R] \tag{2}$$

$$\rho = (3\pi \left(\frac{R}{L}\right)^2 - 8\sqrt{2} \left(\frac{R}{L}\right)^3) \rho_s \tag{3}$$

where ρ_s is the density of the cylinder.

The applicability of this model to the porous glass in the present study can be examined by comparing the estimated surface area, S_g, on this model with the measured value. The estimated S_g of the model cubic cell was 33 m²/g by correlating the experimental data, ρ_0=1.42 g/cm³, ρ_s=2.40 g/cm³ and D=23.4 nm with Eqs. (1) and (3) and substituting into Eq. (4):

$$S_g = \frac{1}{R\rho_s} \left(\frac{6\pi L - 24\sqrt{2} R}{3\pi L - 8\sqrt{2} R}\right) \tag{4}$$

Good agreement between estimated and measured S_g values at 30-40 m²/g assumes that the change ρ/ρ_s, which represents the sintering behavior of the porous glasses, can be expressed using reduced time, $(\gamma \eta L_0)(\rho_s \rho_0)^{1/3}(t - t_0)$. Here, γ is the surface tension, η is the viscosity of the skeletal glass, L_0 is the initial value of L, and t_0 is the fictitious time at which x(=a / l) equals zero.

Figure 8 shows the results for the estimating the change in ρ/ρ_s at various values of viscosity, assuming that γ=0.3 J/m² (300 ergs/cm²). Since ρ/ρ_s increases from an initial value of 0.57 to 0.94 in 20 hrs by isothermal heat-treatment at 750°C, the viscosity of the skeletal glass at that temperature is known to be in the range 1-1.5 x 10⁻¹³ dPa·s. When these values are plotted against the reciprocal of the absolute temperature, the viscosity of the skeleton of the porous glass is estimated to be about 10^{-14} dPa·s at 700 °C. This value is reasonable since the glass transition temperature determined by dilatometry is 726±5°C.

The results for similar estimation at constant viscosity of 10^{-13} dPa·s on the densification time for the material of viscous pore diameters are shown in Fig. 9. It is known from this figure that the densification time of the porous glass with 25 nm pores is about 170 hrs at this viscosity, and that the densification time can be reduced to a half by decreasing the pore diameter from 25 nm to 10 nm. Heat-treatment time of 100 hrs for densification will be within an acceptable range for practical fabrication of quantum dots materials.

Although the phase separated glass with the pore diameter of 10 nm did not maintain integrity under the leaching conditions of present study, it may not be impossible to obtain a porous glass with such a pore structure without disintegration after optimization. Then, the possibility of application to optical and photonic devices will be remarkably increased.

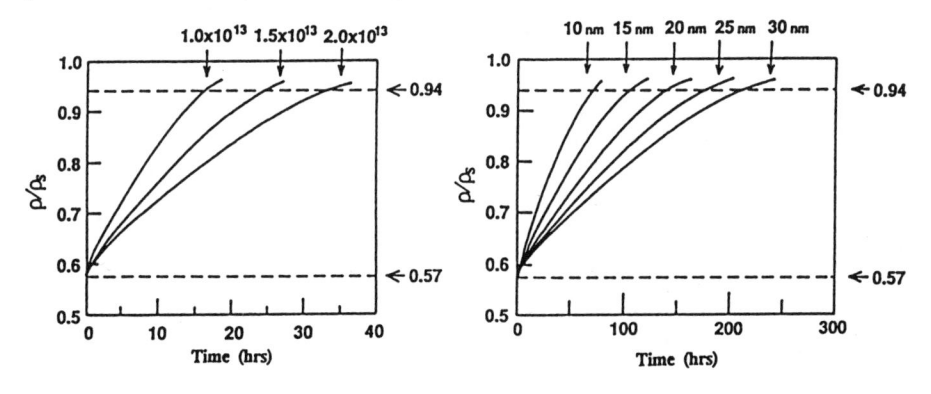

Figure 8 Estimation of densification time at various values of viscosity.

Figure 9 Estimation of desification time for materials with various pore diameters at constant viscosity.

CONCLUSIONS

Preparation of a porous glass having a pore structure similar to that of high silica porous glass of the Vycor-type, but that which densifies at lower temperatures, was studied in order to develop a material applicable to the fabrication of quantum dots materials, optical integrated circuits and other components important for optical and photonic devices.

The following conclusions were obtained:

(1) A porous glass having average pore diameter of 23.4 nm and specific surface area of 40 m^2/g was obtained by the heat-treatment and acid leaching of a Shirasu-type glass with modified composition.
(2) The porous glass showed a good transparency in the UV-Vis region to allow photo-patterning of dopants.
(3) The porous glass completely densified at 750°C and can be applied to the fabrication of quantum dots materials containing non-oxide semiconductors.

REFERENCES

1. T. H. Elmer, "Evaluation of Porous Glass as Desalination Membrane", Amer. Ceram. Soc. Bull., 57, 1051-1053 (1978)
2. S. Hirai, A. Morikawa, Y. Ishinaga, K. Otuka and Y. Wada, "Annihilation of the Catalytic Activity in Butene Isomerization of Porous Vycor Glass by Washing with EDTA Aqueous Solution", Bull. Chem. Soc. Jpn.(in Japanese), 50, 3411-3412 (1977)
3. T. Yazawa, H. Tanaka and K. Eguchi, "Effects of Pore Structure of Porous Glass on Gas Permeation", Nihon-Kagakukai-shi (in Japanese), 2, 201-07 (1986)
4. A. Asahara, H. Sakai, S. Shingaki, S. Ohmi, S. Nakayama, K. Nakagawa and T. Izumitani, "One-Directional Gradient-Index Slab Lens", Appl. Opt., 24, 4312-4315 (1985)
5. T. Takada, N. Nihei, T. Yano, A. Yasumori and M. Yamane, "Preparation of PbS-Doped Glasses by Photochemical Process", Sci. Tech. New Glasses (eds. S. Sakka and S. Soga), Ceram. Soc. Jpn., 418-423 (1991)
6. N. F. Borrelli and D. L. Morse, "Photosensitive impregnated Porous Glass", Appl. Phys. Lett., 43, 992-993 (1983)
7. T. Kokubu and M. Yamane, "Thermal and Chemical Properties of TiO$_2$-SiO$_2$ Porous Glass-Ceramics", J. Mat. Sci., 22, 2583-2588 (1987)
8. T. Nakashima and M. Shimizu, "Porous Glass from Calcium Alumino Boro-Silicate Glass, Bull. Ceram. Soc. Jpn. (in Japanese), 21, 408-412 (1986)
9. T. Nakashima and Y. Kuroki, "Effects of Composition and Heat Treatment of the Phase Separation of Na$_2$O-B$_2$O$_3$-SiO$_2$-Al$_2$O$_3$-CaO Glass", Nihon-Kagakukai-Shi (in Japanese), 8, 1231-1238 (1981)
10. G. W. Scherer, "Sintering of Low-Density Glasses: 1, Theory", J. Am. Ceram. Soc., 60, 236-239 (1977)

Section III. Properties

CHARACTERIZATION OF POROUS SILICA GEL BY MEANS OF ADSORPTION-DESORPTION ISOTHERMS OF WATER VAPOR AND NITROGEN GAS

Hiromitu Naono*, Masako Hakuman*, Hisanao Joh*, Masao Sakurai*, and Kazuyuki Nakai**
*Department of Chemistry, Faculty of Science, Kwansei Gakuin University, Uegahara, Nishinomiya 662, Japan
**Technical Department, Bel Japan, Inc., 1-5, Ebie 6-chome, Fukusima-ku, Osaka 553, Japan

The porous texture and the surface character of porous silica gel were investigated by means of the adsorption-desorption isotherms of water vapor at 25°C and nitrogen gas at 77 K. The adsorption measurements were carried out by the automatic adsorption apparatus constructed by us. Porous silica gel was prepared by hydrolyzing $Si(OC_2H_5)_4$ in the presence of excess water at 70°C. The gel was pretreated *in vacuo* at 100, 500, 700, 900, and 1000°C. On the basis of the H_2O and N_2 isotherms, the pore size distribution was calculated. In the present silica gel, the majority of pores is distributed in the range of 1 to 2 nm in pore radii. The 1st and 2nd isotherms of H_2O vapor were measured for the silica samples pretreated at temperatures described above. The chemisorbed and physisorbed amount of H_2O vapor for the dehydroxylated silica gels were determined by means of the 1st and 2nd isotherms. The reversibility of rehydroxylation on dehydroxylated silica surface was discussed on the basis of the chemisorbed amount of H_2O vapor. The ratio between surface silanol group and physisorbed water at monolayer coverage was found to be one.

INTRODUCTION

The adsorption isotherms of various gases have been widely used to characterize nonporous, mesoporous, and microporous adsorbents. In our previous papers, porous texture and surface character of several adsorbents have been investigated.[1-7] The purpose of the present work is to study the porous texture and the surface character of porous silica gel by means of the adsorption-desorption isotherms of H_2O vapor and N_2 gas.

Porous texture of materials has been mainly studied by analyzing the N_2 isotherms at 77 K. It has been generally accepted that the lower closure point of the hysteresis loop of N_2 at 77 K is situated at a relative pressure close to 0.42.[8] The pore radius at this point could be calculated to be 1.7 nm, which was considered to be the smallest pore radius determined from the N_2 isotherm. It is therefore possible to calculate the pore size distribution curve down to the pore radius of 1.7 nm. On the other hand, as will be shown later, the pore size distribution curve down to the pore radius of 1 nm can be calculated by analyzing the H_2O isotherm at 25°C. In the present work, the H_2O isotherm at 25°C as well as the N_2 isotherm at 77 K were utilized to analyze porous texture of porous silica gel. It has been found that the H_2O isotherm plays an important role for characterizing porous texture of porous silica gel.

Gregg and Sing[9] have pointed out the importance of the H_2O isotherms for characterization of the silica surface. It is well known that H_2O molecules are preferentially physisorbed on the hydrophilic sites (surface silanol groups). Silica surface can be dehydroxylated by thermal treatment at elevated temperatures. When dehydroxylated silica is exposed into water vapor, the silica surface is again rehydroxylated. However, the behavior of rehydroxylation for silica is quite different from that for titania, alumina, and iron oxide. For drastically dehydroxylated silicas, the rehydroxylation is practically inhibited at a low relative pressure, but it occurs slowly at a high relative pressure. As a result, the H_2O isotherm for dehydroxylated silica gives a remarkable low pressure hysteresis in the adsorption and desorption branches.[1] On the other hand, for the oxides mentioned above, the rehydroxylation readily occurs at an extremely low pressure, giving no low pressure hysteresis.

In order to examine the surface character of dehydroxylated silica gel, the adsorption-desorption isotherms of H_2O vapor were measured at 25°C on the silica gels dehydroxylated at the temperatures from 500 to 1000°C *in vacuo*. After the 1st H_2O isotherm had been obtained for dehydroxylated silica, the 2nd H_2O isotherm was determined. As mentioned above, the 1st H_2O isotherm gives the remarkable low pressure hysteresis, but the 2nd isotherm for rehydroxylated silica gel shows no low pressure hysteresis. On the basis of the 1st and 2nd H_2O isotherms, the chemical and physical interaction of H_2O molecules with the dehydroxylated and rehydroxylated silica surfaces will be discussed.

EXPERIMENTAL

Materials

Porous silica gel was prepared by hydrolysis of Si(OC$_2$H$_5$)$_4$ (7.3 g) in the presence of excess water (250 ml). Hydrolysis was carried out at 70°C for 24 h. Porous silica gel thus prepared was separated by centrifuge and dried at 25°C under vacuum. Porous silica gel was pretreated *in vacuo* at the temperatures from 100 to 1000°C. Sample name, pretreatment condition, BET surface areas, and water content are given in Table 1. The BET areas were determined from the N$_2$ isotherms (*cf.* Fig. 1) by assuming the area of a N$_2$ molecule to be 0.162 nm^2. The water content of the present silica gel was measured by the successive-ignition-loss method described previously.[1]

Measurement of Adsorption Isotherm

The adsorption isotherms of H$_2$O at 25°C and N$_2$ at 77 K were measured by means of the computer-controlled automatic adsorption apparatuses selfmade. The standard data of N$_2$ gas and H$_2$O vapor for nonporous reference adsorbent were determined and used for characterization of unknown adsorbents. Analysis of the isotherms were performed by means of the personal computer. Such graphs as BET plot, t plot, V-V plot, and pore size distribution curve can be output by the simple procedures. In the present work, two kinds of the adsorption apparatuses were used; one was the prototype apparatus reported previously[2,3,10] and the other was the commercial type (BELSORP 28SA(N$_2$) and BELSORP 18 (H$_2$O)). Particular attention must be paid on the adsorption measurement of H$_2$O vapor. In measurement of H$_2$O vapor, the adsorption of H$_2$O vapor on the inner wall cannot be avoided. In order to minimize such adsorption, the pressure gauge operated at 100°C and the well polished stainless pipes were used. The adsorption of H$_2$O on the inner wall of the pipes was subtracted from the total adsorbed amount of H$_2$O to give the true isotherms. The method of correction for the adsorption on the wall has been described.[2]

Table 1. Surface Area and Water Content of Silica Gel

Sample	Pretreatment Temp. °C	Time h	S$_{BET}$ m^{2}g^{-1}	Water content OH/nm^2
A-100	100	4	721	5.5
A-500	500	4	681	1.26
A-700	700	4	622	0.61
A-900	900	2	529	0.20
A-1000	1000	2	438	0.10

RESULTS AND DISCUSSION

Porous Texture of Porous Silica Gel

Figure 1 shows the adsorption and desorption isotherms of N_2 and H_2O for A-100. Both of the isotherms are of Type IV (Brunauer's classification), giving an appreciable hysteresis loop in the adsorption and desorption branches. The hysteresis loop has been explained as capillary condensation of the adsorptive molecules into mesopores. The pore size distribution curve was calculated by the method reported by Dollimore and Heal[11]. In calculation, the desorption branches of the isotherms of Fig. 1 were utilized. The standard t curves of N_2 and H_2O on non-porous silica (Aerosil-200) were used to correct the adsorbed thickness.[2,3]

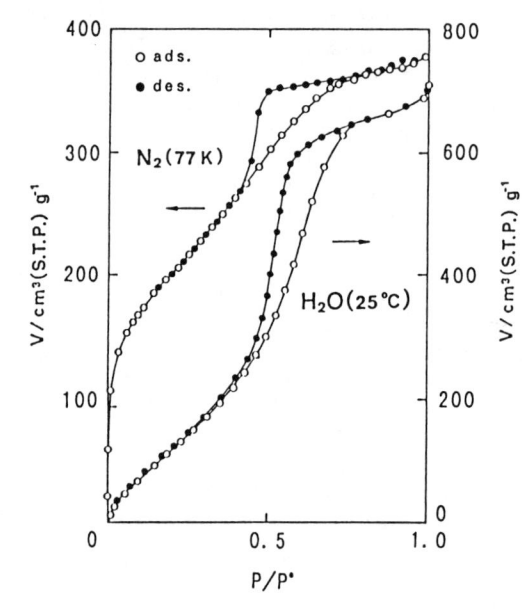

Figure 1 Adsorption-desorption isotherms of N_2 gas and H_2O vapor for porous silica gel (A-100).

Before calculation of pore size distribution, it is very important to determine the lower limit of pore size obtained from the adsorption isotherms of N_2 gas and H_2O vapor. As has been mentioned previously, it has been generally accepted that the minimum pore radius determined from the N_2 isotherm at 77 K is 1.7 nm. On the other hand, few works have been reported on the lower limit of pore size obtained from the H_2O isotherm. We found in several porous adsorbents that the lower closure point in the hysteresis loop of the H_2O isotherm at 25°C lies in the vicinity of $P/P^o=0.27$. One of the examples is shown in Fig. 2, where the H_2O isotherm gives rise to an appreciable hysteresis loop, but no hysteresis is detected in the N_2 isotherm on the same porous adsorbent.[5] The closure point of the H_2O isotherm in Fig. 2 corresponds to the pore radius of 1 nm, which is assumed to be the minimum pore size determined from the H_2O isotherm. It has been clarified from the above finding that the H_2O isotherm enables one to calculate the pore size distribution down to the pore radius of 1 nm. The details on numerical calculation of pore size distribution will be reported elsewhere.

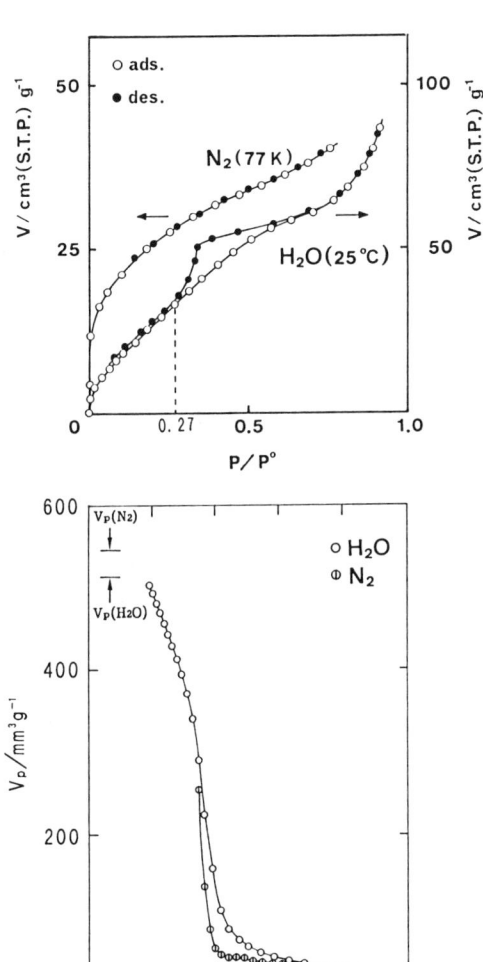

Figure 2
Adsorption-desorption isotherms of N_2 gas and H_2O vapor for porous hematite.

Figure 3
Pore size distribution curves for porous silica gel (A-100).
$V_P(N_2)$:total pore volume obtained from N_2 t plot.
$V_P(H_2O)$:total pore volume obtained from H_2O V-V plot.

Figure 3 shows the cumulative pore size distribution curves of A-100, which were calculated from both of the H₂O and N₂ isotherms. As has been pointed out above, the lower limit of the distribution curve for the H₂O and N₂ isotherms is estimated to be 1 nm and 1.7 nm, respectively. Until 1.7 nm, both the distribution curves are similar to one another. The result of Fig. 3 shows clearly that the majority of pores in A-100 is distributed in the range of 1 nm to 2 nm in pore radii. This means that the N₂ isotherm does not give the satisfactory information on pore size distribution of A-100. On the other hand, the H₂O isotherm is considered to give the satisfactory information on the distribution curve. The cumulative pore volume down to the pore radii of 1 nm is in good agreement with the total pore volume ($V_P(H_2O)$) obtained from the V-V plot for A-100 (cf. Fig. 5). It is clear that the H₂O isotherm plays an important role for evaluating porous texture of the present porous silica gel. The importance of H₂O isotherms for evaluation of porous texture was pointed out by Hagymassy et al.[12]

The pore size distribution curves for A-500, A-700, A-900, and A-1000 were also determined by using the desorption branches of the H₂O isotherms. The pore surface of these silica samples are dehydroxylated by thermal treatment (cf. the water content of Table I). However, as will be shown below, the dehydroxylated silica surface is amply rehydroxylated at the starting point of the desorption branch ($P/P°=1$). Therefore, the distribution curves obtained for A-500, A-700, A-900, and A-1000 corresponds to those for the rehydroxylated silica samples. The pore volume decreases with increasing the pretreatment temperature, but the majority of pores remains in the range of 1-2 nm in pore radius.

Surface Character of Porous Silica Gel

Surface silanol groups on porous silica gel act as the primary sites for adsorption of H₂O molecules. The number of surface silanol groups decreases with an increase in pretreatment temperatures (cf. Table I). As the result, the H₂O isotherm changes from Type IV (A-100) to Type V (adsorption branch of A-500, A-700, A-900, A-1000). When the H₂O isotherm is of Type IV, the BET method can be effectively utilized to determine the monolayer capacity of physisorbed H₂O molecules on silica surface (V_m/H_2O nm^{-2}). From the V_m-values, it is possible to evaluate the number of hydrophilic sites on porous silica gel. On the other hand, when the H₂O isotherm is of Type V, the BET method cannot be applied for the determination of the monolayer value. The BET plot gives rise to the convex curve against the relative pressure. In such a case, it is difficult to determine the linear portion of the BET plots, which means that the V_m-value cannot be evaluated from the BET method. Furthermore, it must be taken into account that the dehydroxylated silica surface can undergo slow rehydroxylation.

One of the H₂O isotherms for dehydroxylated silicas (A-900) is shown in Fig. 4. As will be seen from Fig. 4, the adsorbed amount of H₂O vapor in the adsorption branch of the 1st isotherm is remarkably small at the P/P°=0-0.4. This shows that the dehydroxylated silica surface is highly hydrophobic. From the above result, it is reasonable to conclude that the rehydroxylation or chemisorption of H₂O molecules is practically inhibited at the low P/P° range. The sharp increase in the adsorption branch above P/P°=0.5 is due to the rehydroxylation and capillary condensation. In the 1st desorption branch, the sharp decrease in the isotherm is observed at around P/P°= 0.5, but the 1st desorption isotherm never returns to the origin. It is characteristic of dehydroxylated silica that the 1st H₂O isotherm gives rise to the remarkable low pressure hysteresis. The 2nd H₂O isotherm was measured for the same adsorbent, after the 1st H₂O isotherm had been determined and physisorbed water had been removed by degassing at 25°C for 12 h. As will be seen from Fig. 4, the 2nd H₂O isotherm gives no low pressure hysteresis. The adsorption and desorption branches in the 2nd H₂O isotherm coincide, except the hysteresis loop due to capillary condensation.

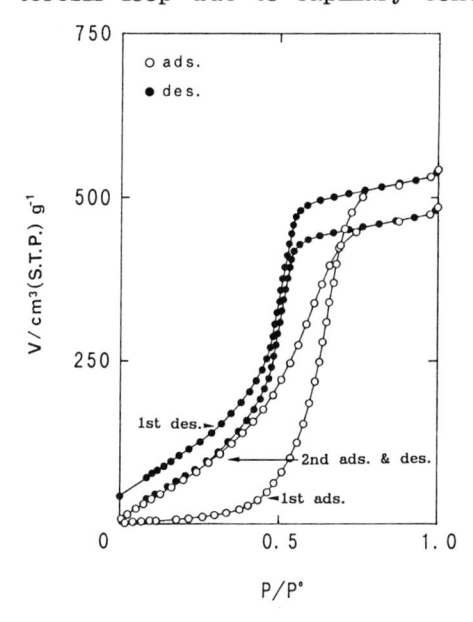

Figure 4
1st and 2nd adsorption-desorption isotherms of H₂O vapor at 25°C for porous silica gel(A-900).

For the dehydroxylated samples of A-500, A-700, and A-1000, the 1st isotherms show a remarkable low pressure hysteresis, while the 2nd isotherms give no low pressure hysteresis. In the present paper, we propose the tentative method to determine the monolayer capacity of physisorbed H_2O molecules (V_m) for both of dehydroxylated and rehydroxylated silica surfaces. The method is based on the comparison plot (V-V plot)[13];*i.e.*, the amount of H_2O adsorbed on the adsorbent under examination is plotted against the number of H_2O adsorbed on the reference adsorbent. As the reference adsorbent, nonporous silica gel (Aerosil-200) was selected in the present work. The monolayer value of H_2O vapor was found to be $V_m = 3.5$ H_2O/nm^2 from the BET method.[2]

The V-V plot for A-900 is shown in Fig. 5, which was constructed from both of the H_2O isotherms of Fig. 4 and the H_2O isotherm for reference adsorbent previously reported.[2] In Fig. 5, the ordinate is expressed as the adsorbed amount of H_2O (mm^3(liq.)/g) on A-900 and the abscissa is expressed as the number of H_2O molecules per unit area (H_2O/nm^2) on Aerosil-200. Since V_m for Aerosil-200 is known, two kinds of V_m-values for A-900 can be determined from the initial slope of V-V plots for the 1st and 2nd adsorption isotherms. The V_m-values for dehydroxylated and rehydroxylated silicas were calculated by the following equation.

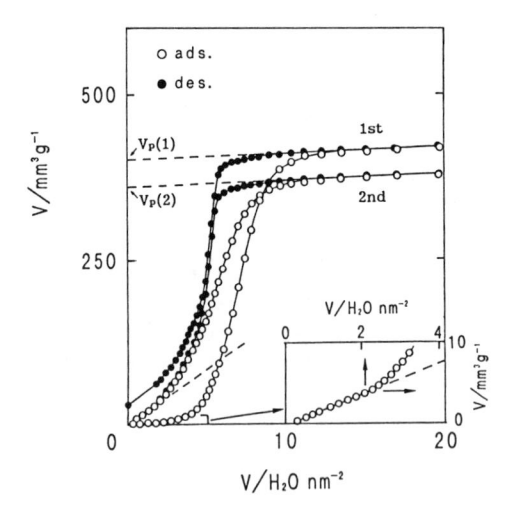

$$V_m(H_2O\ nm^{-2})\ \text{for A-900} = k \times (\text{initial slope}) \times 3.5$$

Figure 5
V-V plots of 1st and 2nd H_2O isotherms for porous silica gel(A-900).
Reference isotherm: H_2O isotherm for nonporous silica(Aerosil-200).

where $k = (6.02 \times 10^{23})/(18 \times 10^{21}) S_{BET(A-900)}$. The V_m-values thus calculated are listed in the 2nd and 3rd columns of Table 2. As shown in the 2nd and 3rd columns of Table 2, the hydrophilic sites for dehydroxylated silica decrease drastically with an increase in pretreatment temperature and the hydrophilic sites for the rehydroxylated silica decreases gradually above 500°C.

In addition to the V_m-value, it is very important to determine the chemisorbed amounts of H_2O vapor. It is evident from Fig. 4 that the desorption branch in the 1st and 2nd isotherms run parallel to one another at the $P/P°=0-0.3$. This difference in the isotherms is equal to the irreversible adsorbed amount of H_2O vapor (the last point of the 2nd desorption isotherm).[1] The chemisorbed waters determined from the difference between the 1st and 2nd desorption isotherms are given in the 4th column of Table 2. The sum of water content (Table 1) and chemisorbed water is considered to be the total surface silanol content (5th column of Table 2), which decreases gradually with an increase in temperatures. This result shows that the rehydroxylation is not reversible on the silica surface pretreated at the elevated temperatures higher than 500°C. In other words, silica surface may be stabilized by thermal treatment at elevated temperatures.

Finally, the relation between the surface hydroxyl group and the physisorbed water was discussed on the basis of the data in Table 2. The ratio of V_m (3rd column) to the total hydroxyl group (5th column) is listed in the 6th column of Table II. The ratio is found to be 0.98 ± 0.04 for the five samples. This result indicates that one H_2O molecule interacts with one OH group at monolayer coverage.

Table 2. Physisorbed and Chemisorbed Waters on Silica Gel

Sample	V_m H_2O/nm^2 (1st ads.)	(2nd ads.)	Chemisorbed Water OH/nm^2	Total OH OH/nm^2	Ratio H_2O/OH
A-100	5.6	5.6	0.0	5.5	1.0₂
A-500	0.70	5.0	4.0	5.3	0.9₄
A-700	0.56	4.9	4.8	5.0	0.9₈
A-900	0.40	3.9	3.9	4.1	0.9₅
A-1000	0.38	3.7	3.6	3.7	1.0₀

REFERENCES

1) H. Naono, R. Fujiwara, and M. Yagi, "Determination of Physisorbed and Chemisorbed Waters on Silica Gel and Porous Silica Glass by Means of Desorption Isotherms of Water Vapor", *J. Colloid Interface Sci.*, 76, 74–82 (1980).
2) H. Naono and M. Hakuman, "Analysis of Adsorption Isotherms of Water Vapor for Nonporous and Porous Adsorbents", *J. Colloid Interface Sci.*, 145, 405–412 (1991).
3) H. Naono and M. Hakuman, "Evaluation of Surface Character and Porous Texture (in Japanese)", *Hyomen (Surface)*, 29, 362–373 (1991).
4) H. Naono and K. Nakai, "Thermal Decomposition of γ-FeOOH Fine Particles", *J. Colloid Interface Sci.*, 128, 146–156 (1989).
5) H. Naono and R. Fujiwara, "Micropore Formation due to Thermal Decomposition of Acicular Microcrystals of α-FeOOH", *J. Colloid Interface Sci.*, 73, 406–415 (1980).
6) H. Naono, R. Fujiwara, H. Sugioka, K. Sumiya, and H. Yanazawa, "Micropore Formation due to Thermal Decomposition of Acicular Microcrystals of β-FeOOH", *J. Colloid Interface Sci.*, 87, 317–332 (1982).
7) H. Naono, "Micropore Formation due to Thermal Decomposition of Magnesium Hydroxide", *Colloids Surfaces*, 37, 55–70 (1989).
8) S. J. Gregg and K. S. W. Sing, "The Physical Adsorption of Gases by Mesoporous Solid: The Type IV Isotherm (Chapter 3, p. 154)" in "Adsorption, Surface Area and Porosity (2nd Ed.)", Academic Press, London (1982).
9) S. J. Gregg and K. S. W. Sing, "Type III and Type V Isotherms – The Special Behavior of Water (Chapter 5, p.269)" in "Adsorption, Surface Area and Porosity (2nd Ed.)", Academic Press, London (1982).
10) N. Naono, K. Nakai, and M. Hakuman, "Application of Computer Controlled Adsorption Apparatus I. Tentative Method of Separation of Composite Isotherm into Type I, II, and IV Isotherms", *International Symposium on Adsorption* (Kyoto), Abstract, p.121–123 (1988).
11) D. Dollimore and G. R. Heal, "Pore-Size Distribution in Typical Adsorbents Systems", *J. Colloid Interface Sci.*, 33, 508–519 (1970); "An Improved Method for the Calculation of Pore Size Distribution from Adsorption Data", *J. Appl. Chem.*, 14, 109–114 (1964).
12) J. Hagymassy, Jr., S. Brunauer, and R. Sh. Mikhail, "Pore Structure Analysis by Water Vapor Adsorption I. t Curves of Water Vapor", *J. Colloid Interface Sci.*, 29, 485–491 (1969).
13) S. J. Gregg and K. S. W. Sing, "The Physical Adsorption of Gases by Nonporous Solids: The Type II Isotherm (Chapter 2, p.100)" in "Adsorption, Surface Area and Porosity (2nd Ed.)", Academic Press, London (1982)

PERMEATION CHARACTERISTICS OF CO_2 THROUGH SURFACE-MODIFIED POROUS GLASS MEMBRANE

Tetsuo YAZAWA and Hiroshi TANAKA
Government Industrial Research Institute, Osaka
Midorigaoka, Ikeda City, Osaka 563, Japan

The permeation characteristics of CO_2 through a surface-modified porous glass membrane were investigated. A novel porous glass membrane was prepared from the sodium borosilicate glass containing ZrO_2 and CaO (Composition: $53SiO_2$ $24.5B_2O_3$ $5.5Na_2O$ 8CaO $6ZrO_2$ $3Al_2O_3$ wt%). The borate rich phase was leached with an aqueous solution of 1N HNO_3 for 48 hours at 98°C. The median pore diameter of the porous glass membrane was 3.0-3.5nm. The surface modification was performed by the following two methods. 1. The membrane was dipped in an aqueous nitrate solution of K, Mg, Ag and Cr and subsequently heated at the decomposition temperatures of the nitrate salts. 2. The membrane was dipped in xylene and subsequently heated under a reductive atmosphere.

The permeability of CO_2 and N_2 and the separation factor of CO_2 for CO_2/N_2(50/50) gaseous mixture were measured. The measurement was performed at 30-400°C. The permeability and separation factor decreased with an increase in temperature. However, the surface-modified membrane prepared by second method maintained a high separation factor value even at a high temperature region.

INTRODUCTION

Global warming has been a serious issue because of its potential threat to human life. Of the gases that cause global warming, CO_2 is of the most serious concern because it is produced in large quantities by many industries. In order to reduce CO_2 emitted to the atmosphere, it is necessary to separate and withdraw CO_2 from the waste gas of electric power generation plants, cement production plants, etc.

In general, CO_2 is separated by two methods. One method is based on absorption by an alkali solution or adsorption by a zeolite. The other method is the membrane separation. The latter has advantages of being energy efficient and being a continuous operation. Though some excellent organic membranes for CO_2 separation have been developed[1], these membranes cannot be used at more than ca. 100°C. On the other hand, inorganic membranes can be used at these temperatures[2] and therefore can be directly applied to the flue gas. Porous glass membranes are widely applicable for their narrow pore size distribution and good shaping capabilities.

EXPERIMENTAL

Porous Glass Membrane

The porous glass was prepared from a glass of the following composition: $53SiO_2$- $25B_2O_3$- $5.5Na_2O$- $8.0CaO$- $6.0ZrO_2$- $3.0Al_2O_3$(wt%). The phase separation was carried out at 750°C for 10 hours. The borate rich phase was leached with an aqueous solution of 1N HNO_3 for 48 hours at 98°C[3]. The pore size distribution of the porous glass was calculated from the nitrogen adsorption isotherms based on the Cranston-Inkley method. Figure 1 shows the pore size distribution of the porous glass membrane. The composition of the porous glass membrane was $72SiO_2$- $10B_2O_3$- $2.7Na_2O$- $1.6CaO$- $8.1ZrO_2$- $5.8Al_2O_3$ 5.8(wt%).

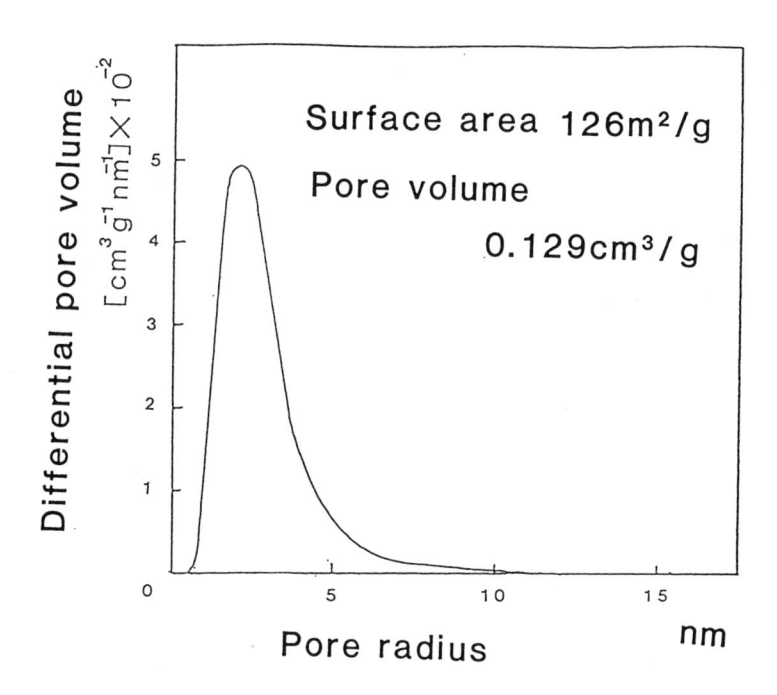

Figure 1 Pore size distribution of the membrane

Surface Modification

The porous glass membrane was dipped in an aqueous

nitrate solution of Mg, K, Ag and Cr. After drying, the membrane was heated at temperatures slightly above the decomposition temperatures of nitrate salts.

Another surface modification was as follows. The membrane was dipped in xylene and subsequently heated at 400°C under N_2. Xylene was pyrolyzed, resulting in a deposition of carbon onto the surface of the pore.

Membrane Module

A cylindrical porous glass membrane (one end was sealed by melting), with an outer diameter of 5mm and an inner diameter of 4mm, was connected to a Pyrex glass tube as shown in Figure 2, and placed in a stainless steel casing. The effective surface area of the membrane for the separation experiment ranged from 30-60 cm^2.

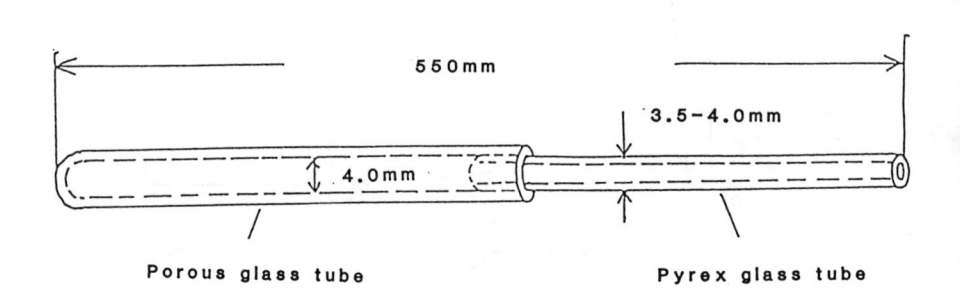

Figure 2 Membrane module

Experimental Apparatus

The schematic diagram of the experimental apparatus is shown in Figure 3. The arrows show the direction of gas flow.

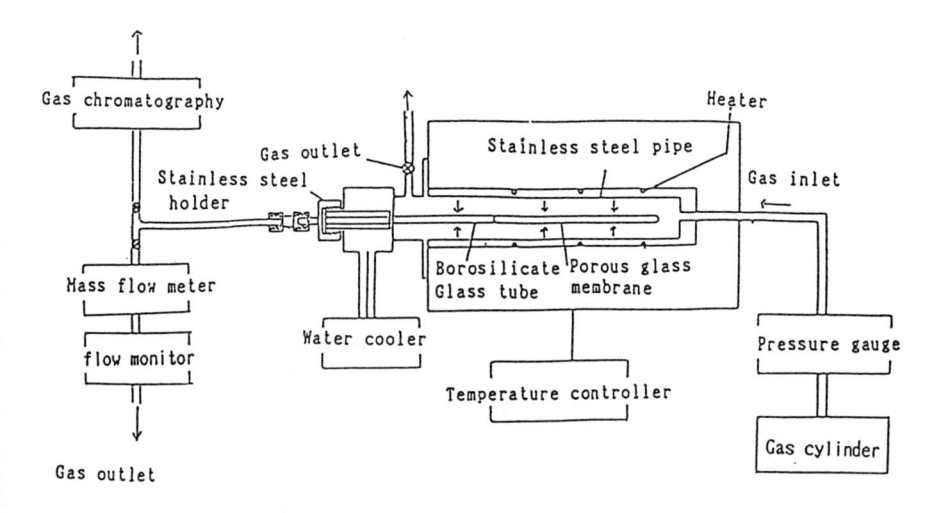

Figure 3 Schematic diagram of the apparatus

Measurement of Permeability and Separation Factor

The permeability was measured by a mass flow meter. The feed gas mixture used for the determination of the separation factor was 50 mol% CO_2 and 50 mol% N_2. The pressure of the feed gas was kept in the range of 200-400kPa. The composition of the permeate gas was analyzed by gas chromatography. The experiment was performed at temperatures ranging from 30-400°C.

RESULTS

Table 1 shows the permeability of N_2 and CO_2 for various kinds of porous glass membranes. The separation factor analyzed by gas chromatography is also shown in Table 1. The separation factor was calculated by the following equation by considering the effect of back diffusion[4].

$$f=\frac{X_1}{1-X_1}\cdot\frac{(1-X_h)-(1-X_1)\cdot k}{X_h-k\cdot X_1} \tag{1}$$

$$k=\frac{P_1}{P_h}$$

where f=separation factor of CO_2, X_1=CO_2 concentration of permeate gas, X_h: CO_2 concentration of feed gas, P_1=pressure of permeate gas and P_h=pressure of feed gas.

Table 1. Permeation Characteristics of CO_2 through Surface-modified Porous Glass Membrane

Temp.(°C)	$J(CO_2)$[a]	$J(N_2)$[a]	$J(CO_2)/J(N_2)$	f[b]
Without surface modification				
30	1.80×10⁻⁵	1.43×10⁻⁵	1.26	1.33
200	2.09	2.12	0.99	1.02
300	2.34	2.73	0.86	0.87
400	2.67	3.19	0.84	0.88

Surface modified with $Mg(NO_3)_2$

30	4.71	3.12	1.51	1.12
200	3.94	4.64	0.85	0.91
400	3.81	4.81	0.79	0.85

Surface modified with KNO_3

30	0.99	0.84	1.16	1.28
200	2.05	2.15	0.96	1.06
300	1.92	2.13	0.90	0.93
400	4.16	5.19	0.80	0.91

Surface modified with $AgNO_3$

30	13.0	12.9	1.01	1.03
200	14.0	17.4	0.80	0.90
300	12.4	15.4	0.81	0.87
400	12.4	16.3	0.76	0.85

Surface modified with $Cr_2(NO_3)_3$

30	15.0	16.0	0.94	1.05
200	18.4	10.3	0.82	0.90
300	9.5	12.2	0.78	0.85
400	9.1	11.6	0.78	0.84

Surface modified with Xylene

30	5.19	3.41	1.51	1.76
200	3.77	3.24	1.16	1.22
300	3.12	3.28	0.95	1.05
400	2.81	2.98	0.94	0.98

a) $cm^3(STP)\ cm^{-2}\ s^{-1}\ cmHg^{-1}$, b) Separation fctor of CO_2 determined by mixed gas separation

DISCUSSION

The separative flow through porous glass membranes is composed of Knudsen flow and surface flow[5]. Knudsen flow predominates when the pore size is similar to or smaller than the mean free path of gas molecules and obeys the equation (2).

$$J = K \cdot \frac{\Delta P}{\sqrt{\pi M R T} \cdot l} \tag{2}$$

where J=Permeability, k=characteristic constant of the membrane, M=molecular weight of gases, T=absolute temperature, R=gas constant, P=pressure drop and l=membrane thickness

If CO_2 and N_2 permeate the membrane by Knudsen flow, the ideal separation factor is 0.80 calculated from eq.(2); therefore, this mechanism of separation can not be applied to the CO_2 separation. On the other hand, surface flow is controlled by the interaction between pore surface and gas molecules. If one of the gases adsorbs on the pore surface, it will concentrate on the pore surface and permeate along the concentration gradient on the surface. Thus, adsorbed gas flows preferentially (Figure 4)[5]. In this case a higher separation factor of CO_2 can be expected because CO_2 is adsorbed at a much higher rate than N_2.

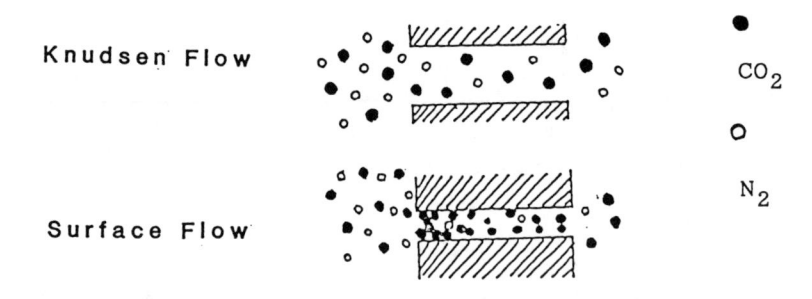

Figure 4 Shematic representation of flow mechanisms

If the flow through the porous glass membrane obeys Knudsen flow, the separation factor is 0.80 regardless of the temperature. From Table 1, it is clear that at the high temperature region the flow through the membrane approaches Knudsen flow, but at room temperature, the flow contains some surface flow.

Fundamentally, surface-modified membranes indicate the same tendency with the unmodified membrane. From these results, CO_2 seems to adsorb on the porous glass wall mainly by physical adsorption[6]. However, the separation factor of the membrane modified with xylene maintains a high separation factor value even in the high temperature region. In this membrane, the CO_2 adsorption might have somewhat of a chemical characteristic, because chemical adsorption of the CO_2 molecule is much stronger than that of physical adsorption in a high temperature region.

CONCLUSION

A novel porous glass membrane was prepared from the sodium borosilicate glass containing ZrO_2 and CaO. The median pore diameter of this membrane was 3.0-3.5nm. The permeability of CO_2 and N_2 and the separation factor of CO_2 for CO_2/N_2(50/50) decreased with an increase in temperature. Surface-modified membranes indicate the same tendency with the unmodified membrane, however, the separation factor of the membrane modified with xylene maintained a high separation factor value even in the high temperature region.

REFERENCES

1)T.H.Kim, W.J.Koros, G.R.Husk and K.C.O'brien, Relationship between gas separation properties and chemical structure in a series of aromatic polyimides, J.Memb.Sci., 37,45(1988).
2)T.Yazawa and K.Eguchi, Application of porous glass for membrane separation technology, New Ceramics, 1,49(1988).
3)H.Tanaka, T.Yazawa, K.Eguchi, T.Yamaguro, T.Einishi, H.Nagasawa and N.matusda, Precipitation of colloidal silica gel in high silica porous glass, J.Non-cryst.Solids, 65,301(1984).
4)R.W.Tock and K.Kammermeyer, Temperature-separation factor relationships in gaseous diffusion, AIChE J., 15,715(1969).
5)S.T.Hwang and K.Kammermeyer, "Membrane in Separation", John Wiley & Sons, New York(1975) P.57-67.
6)S.T.Hwang, Interaction energy in surface diffusion, AIChE J., 14,809(1968).

FRACTAL ANALYSIS ON PORE STRUCTURE IN RELATION TO FROST DURABILITY OF FIBER-REINFORCED BUILDING CEMENT MATERIALS

Masahiko NAKAMURA, Shinji URANO and Takahiro FUKUSHIMA
Kyoto Institute of Technology, Faculty of Engineering and Design,
Department of Chemistry and Materials Technology
Matsugasaki, Sakyo-ku, Kyoto 606 Japan

The high correlation between frost durability and fractal dimension calculated from pore size distribution of several fiber reinforced cement boards was explained by taking the frost mechanism into consideration.

INTRODUCTION

The use of asbestos has been severely restricted because of its negative health effects. Asbestos is widely used as a fiber reinforcement in portland cement building materials. Therefore, development of highly-durable building cement materials reinforced with non-asbestos fibers is needed world wide. For such materials, "frost durability" is one of the important factors which determine their service lives.

Improvements of frost susceptibility have been realized to some extent by a control of pore size distribution for a wide range of inorganic building materials. However, much higher frost durabilities are still needed, as well as reliable frost evaluation methods than those currently in use for building cement boards.

The present investigation was undertaken to discuss the relation between the distribution structure of pores and frost durability of building cement boards reinforced with different kind of fibers. The validity of an argument about the distribution structure of pores on the basis of "fractal" geometry has previously been reported[1].

As a numerical expression for the structural complexity of pore distribution, a fractal dimension (D) was calculated from an accumulated pore size distribution curve measured by the mercury intrusionmethod which is widely used.

EXPERIMENTAL PROCEDURE

Samples

Samples used in the present investigation are listed in Table 1. In order to prepare samples of fiber-reinforced cement boards having widely scattered physical properties, the samples were chosen partly from the commercial products and others from specimens which were made in the authors' laboratory.

Table 1. Specification of Samples used.

Sample symbol	Reinf. fiber	Forming method	Curing method	Additive aggregat.	Absorp. water(%)	Frost durab.(FD) (%)n=300cycles
P1	Viny.	Press	Steam	Fly ash	13.0	81.1

P2	Asbe.	Press	A.C.	—	13.2	59.3
P3	Pulp	Press	A.C.	—	17.5	66.6
E1	P.P.	Extr.	Steam	Fly ash	2.8	100.0
E2	P.P.	Extr.	A.C.	—	9.4	86.7
F1	P.P+Pulp	Filt.	A.C.	—	16.3	53.3
F2	Asbe.	Filt.	A.C.	—	19.6	39.0
F3	Viny.	Filt.	Steam	—	14.3	36.3

Viny.= Vinylon, Asbe.= Asbestos, A.C.=Autoclave, Extr.= Extrusion, P.P.= Polypropylen, Filt.= Filtration, FD(%)= Calcul.from Eq.(1).

Frost Durability Tests

The ASTM (C-666 A) test of frost susceptibility for concrete was applied to the samples in Table 1. Up to 300 freez-thaw cycles between +20 and -20°C were completed. All the samples for the freez-thaw test were prepared in a rectangular shape of 80x40 mm^2 and 4~8 mm in thickness. After accurate weighing of an oven-dried sample, water absorption of the respective samples was calculated from an amount of absorbed water after dipping it in distilled water for 24 hrs at 20°C. Water absorption for each starting sample for the freez-thaw test was carried out in the same way as stated above. Frost durability of each sample was evaluated by an index (FD) defined from the following equation:

$$FD = (f_n/f_0)^2 \times 100(\%) \quad \cdots\cdots(1)$$

where f_0 and f_n are respectively a primary resonant frequency (f) of a starting wet sample under a condition of deflecting vibration and of the same sample being tested after the (n) cycles of freezing-thawing. A dynamic modulus of elasticity (E_D) can be expressed by the following equation:

$$E_D = C \times W \times f^2 \quad \cdots\cdots(2)$$

where W is the weight of the specimen, and C is a factor which depends upon the shape, size and Poisson's ratio of the specimen, as well as the mode of vibration.

Pore Size Distribution Measurements

Pore size distributions of all samples were obtained by a mercury porosimetry, using a pressure up to 405 MPa. A contact angle of 140° was used to calculate pore radius. Samples for the porosimetry were prepared by breaking a sample into small pieces and drying them for 24 hrs at 105°C in oven.

Fractal Dimension Analysis

The authors have reported in the previous paper[2] that there was a close correlation between frost durability of clay roofing tiles and their fractal dimensions (D), calculated from the pore size distribution by the following equation:[1]

$$(V - V_{min})/(V_{max} - V) = \{(R - R_{min})/(R_{max} - R)\}^{D-3} \quad \cdots\cdots(3)$$

where R is a pore radius, V an accumulated pore volume, and suffixes $_{min}$ and $_{max}$ imply respectively minimum and maximum, when the logarithmic plot of $(V-V_{min})/(V_{max}-V)$ vs $(R-R_{min})/(R_{max}-R)$ shows the highest linearity.

RESULTS AND DISCUSSION

The curves of an accumulated pore size distribution were compared in Fig.1 to three different forming methods. The finaly-determined points of the co-ordinates of (V_{min}, R_{min}) and (V_{max}, R_{max}) after confirming the criterion mentioned above are indicated as arrows on the respective curves in Fig.1.

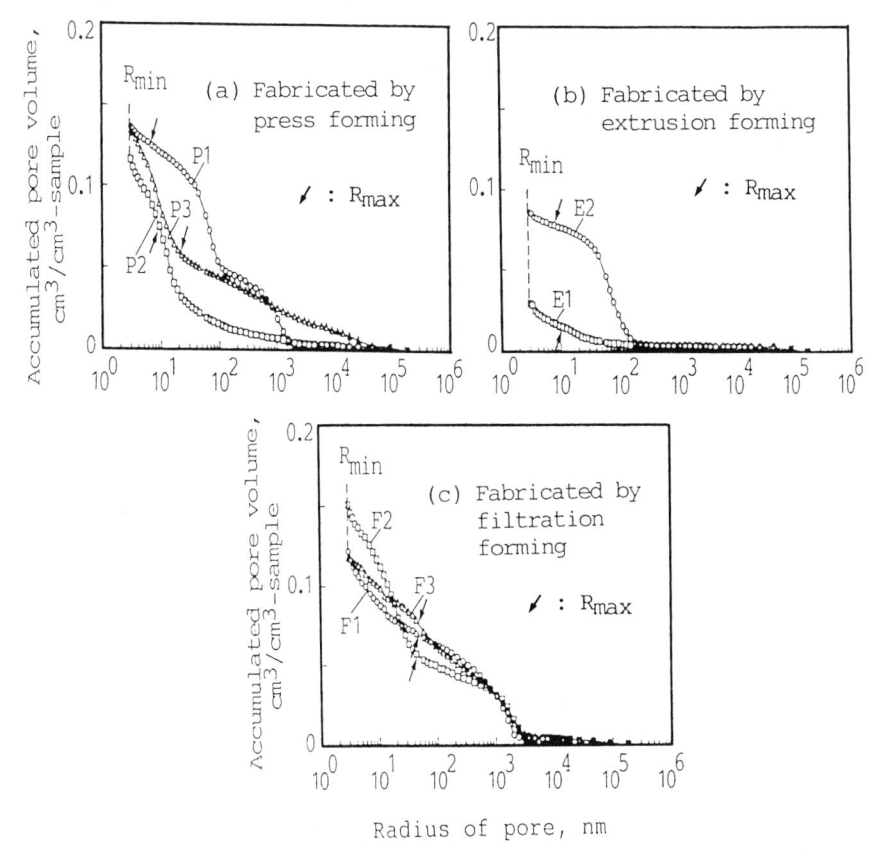

Figure 1 Comparison of Pore Size Distribution Curves. Symboles used for Respective Samples are the Same in Table 1.

Porous Materials 227

By substituting several co-ordinates existing between (V_{min}, R_{min}) and (V_{max}, R_{max}) into Eq.(3), the gradient of the highest linearity was obtained as shown in Fig.2. The correlation coefficient (r) value that best fits each line is also indicated in Fig.2.

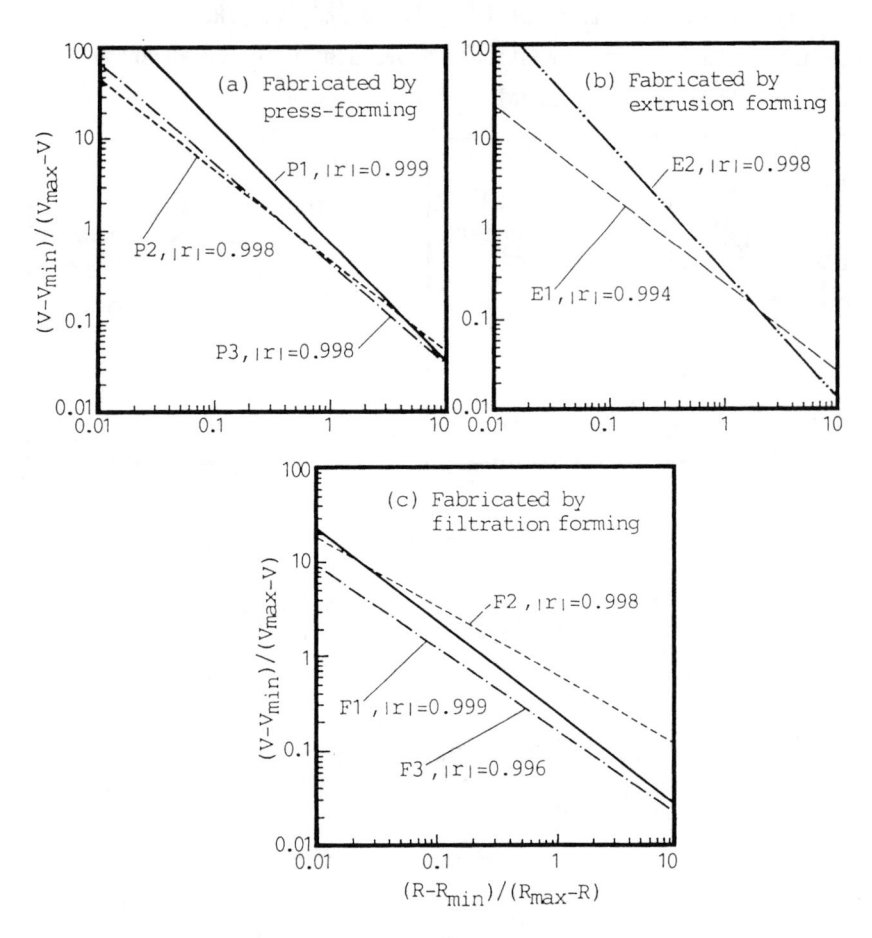

Figure 2 Logarithmic Plots between $(V-V_{min})/(V_{max}-V)$ and $(R-R_{min})/(R_{max}-R)$. Correlation Coefficient, r, for each Line are Indicated Following the Names of Respective Samples.

The fractal dimension (D) indicating a geometrical complexity of distribution or three dimensional connectivity of pores can be obtained from the gradient of (D-3) from Eq.(3), i.e., $\log\{(V-V_{min})/(V_{max}-V)\}/\log\{(R-R_{min})/(R_{max}-R)\}$ value of the respective line in Fig. 2.

By plotting (D) against the frost durability index (FD), which can be calculated from Eq.(1), a close correlation is observed, as shown in Fig. 3.

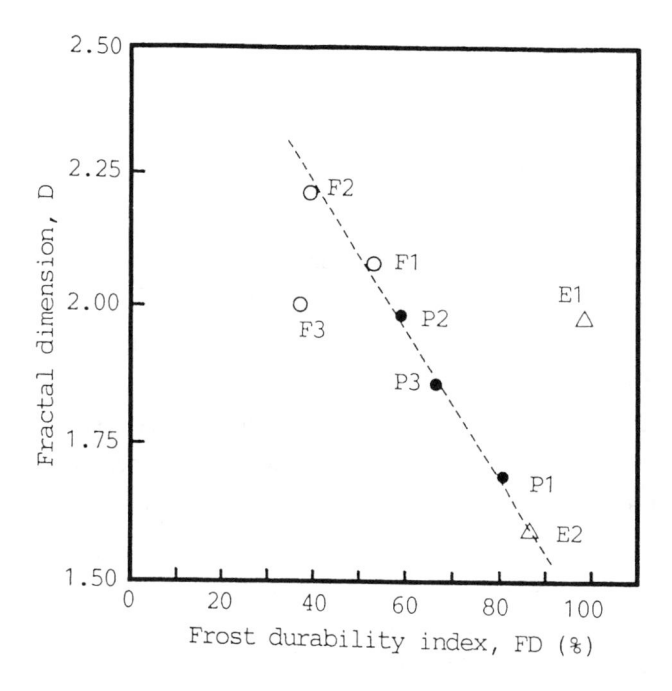

Figure 3 Relationship between Fractal Dimension and Frost Durability Index.

Investigating the relation between (D) and (FD) of the fired clay roofing tiles in the previous paper[2], the authors pointed out that the higher interconnectivity results in an increase in frost damage in those samples.

The authors also have previously explained such close dependence between (FD) and (D), based on the mechanism of frost action. The frost action is generally believed to be caused by two different driving forces; one is the hydraulic pressure genarated by freezing of water which is confined in interconnected pores, the other is the heaving pressure arising perpendicularly to the freezing-plane at an ice-water interface, due to the chemical potential differences of water molecules in those two phases under a temperature gradient.

Considering from the freezing condition resulting from the freez-thaw test (ASTMC-666A), the hydraulic pressure is believed to cause more damage. In a sample having a higher interconnected pore structure, the amount of unfrozen water entrapped in fine pores during cooling increases with an increase in pore interconnectivity. Such an increased amount of entrapped water will genarate more destructive force due to the hydraulic pressure during freezing of the water. When the rate of cooling increases, or the degree of branching in a connecting network of pores becomes more complex, the porous structure will become significantly damaged. This is due to an accumulated fracture pressure that cannot be released by the hydraulic flow of water and/or the plastic deformation of ice through the interconnected fine pores.

In the samples of fiber-reinforced cement boards used in this research, the frost damage may occur in the same manner as in the case of the fired clay roofing tile samples reported in the previous paper[2]. In the samples used in this research, the pore size ranges (i.e., 3~30 nm) indicating the highest linearity of lines in Fig.2 are much smaller than those (i.e., 5~100 nm) in the samples of the fired clay roofing tile. This means that the harmful pore size for frost damage of the fiber-reinforced cement boards is significantly smaller than that of fired clay roofing tiles. Such a discrepancy of the harmful pore size ranges between the two kinds of samples may come from the difference in the formation process of pores. In the sample of clay roofing tiles, the small pores easily disappear due to the mutual coalescece during firing. But in the sample of cement materials, much finer micro- or meso- pores are progressively developed by CSH-gel formation during curing and hardening.

No investigation has previously been reported on the frost problems of inorganic porous building materials. By considering the results obtained here together with those of the previous report[2], the fractal analysis of the geometrical complexity of porosity can be reliably used as an indirect evaluation of the frost durability or for designing building materials with high frost durability.

CONCLUSIONS

The present investigation was undertaken to discuss the relationship between a pore distribution structure expressed by a fractal dimension (D) and a frost durability (FD), in order to develop fiber-reinforced cement boards with improved frost durability, and to establish a reliable indirect method for evaluation of frost durability. The results obtained are summarized as follows;
(1) It was found that (FD) of fiber-reinforced cement boards was strongly dependent on (D), which was calculated from an

accumulated pore size distribution. This means that the samples having a higher complexity in the distribution or connection of pores, i.e., the higher value of (D), show the higher frost damage, as in the case of the fired clay roofing tiles.

(2) In the case of samples of fiber-reinforced cement boards, the pore size range indicating the highest correlation coefficient between $\log\{(V-V_{min})/(V_{max}-V)\}$ and $\log\{(R-R_{min})/(R_{max}-R)\}$ was very smaller than that of samples of fired clay roofing tiles.

(3) The fractal analysis used in the present investigation on the basis of pore size distribution curves has potential as indirect evaluation method for the frost durability of inorganic porous building materials.

ACKNOWLEDGEMENTS

This work was partially supported by the Science Research Grant of Ministry of Education in Fiscal year of 1990 (Theme No. 02555161; representative, M. Nakamura). The support is gratefully acknowledged. We wish to thank to Mr. M. Kamitani at Sekisui Chemical Co. Ltd., for donation of commercial cement boards and for measurement of pore size distritution of the samples used.

REFERENCES

1) F. Ehrburger and R. Jullien,"Characterization of Porous Solids"(K. K. Unger et al., Eds.), p.441 Stud. Surf. Sci. Catal. Vol.39, Elsevier, Amsterdam, 1988.
2) M. Nakamura,"Quantitative Analysis of Pore Structure on Frost Durability of Inorganic Building Materials", J. Ceram. Soc. Japan, 99, [11], 1114-1119 (1991).

MECHANICAL PROPERTIES OF HIPED POROUS COPPER

Atsushi TAKATA* and Kozo ISHIZAKI
Nagaoka University of Technology
Nagaoka, Niigata 940-21, JAPAN
*Present address: Japan Grain Institute Co., Ltd., Kokubunji, Kagawa 769-01 JAPAN

Open porous copper metals, which have high strength, high open porosity and well controlled pore size distribution, were produced by a hot isostatic press (HIP) process. They were sintered at different temperatures from 973 to 1273 K under various HIPing pressures up to 200 MPa. Young's modulus and pore size distribution of sintered samples were analyzed. Both strength and open porosity were improved by the HIPing pressure as compared with normally sintered product. The internal structural parameters such as pore size distribution were controlled by changing the HIPing pressure.

INTRODUCTION

Porous metals, especially with open pores, are important to industrial products, such as filters[1], oil containing bearings[2], and catalytic converter filters for exhaust gases[3]. High mechanical strength and high porosity are required for these applications. There is a common shortcoming which any conventional process can not overcome. The reason for this shortcoming is the contradiction between porosity and strength of materials, and increasing porosity inevitably leads to the strength reduction. According to the conventional methods, the designed porosity of expected strength can be hardly obtained within a reasonable error tolerance because of various uncontrollable pore producing factors.

Several researchers have investigated the relationship between the strength and the porosity of porous materials[4, 5]. The configuration of pores, flaws and defects play sensitive roles in the mechanical properties of porous materials. Almost all commercialized porous materials are produced at various sintering temperatures, holding times, compressive pressures and particle sizes, in various kinds of surrounding atmospheres, but sintering pressure of 0.1 MPa has never been changed.

The authors have developed a HIP process, in which no capsule was used, to improve the mechanical properties of porous ceramics while retaining adequate porosity[6, 7]. Previously HIP has only been used to produce densified materials[8, 9], and is considered to be unsuitable for production of porous materials, to the best knowledge of the authors. In this study, a porous copper metal with high strength and high open porosity was obtained by the HIP process. The effects of HIPing temperatures and pressures on porosity, pore size distribution and mechanical properties were investigated.

EXPERIMENTAL PROCEDURE

Copper powder (Kojundo Kagaku Kenkyujo, purity 99.9%) with an average particle size of 15 μm was used to obtain a porous metal. Twelve grams of powder were uniaxially pressed in a stainless steel mold at a pressure of 5 MPa for 60 s and then cold isostatically pressed (CIP) under 10 MPa for 300 s, to form green bodies. Three samples were sintered for each HIP process.

Copper samples were hot isostatically pressed (Kobe Steel, Dr-HIP) for 1 hour in an argon gas atmosphere under different pressures (0.1–200 MPa) and temperatures (973–1273 K) without capsules. At first, the HIP chamber was pressurized to half of the holding pressure, and then heated at a heating rate of 900 K/h. The pressure increased by heating, and reached the final pressure at 923 K. Beyond this temperature, the pressure was maintained by means of degassing. The normal sintering (under 0.1 MPa of argon) was also carried out in a graphite furnace under the same heating conditions.
The density and porosity of green and sintered bodies were measured by the Archimedean method in toluene. The pore size and the pore size distribution were measured by a mercury pressure porosimetry (Carlo Erba Strumentazione, Macropore 120 and Porosimeter 2000, courtesy of Kagawa Industrial Center), and Young's modulus by the acoustic method[10].

RESULTS

Figure 1 shows the changes of relative density, open and closed porosity of the normal and the HIPed copper samples as a function of the sintering temperature. The relative density of the normally sintered samples rapidly increases from 56 to 85% and reaches the highest value at temperatures higher than 1173 K. The open porosity of the samples decreases with increasing temperature. The closed porosity increases at the temperatures higher than 1073 K. The increasing rate of the relative density of the HIP sintered samples is lower than that of the normally sintered ones. The open porosity decreases with increasing sintering temperature. The open porosity of the HIP sintered samples exceeds the normal sintered samples about 30% at sintering temperatures of 1123 and 1173 K. The increase of HIPing pressure from 20 to 200 MPa results in increased open porosity but decreased closed porosity.

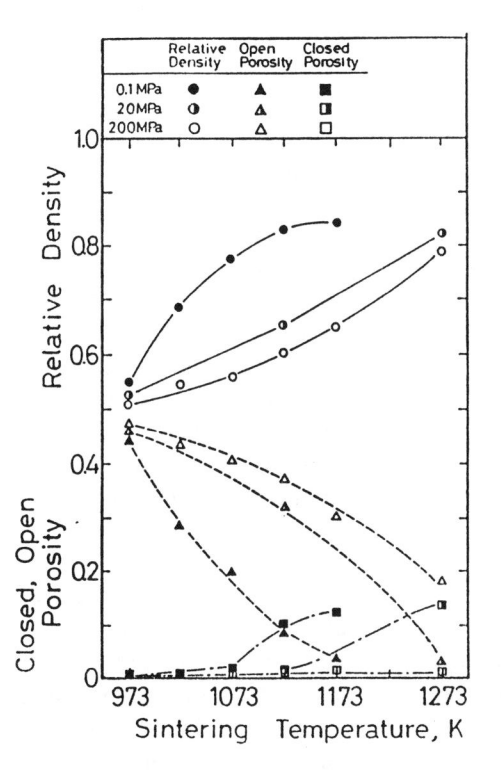

Figure 1 Changes of relative density, as well as closed and open porosities with the sintering temperature. The circle, triangle and square marks correspond to the relative density, the open and the closed porosity, respectively. The solid, half solid and open marks mean normal sintering (i.e. 0.1 MPa) and HIPing under 20 and 200 MPa respectively. The open porosity of the samples HIPed under 200 MPa was much higher about 30%, than the sintered one under 0.1 MPa at 1123 K.

Figure 2 shows the effects of the HIPing pressure on the total and open porosity of copper samples sintered at different temperatures between 973 and 1273 K. For all sintering temperatures, especially at intermediate temperatures between 1073–1173 K, the amount of total porosity increases

porosity was observed at a sintering temperature of 973 K, which is lower than the sintering temperature. At 1273 K the total porosity increased slightly, but the open porosity increased from 2% to 18% by increasing the HIPing pressure from 20 MPa to 200 MPa.

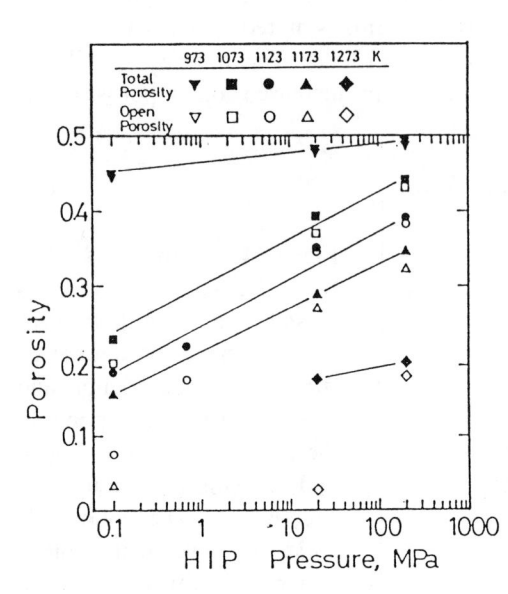

Figure 2 Effect of HIPing pressure on total(solid marks) and open(open marks) porosity at different sintering temperature. ; at 973 K (inverse triangles), 1073 K (squares), 1123 K (circles), 1173 K (triangles) and 1273 K (diamonds). The treatment under 0.1 MPa implies the normal sintering under argon atmosphere. The total porosity increased with the increasing treatment pressure. The samples sintered under low HIPing pressure had more closed porosity.

A photograph of porous copper samples sintered at 1123 K under different pressure and pre-sintering green body (i. e. the CIPed sample) is shown in Fig. 3. With increasing pressure, the shrinkage of samples decreases due to increasing porosity. The microstructures of the copper sample sintered under different pressures are shown in Fig. 4. The pores are almost invisible in the sample sintered under 0.1 MPa. The number of pores increases with increasing pressure.

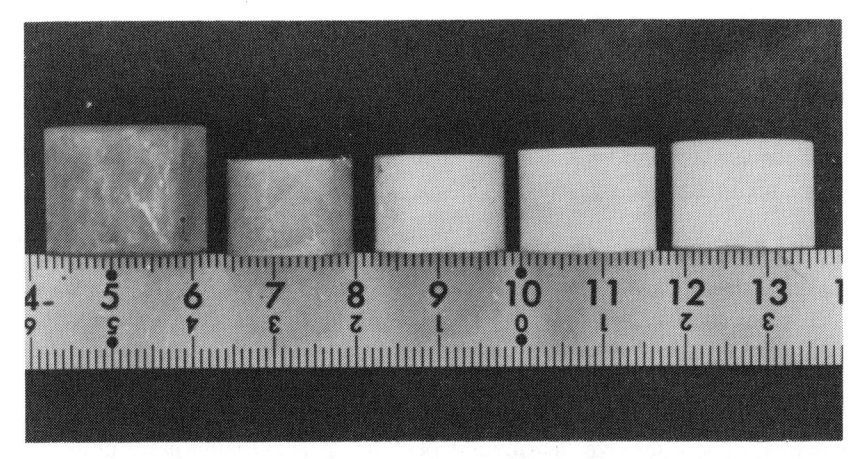

Figure 3 Photograph of porous copper samples. CIPed green body, and sintered at 1123 K under different pressures(0.1, 1, 20 and 200 MPa) from left to right.

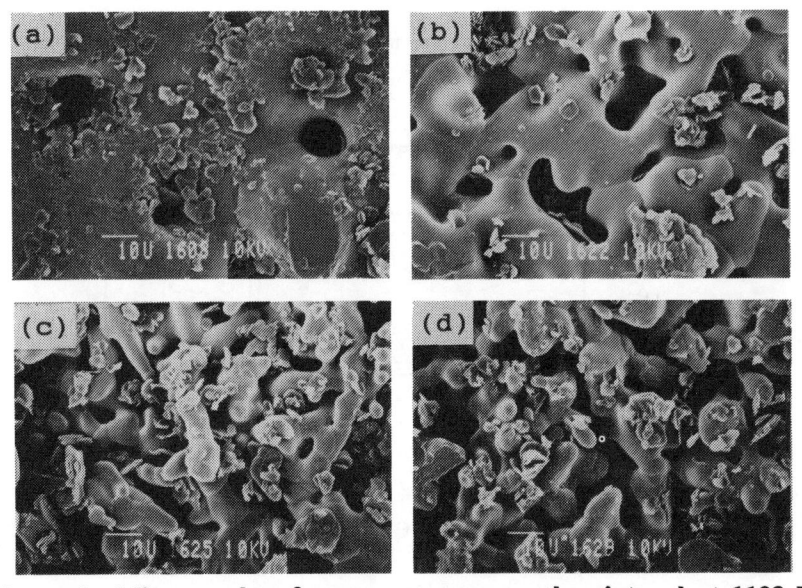

Figure 4 Micrographs of porous copper samples sintered at 1123 K under (a) 0.1 MPa, (b) 1 MPa, (c) 20 MPa, (d) 200 MPa. The pores are almost invisible in the sample sintered under 0.1 MPa. The number of pores increases with increasing pressure.

Porous Materials

Relationships between the specific pore volume and the pore size of copper samples sintered at 1123 K under different pressures are shown in Fig. 5. The specific pore volume increases with increasing pressure. This coincides with the fact that the porosity of the samples increases for the higher HIP pressures as shown in Figs. 1 and 2. The mean pore size, however, does not change significantly.

Figure 6 shows the relationship between the mean pore size and the HIPing pressures of copper samples sintered at different sintering temperatures of 1073, 1123 and 1173 K. The top and the bottom bars and the middle mark in the bars correspond to the pore size of 90, 10 and 50% of the specific pore volume respectively. The circle, triangle and square marks move to the middle position of both top and bottom bars, as the HIPing pressure increases. The pore size distribution becomes narrower and more uniform with increasing HIPing pressure.

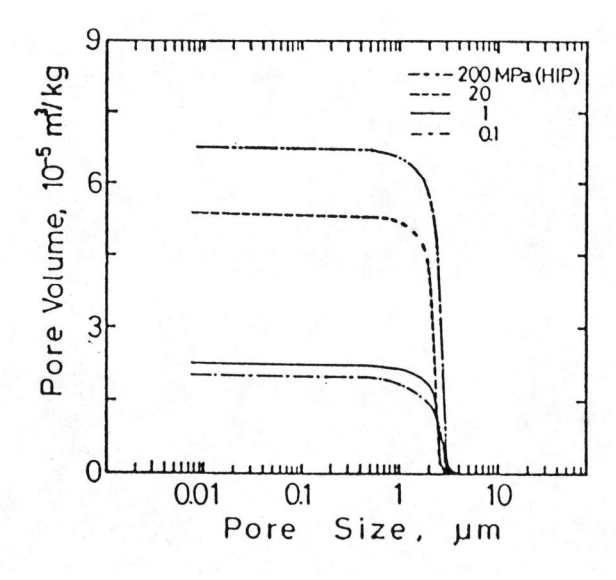

Figure 5 Relationships between specific pore volume and pore size of copper samples sintered at 1123 K under different pressure.

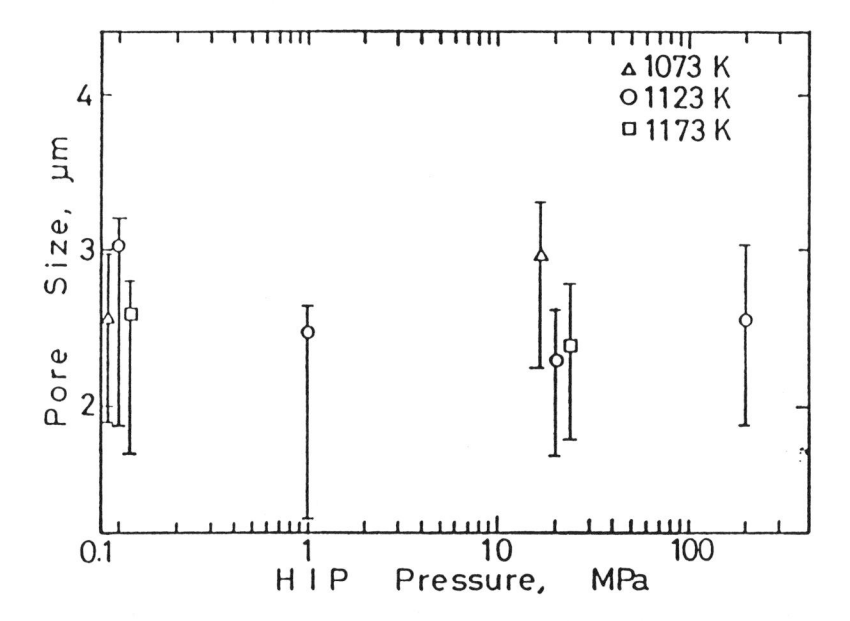

Figure 6 Effect of HIPing pressure on pore size. The triangle, circle and square marks correspond to samples HIPed at 1073, 1123 and 1173 K, respectively. The mean values imply the pore size of 50% pore volume. The upper and bottom lines of the marks correspond to 90 and 10% amount of pore volume, respectively. By increasing the HIPing pressure, the pore size distribution became narrower.

Normalized Young's modulus (E/E_0) are plotted as a function of the relative density or the total porosity as shown in Fig. 7, where Young's modulus of dense copper (E_0) is 123 GPa. The solid and dashed lines show the change of the normalized Young's modulus for the normally (0.1 MPa) sintered samples and the HIP (200 MPa) sintered ones, respectively. The slopes (calculated by the least square method) of the curves were 3.23, 3.04 and 3.03 for the normal sintered samples, and the HIPed ones at 20 MPa and at 200 MPa respectively. Figure 7 reveals that there is a considerable difference in normalized Young's modulus between HIPed samples and normal sintered ones.

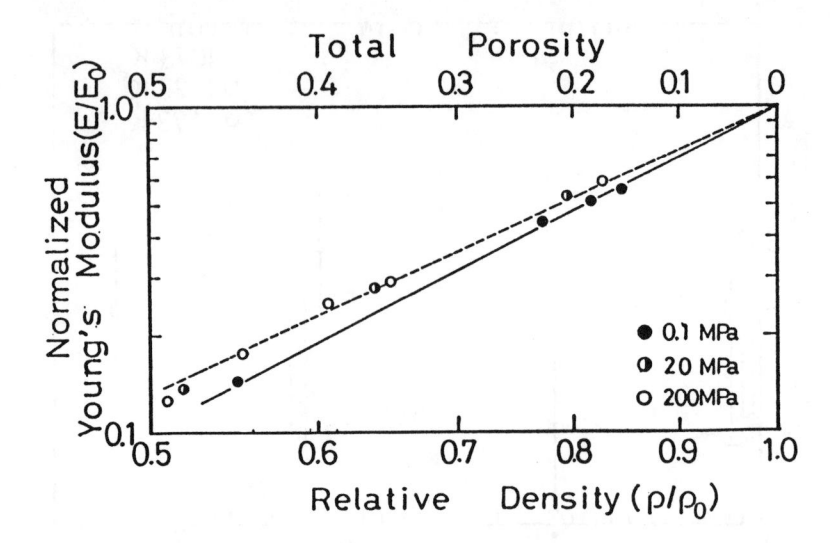

Figure 7 **Normalized Young's modulus as a function of the relative density or the total porosity.** The solid, half solid and open marks mean normal sintering (i. e. 0.1 MPa) and HIPing under 20 and 200 MPa, respectively. The solid and dashed lines are approximate lines of the normally sintered samples and the HIPed ones at various sintering temperature respectively.

DISCUSSION

In the normal sintering (i. e. the sintering pressure (0.1 MPa)), the open pores began to close and the closed pores increased rapidly at temperatures above 1073 K. An increase of pressure to 20 MPa shifted to the temperature, at which the samples started to shrink, to 1123 K (Fig. 1). Increasing closed porosity did not take place in the whole investigated temperature range under 200 MPa of HIPing pressure (Fig. 2).

The pore distribution obviously changed as shown in Fig. 6. By increasing the HIPing pressure, the distribution bar became smaller. This means that the small pores of the green body were removed by HIP sintering. Almost all commercialized porous metals are produced without varying the sintering pressure. The effect of the porosity on Young's modulus of porous material[11] is represented by the following equation,

$$\frac{E}{E_0} = (1-P)^\alpha = \left(\frac{\rho}{\rho_0}\right)^\alpha \tag{1}$$

where E, ρ, P and α are respectively Young's modulus, density, porosity and a constant, and the subscript 0 means at $P=0$. The relationship defined by Eq. (1) is valid for $P<0.5$. Usually, Young's moduli of porous materials are represented by Eq. (1). The value of α (theoretically calculated to be 3 [11]) is depend on pore configuration and bridge strength, hence is constant for different sintering and molding conditions.

The porosity of the open porous materials was controlled by HIPing pressure even in the case of the same manufacturing condition. The enhancement of mechanical properties of porous materials by HIPing effects may be expressed by the change of α. The value of α as a function of the HIPing pressure is shown in Fig. 8, and decreases with increasing HIPing pressure. The Young's modulus of the HIPed porous copper samples was higher than that of the normally (0.1 MPa) sintered ones as shown in Fig. 7. The increase of the Young's modulus by applying the HIPing pressure is due to the conventional HIP effects on the microscopical pore bridge parts, and the porosity of increasing temperature. Test results suggest that the high pressure collapses the closed pores of the particles and high sintering temperatures accelerate the diffusion. The HIP method removes flaws and defects inside the metal, and enhances bridging of particles.

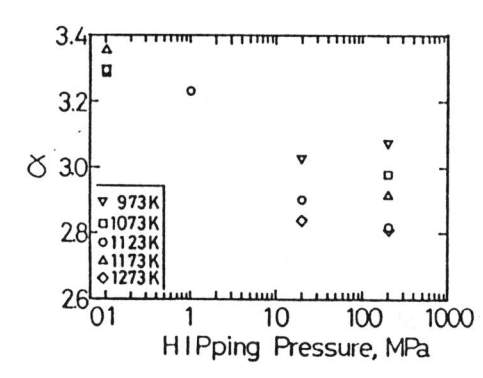

Figure 8 Relationship between HIPing pressure and value of α. The inverse triangle, square, circle, triangle and diamond marks correspond to samples sintered at 973, 1073, 1123, 1173 and 1273 K respectively.

CONCLUSIONS

Porous copper containing a large quantity of open pores was successfully produced by using the modified HIP process. The effects of HIP pressure on porosity, pore size distribution and Young's modulus of copper were concluded as follows:
1) The porous metal becomes stronger by increasing the HIPing pressure.
2) The modified HIP process provides higher open porosity than the normal sintering process. Open porosity increases with increasing HIPing pressure, which also leads to elimination of closed pore.
3) Pore size distribution becomes narrower by increasing the HIPing pressure. The change of the mean pore size was not observed.

REFERENCES

1. N. Kato, Ceramic Filter for Microfiltration and Ultrafiltration, Bull. Ceram. Soc. Jpn., **23**, 726-730(1988).
2. H. Shikata, N. Funabashi, Y. Ebine and T. Hayasaka, Modern Development in Powder Met., **17**, 497 (1984).
3. T. Hiki, Ceramic Porous Media and Its Application, Bull. Ceram. Soc. Jpn., **20**, 168-174 (1985).
4. J. G. J. Peelen, B. V. Rejda and J. P. W. Vermeiden, Sintered Hydroxylapatite as a Bioceramic, Philips Tech. Rev., **37**, 234-236(1977).
5. R. M. Spriggs, L. A. Brissette and T. Vasilos, Effect of Porosity on Elastic and Shear Moduli of Porycrystalline Magnesium Oxide, J.Amer. Ceram. Soc., **45**, 400(1962)
6. K. Ishizaki, A. Takata and S. Okada, Mechanically Enhanced Open Porous Materials by HIP Process, J. Jpn. Ceram. Soc., **98**, 533-540 (1990).
7. K. Ishizaki, S. Okada, T. Fujikawa and A. Takata, "Process for Modifying Porous Materials Having Open Cells", US Pat., No.5, 126, 103(1992) and Germany Pat., No.P4091346.5(1991), Patent Pending (Japan).
8. P. E. Price and S. P. Kohler, Hot Isostatic Pressing of Metal Powders, in Metal Handbook 9th ed. vol 7 Powder Metallurgy, (ASM, Metal Park, Ohio, 1984), pp. 418-443.
9. K. Ishizaki and K.Tanaka, Sintering of Advanced Materials -Applications of Hot Isostatic Pressing-, (Uchida Rokakuho, Tokyo, 1987).
10. W. R. Davies, Trans. Br. Ceram. Soc., **67**, 515-541 (1968).
11. M. Yu. Bal'shin, Relationship between Porosity Contact Cross -Section and Properties of Powders, Dokl. Akad. Nauk. SSSR, **154**, 80 (1964).

BENDING STRENGTH AND ELECTRIC RESISTIVITY OF POROUS Si-SiC CERAMICS

Haruhiro Osada, Akira Kani, Shoji Katayama and Tadashi Koga
Eagle Industry Co.,Ltd., Katayanagi, Sakado-shi 350-02, Japan

It is well known that Si-SiC materials have high resistance to heat and corrosion and can be produced with a high purity. Therefore, Si-SiC porous products are recently applied in dispersion boards for IC etching. The physical properties of Si-SiC were found to depend very much on the amount of free Si, that filled up the pores in the SiC matrix. According to the order of wettability, molten Si filled up the small pores which had sharp corners, first. Great improvement in bending strength was achieved, because the number of sharp notches causing crack initiation was diminished by the free Si. Based on the microstructure study, these phenomena and the change of electric resistivity were experimentally as well as theoretically explained.

INTRODUCTION

Porous ceramics with high corrosion and heat resistance are being promoted for wide application in filtration and separation[1,2]. Si-SiC ceramics have a high corrosion resistance and can be produced with a high purity; therefore, Si-SiC porous products are used as dispersion boards and electrodes for etching gas in IC manufacture. The bending strength, the electric resistivity, and the porosity are important factors for the applications of porous Si-SiC. The effects of the free Si content in the reaction sintering on these properties were investigated. The relationship between these properties and the microstructure of the Si-SiC ceramics was also examined. A conceptual model was used to calculate the electric resistivity.

EXPERIMENTS

Manufacture of Porous Ceramics

The tested porous ceramics were manufactured as follows. SiC abrasive powder GC#150 (mean particle diameter:69 μm) and carbon powder (mean particle diameter: 4 μm) in a weight ratio of 95:5 were mixed in a solvent (acetone) together with 5 parts of stearic acid as an organic binder. The slurry was then dried. The material was molded (uniaxial molding pressure: 90 MPa) into green disks of 37 mm in diameter and about 5 mm thick. The disks were contacted with the appropriate amount of Si, and heat treated in a vacuum furnace at 1500°C for 1 h.

Measurement

The bulk density was obtained by dividing the weight by the volume derived from the dimensions of the specimens. The apparent density was derived from a water absorption method(JIS R 2205). The porosity was obtained by comparing the bulk density with the apparent density. The free Si contents were obtained by subtracting the amount of Si reacted with C to form SiC from the total amount of metallic Si pasted on the sample. Bending strengths were obtained from a 3-point loading test at a crosshead speed of 0.5 mm/min by using an autograph AG-5000A made by Shimadzu. The pore diameter distributions were measured by a mercury penetration porosimetry method. The specimen (5 mm X 8 mm X 25 mm) was taken out of the Si-SiC disk, and the electric resistivity was measured by a four-probe dc method.

EXPERIMENTAL RESULTS AND REVIEW

Bending Strength

The porosity and the free Si contents of the sintered samples were changed by varying the amount of Si brought into contact. In Fig. 1 the optical micrographs show that the porosity was reduced gradually with increasing free Si level. The microstructure was composed of SiC (gray), free Si(white), and pores (black). From Fig. 1(a),(b) and (c),it became evident that free Si preferably entered small pores. However ,free Si also entered large pores in the samples with a free Si content of more than 8.5 wt%. Figure 2

(a) ε = 0.31
free Si = 1.0 wt%

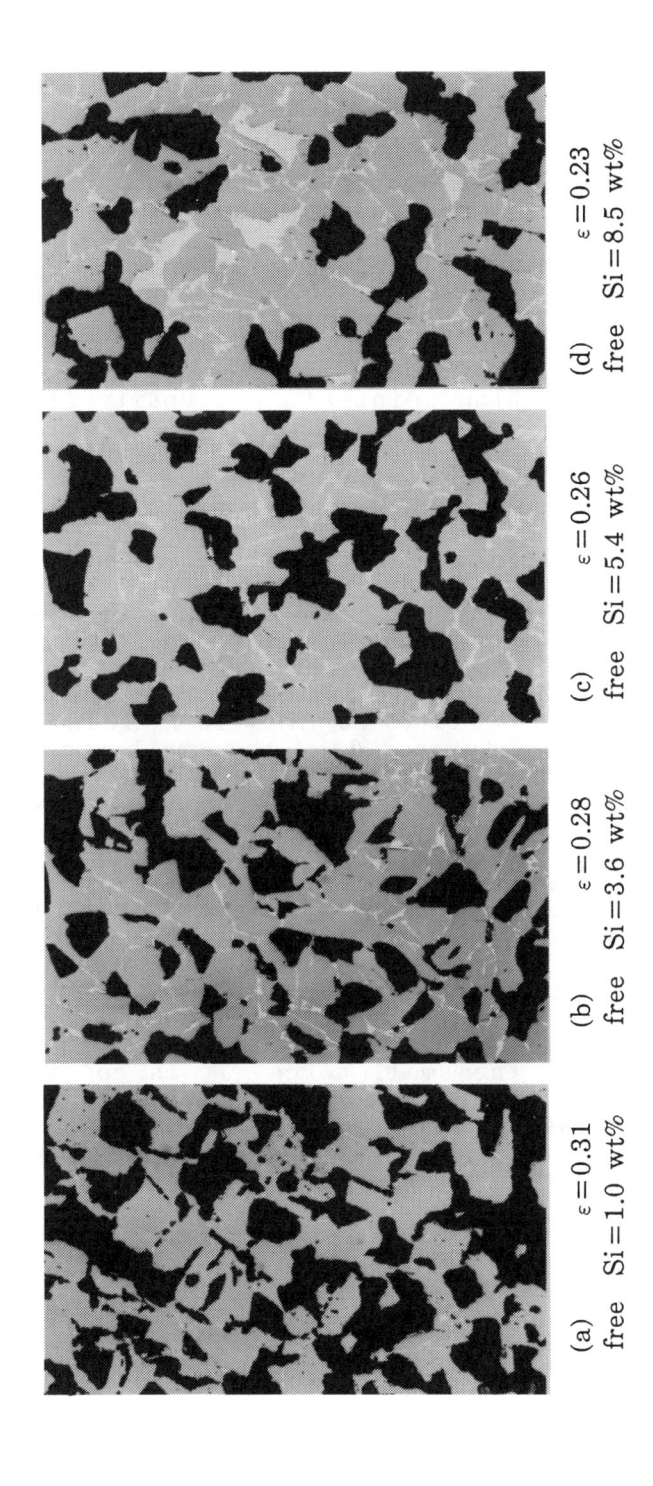

(b) ε = 0.28
free Si = 3.6 wt%

(c) ε = 0.26
free Si = 5.4 wt%

(d) ε = 0.23
free Si = 8.5 wt%

Figure 1 Microstructure of porous SiC ceramics with different free Si contents. (ε is the porosity)

shows the effect of the porosity on the bending strength. Figure 2 shows that decreases in porosity due to increases in free Si level resulted in the rise of the bending strength. However, significant increases in bending strength were not obtained below a porosity of 0.26. Evaluating the photos (a),(b),and (c)in Fig. 1, it can be seen that the number of sharp edges at the corner of small pores decreased in the porosity range from 0.31 to 0.26 by increasing the free Si level. This decrease in the number of sharp edges caused the increase in the bending strength, because sharp edges acted as notches in the bending test. The inflection point at a porosity of 0.26 corresponds to Fig. 1(c): showing the sample for which all the smaller pores are completely filled with free Si. The above results suggested that at a porosity of about 0.26 a big change in the pore diameter distributions occurred. The relationships between the cumulative porosity and pore diameter for varying free Si content are plotted in Fig. 3. Porous specimens without free Si have 2 types of pores: smaller pores 2 to 6 μm and larger pores 10 to 20 μm. As the free Si level increased, free Si filled the small pores(2 to 6 μm) first and after they were filled, large pores (10 to 20 μm) were filled. Figure 4 shows the porosity caused by the small pores and the porosity caused by the large pores. In the total porosity range above 0.26, the number of small pores decreased rapidly and in the lower total porosity range the large pores decreased with increasing free Si content. The inflection point occurring at a porosity of 0.26 agreed with the porosity at which bending strength variation occurred. The above data proved that the strength was greatly affected by the 2 to 6 μm pores.The strength increased significantly as the small pores were filled with free Si. When the disks were contacted with a small amount of Si, free Si first filled the small pores. It is considered that this is caused by the fact that molten Si and SiC have high wettability[3].

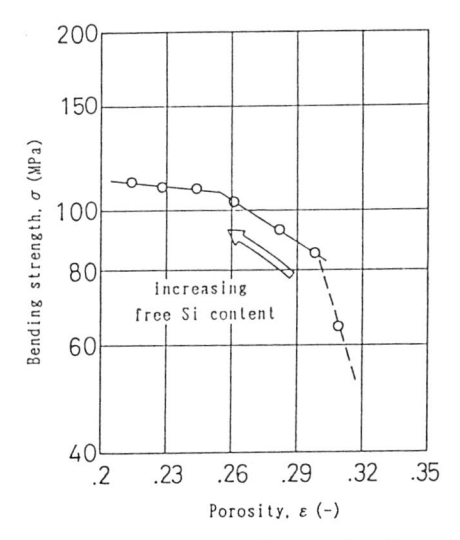

Figure 2 Effect of porosity on bending strength for increasing free Si content.

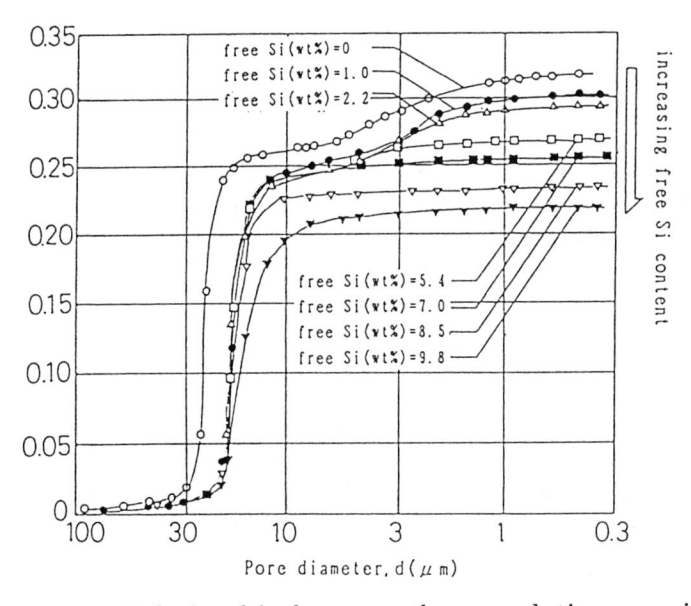

Figure 3 Relationship between the cumulative porosity and pore diameter for various free Si contents.

Porous Materials

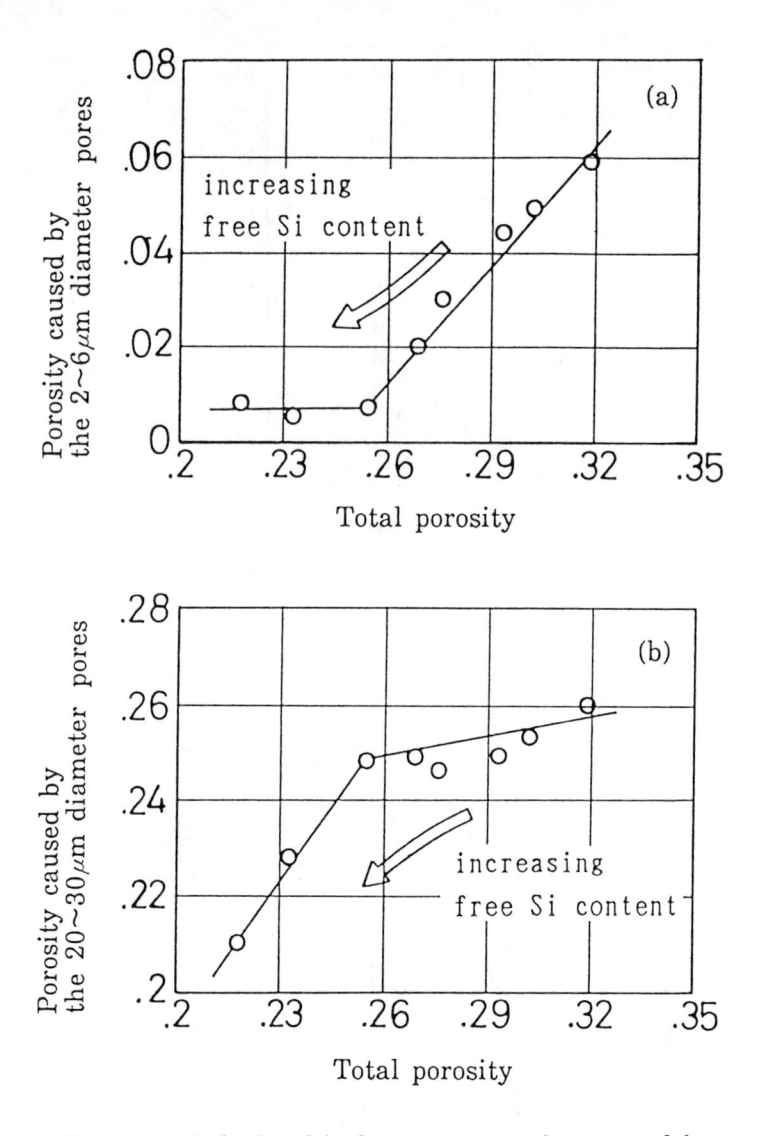

Figure 4 Relationship between porosity caused by the small pores and the total porosity (a) and between the porosity caused by the large pores and the total porosity (b).

Electric Resistivity

Measured and calculated electric resistivities are shown in Fig. 5. A remarkable increase of the electric resistivity was found when the porosity was larger than about 0.30. To calculate the electric conductivity of binary mixtures in non porous structures, the basic equation is used[4]. To derive this formula, it was assumed that material 2 was present as spherical particles embedded in the matrix 1, as shown in Fig. 6.

$$k_{mix} = k_1 + \frac{3k_1(k_2-k_1)x_2}{2k_1+k_2-(k_2-k_1)x_2} \qquad (1)$$

where k_{mix}, k_1, and k_2: are the electric conductivity for the mixture, material 1 and material 2, respectively. x_2 is the volume fractions of material 2.

The electric conductivity of non porous Si-SiC ceramics was first calculated by formula (1), assuming that material 1 and 2 were Si and SiC, and using 2×10^2 and $1\times10^{-3}(\Omega\cdot cm)^{-1}$ as the electric conductivity of Si and SiC, respectively. The same formula was then used again, considering that material 1 was Si-SiC ceramic and material 2 was the pores. Consequently, the electric conductivity of the Si-SiC ceramic(k_1), pores(k_2), and the volumetric fraction of pores(x_2) were assumed as k_{mix}, zero, and porosity(ε), respectively. Then equation (1) was rewritten as follows.

$$k = k_{mix} \times \frac{2(1-\varepsilon)}{2+\varepsilon} \qquad (2)$$

Eventually the electric resistivity was calculated and plotted in Fig. 5. It was in good agreement with the measured electric resistivity.

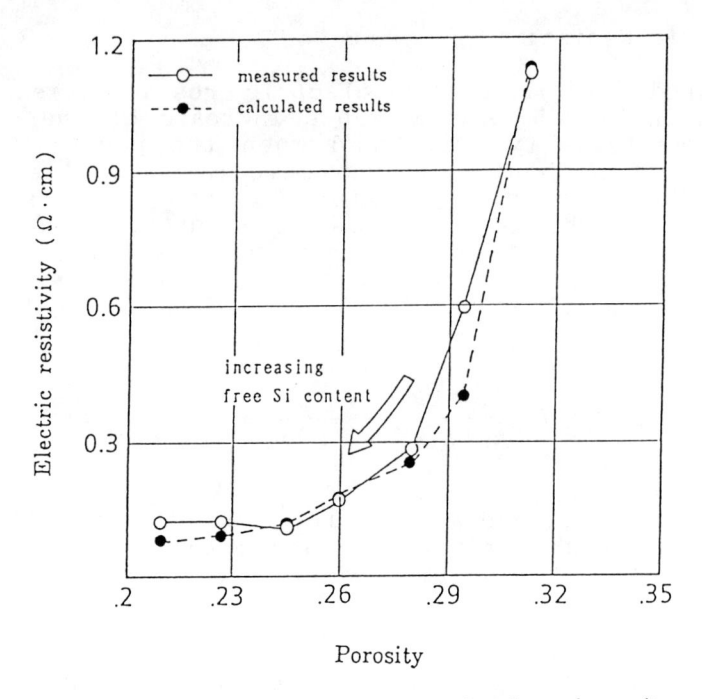

Figure 5 Relationship between electric resistivity and porosity.

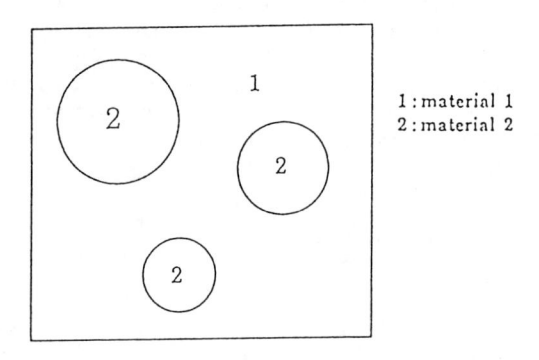

Figure 6 Model of binary mixtures ; material 2 was assumed to be spherical.

CONCLUSIONS

The effects of the free Si level of the porous Si-SiC ceramics on the bending strength and the electric resistivity were investigated. The conclusions are as follows:
1) Increasing the free Si level reduced the porosity ( ) and increased the bending strength. A steep rise of the strength was obtained in the region where < 0.26. In this region, the free Si first filled the small pores, thereby decreasing the number of crack initiation points.
2) By increasing the free Si level further, the molten Si filled also the large pores. However the bending strength did not improve significantly, because a large amount of free Si was required to fill the large pores.
3) Increases in porosity due to decreases in free Si level increased the electric resistivity. Especially in the region with a porosity above 0.30, the increase was remarkable. The calculated electric resistivity was in good agreement with the measured values.

REFERENCES

1) T.Hioki,"Ceramic Porous Media and Its Application," Ceramics Japan, **20**[3]168-174(1985)
2) T.Kato,"Ceramic Filter for Microfiltration and Ultrafiltration,"Ceramics Japan,**23**[8]726-730(1988)
3) T.Cho and T.Oki,"Wettability for Fabrication of Metal Matrix Composite,"Bulletin of the Japan institute of metals,**28**[4]285-293(1989)
4) M.Koiwa and J.Takada,"Physical Properties of Heterogeneous System,"Bulletin of the Japan institute of metals,**27**[7]525-531(1988)

ELECTRICAL AND ELECTROCHEMICAL PROPERTIES OF TiO$_2$-SiO$_2$ POROUS GLASS-CERAMICS

Toshinori Kokubu and Masayuki Yamane†
Miyakonojo National College of Technology, Yoshio-cho, Miyakonojo, Miyazaki 885, Japan,   TEL 0986-38-1010   FAX 0985-38-1508
† Tokyo Institute of Technology,   Ookayama, Meguro-ku, Tokyo 152, TEL 03-3726-1111   FAX 03-3726-0393

Platelet TiO$_2$-SiO$_2$ porous glass-ceramics (TPGC) were made from the mother glass of the composition 23.0 TiO$_2$ - 28.0 SiO$_2$ - 14.0 Al$_2$O$_3$ - 3.5 P$_2$O$_5$ - 28.0 CaO - 1.5 MgO - 1.0 ZrO$_2$ - 1.0 Ni$_2$O$_3$ (mol%). The electric conductivity of the TPGC near room temperature was about 10^{-7} S m^{-1}. TPGC humidity sensor worked well sensing the humidity, especially in the low humidity region below 20%. The sensitivity in the high humidity region improved with impregnation of KCl. Electrochemical property of Pd-TPGC electrode was surveyed by cyclic voltammetry. Current peaks due to hydrogen adsorption and desorption, oxidation and reduction of the electrode, and O$_2$ and H$_2$ generations were observed in the cyclic voltammogram. The peak currents of hydrogen adsorption and desorption increased gradually with scan cycles. With irradiation of Xe-light to the Pd-TPGC electrode, photo-induced current which was higher about 120 mA than the current under dark was observed at the voltage 1.5 V. The Pd-TPGC electrode worked successfully as a photo-electrochemical catalyst. The electrode reacted with phenol which was added in the electrolyte solution and a photo-decomposition of the phenol occurred when the electric potential was held at the voltage 1.0 V or 1.7 V for 1 h under irradiation.

INTRODUCTION

The porous glass-ceramics (TPGC) of TiO$_2$-SiO$_2$ system containing TiO$_2$ more than 50 mol% is obtained by a similar process for the fabrication of porous Vycor glass. This consists of the phase separation of the mother glass and subsequent acid treatment of the phase separated glass[1-3]. The average pore diameter of TPGC is in the range between 2 and 25 nm and the specific surface area is between 50 and 400 m^2 g^{-1}[1-3] . These values are comparable with those of porous Vycor glass[4-5]. TPGC contains crystallites of rutile and/or anatase. Some TPGC have a color originated from a transition metal like Ni, Co, Fe or Mo that was added in mother glass and incorporated into the skeletal part[3]. The thermal and chemical properties[6] and uranium adsorption capacity[7] of TPGC have been already reported. The other properties such as electrical and electrochemical properties were investigated in the present study.

TiO_2 is an n-type semiconductor having an electronic conductivity, and have been applied to catalysts[8-10], photo-catalysts[11-13] and gas sensors[14-15]. If the TPGC has electronic conduction based on constituent TiO_2, it may be applied to humidity sensors or the other chemical sensors. If the TiO_2 contained in TPGC has photo-catalytic activities like TiO_2 crystal, the large specific surface area, relatively uniform pore and many surface hydroxyls on micro-pore wall might be advantageous in the catalytic application.

This paper reports on electric properties of TPGC as a humidity sensor, the electrochemical and photo-electrochemical properties of a TPGC electrode surveyed by cyclic voltammetry, and photo-electrochemical decomposition of phenol on Pd-TPGC electrode.

EXPERIMENTS

The platelet TPGC employed in this work was made from the mother glass of the composition 23.0 TiO_2 - 28.0 SiO_2 - 14.0 Al_2O_3 - 3.5 P_2O_5 - 28.0 CaO - 1.5 MgO -1.0 ZrO_2 - 1.0 Ni_2O_3 (mol%) as previously reported[1-3]. Chemical composition of the TPGC was 51.4 TiO_2 - 36.9 SiO_2 - 4.6 Al_2O_3 - 2.4 P_2O_5 - 1.6 CaO - 0.7 MgO - 1.0 ZrO_2 - 1.3 Ni_2O_3 (mol %) . The pore volume and the specific surface area were 0.29 ml g^{-1} and 84.5 m^2 g^{-1} respectively. The pore diameter distribution was in the range between 8 and 20 nm and the average pore diameter was 13.7 nm.

Electric conductivity measurement was carried out on the platelet TPGC of 0.8 mm x 10 mm x 10 mm. The electrodes of 5 mm in diameter were formed on both surfaces of the plate by vacuum evaporation of gold and connected to lead wire of Pt by silver paste. The change in DC current with temperature was measured under the voltage of 10 V in a cryostat from 200 K to room temperature. The calculation method of conductivity; σ (S m^{-1}), was described elsewhere[21].

Humidity sensing was carried out at 50°C in a vacuum desiccator of the volume 10.7 dm^3 equipped with a thermostat and a fan. A TPGC sensor of the size 5 mm x 10 mm x 0.4 mm having a gold mesh electrode on the upper surface and a gold plane electrode on the other side was tightened between two Cu plates. The plate on the surface side had an opening of 3 mm x 4 mm to allow the sensor contact with the atmosphere. The sensor was placed in the center of the desiccator. The humidity in the desiccator was adjusted to various levels by injecting and vaporizing water. Details for the measurement were described elsewhere[23]. Impedance (Ω) of the sensor in respective levels of humidity was measured using 4192 LF type IMPEDANCE ANALYZER made of YOKOGAWA HEWLET PACKARD Co. A sensor which was impregnated with KCl of about 0.1 wt% to the weight of TPGC was then fabricated and humidity sensing was carried out in the same way.

Electrochemical and photo-electrochemical properties were studied by means of cyclic voltammetry with a potentiostat on a Pd-TPGC electrode. About 0.1 wt% of Pd was impregnated to a platelet TPGC with 0.05 mol l^{-1} PdCl$_2$. After decomposition of PdCl$_2$ at 600°C, the TPGC plate was subjected to reduction at 300°C under H$_2$ gas flow. The Pd-TPGC of the size of 10 mm x 5mm x 0.8 mm was then fabricated into an electrode. The details of the fabrication method of TPGC electrode and the construction of measurement system were described in a previous report[23]. A surface sealing tape which was masking the surface of electrode was cut out before measurement into a size 2 x 2 mm square opening. Cyclic voltammetry was measured in 0.5 mol l^{-1} H$_2$SO$_4$ aqueous solution using Pd-TPGC electrode as the working electrode. A Pt wire electrode and saturated calomel electrode (SCE) were used as the counter electrode and the reference electrode, respectively. The instrument employed was a YANACO V-11 POTENTIOSTAT and the measurement was carried out with a voltage scan speed of 50 mV s^{-1} at 25°C. For the photo-electrochemical properties, a quartz electrolytic cell was employed. A cyclic voltammogram was measured under irradiation of 500 W Xe-light in 0.05 mol l^{-1} H$_2$SO$_4$ aqueous solution.

The photo-electrochemical response with phenol was measured in 10 ml of 0.5 mol l^{-1} H$_2$SO$_4$ aqueous solution containing 2 ml of 1.0 mol l^{-1} phenol solution. The surface of TPGC electrode was brought into contact with the electrolyte solution through an opening of 4 mm x 5 mm and irradiated by Xe-light. The applied voltage was held constant at 0 V, 1.0 V and 1.7 V for 1 h respectively. Sample electrolyte solutions from these respective voltages were then extracted by ethyl acetate and the extracts were analyzed by gas chromatography with temperature rise 7.5°C min^{-1} from 80°C. A SHIMADZU GC 14-A gas chromatograph equipped 2 m glass column packing OV-1 was used.

RESULTS

Electrical Conductivity and Humidity Sensing

A relation between electrical conductivity of TPGC measured in cryostat and the reciprocal of temperature (K^{-1}) is shown in Fig. 1. The conductivity of the TPGC near room temperature was about 10^{-7} S m^{-1}.
A relation between impedance (Ω) and relative humidity at 50°C is shown in Fig. 2. The response time was 5 min after the injection of water. The impedance was 10^9 Ω at 0% and decreased gradually to the order of 10^5 Ω at 80%. With impregnation of KCl the impedance decreased to the order of 10^4 Ω at above 70%.

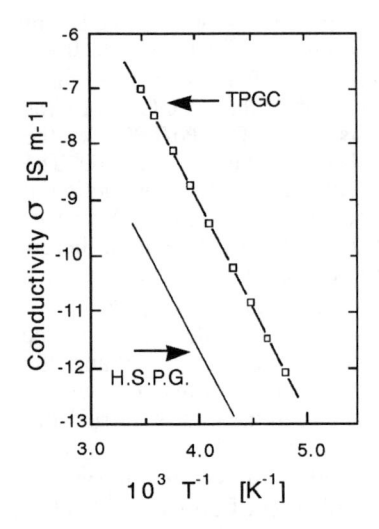

Figure 1 Electrical conductivity plot of the TPGC

Figure 2 Relation between relative humidity and logarithm of impedance

Electrochemical and Photo-electrochemical Properties

Cyclic voltammograms of Pd-TPGC electrode between 5 and 12 cycles since initiation of scanning voltage under dark is shown in Fig. 3. In the voltammogram, currents attributable to desorption of hydrogen (region A), oxidation of electrode (region B) and oxygen generation (region C) were observed in the anodic scan, and currents attributable to reduction of electrode (region D), hydrogen adsorption (region E) and hydrogen generation (region F) were observed in the cathodic scan. Both of the currents in region A and E increased gradually with increasing scan cycles and finally a cyclic voltammogram showing only hydrogen adsorption and desorption with a large current of mA order was obtained.

A cyclic voltammogram of Pd-TPGC under the irradiation of Xe-light to the electrode is shown in Fig. 4. After 10 min of irradiation, the photo-induced current, which was higher about 120 mA than the current under dark, was observed. When the irradiation was stopped, the photo-induced current decreased gradually to the current under dark.

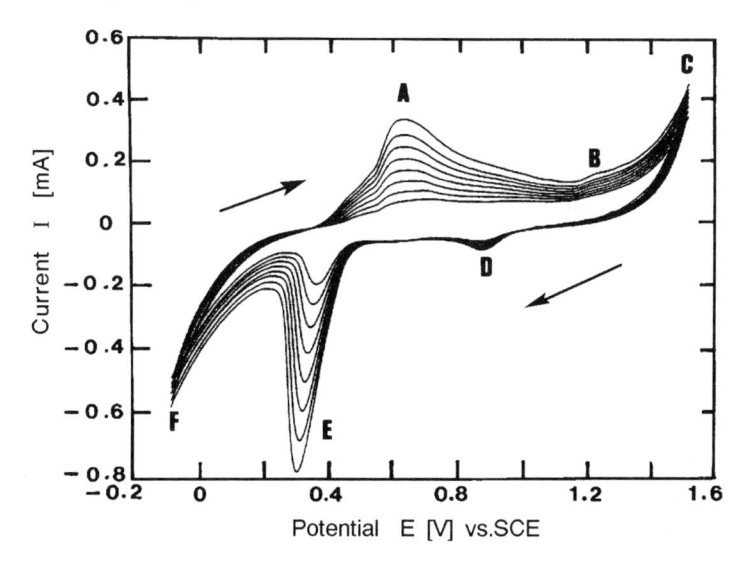

Figure 3 Cyclic voltammogram of Pd-TPGC electrode

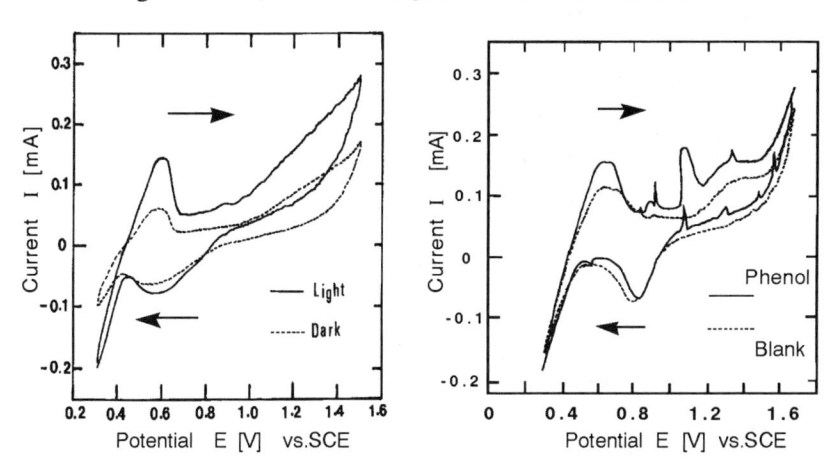

Figure 4 Cyclic voltammogram of Pd-TPGC electrode under irradiation of Xe-light

Figure 5 Phenol response in cyclic voltammogram under irradiation to the electrode

Porous Materials 257

A response with phenol in cyclic voltammogram under irradiation to the Pd-TPGC electrode is shown in Fig. 5. The photo-induced current became larger and more unstable than the voltammogram in blank i.e., when phenol is absent.

Gas chromatograms of the extracts from electrolyte solutions, which were held at 0 V, 1.0 V and 1.7 V for 1 h, are shown in Fig. 6. A peak at the retention time 4.5 min is the peak of phenol and the peak at 6.5 min is due to an impurity which had been contained in phenol reagent. There was no peak except for two respective peaks of phenol and an impurity in the gas chromatogram of the sample held at 0 V. When the solution was held at 1.0 V, however, a few unknown peaks were observed and when held at 1.7 V the peak of phenol almost disappeared.

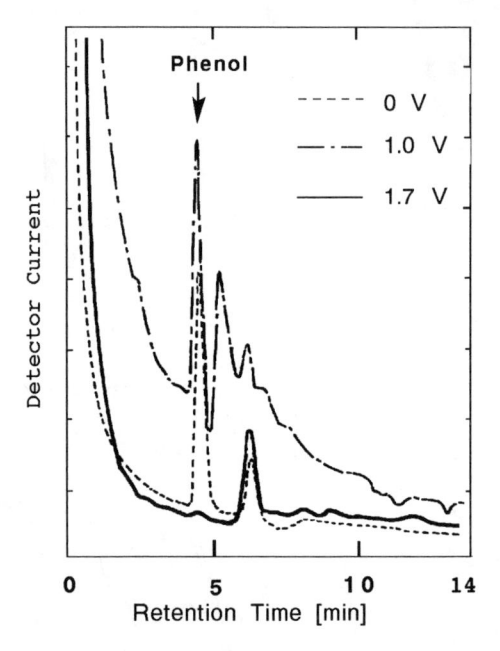

Figure 6 Gas chromatograms of extracts from respective samples held at the voltage 0 V, 1.0 V and 1.7 V for 1 h under irradiation

DISCUSSION

Electrical Conductivity and Humidity Sensing

Among the humidity sensors developed up-to-date, ceramic humidity sensors are excellent because of their mechanical strength, thermal stability and chemical durability[16,17]. In the ceramic humidity sensor, two sensing mechanisms have been suggested[16-20]. One is a proton conduction type which senses a change of conductivity due to physically adsorbed water or capillary condensation water. The other is a semiconductor type which senses a change of conductivity based on an electron donative adsorption of water on the surface of an oxide semiconductor at high temperature above 400°C. The former ceramic sensor, which is made by sintering of a compound metal oxide like $MgCr_2O_4$-TiO_2, is now commonly used[16]. The sensitivity of the sensor of this type, however, decreases in the low humidity region below 20 %[18]. For the sensing in such a low humidity region it has been suggested that a material of the pore size below 30 nm is essential [17,18]. Porous glasses of Vycor type[4-5] could have a regulating pore below 30 nm, with many surface hydroxyls which adsorb water on the micropore wall.

However, the conductivity of high silica porous glass (H.S.P.G.) is very low and is on the order of 10^{-9} S m^{-1} near room temperature, whereas TPGC had a conductivity on the order of 10^{-7} S m^{-1} at the same temperature, as shown in Fig. 1. Since TPGC had penetrating pores, which were ascertained by permeation of a gas[21], and many surface hydroxyls on the micropore wall[21], a proton conduction via surface hydroxyls will be contributing to the conductivity significantly. The difference in the conductivity between TPGC and high silica porous glass by 10^2 Ω may be attributable to the electronic conductivity of TPGC based on TiO_2.

As shown in Fig. 2, TPGC had an impedance of 10^9 Ω at 0% and showed gradual decrease with increasing humidity to the order of 10^5 Ω at 50°C. The relation between impedance and the square root of relative humidity was linear[21]. TPGC sensor, therefore, might be a sensor of typical proton conduction type. The relation between impedance and square root of relative humidity showed good linearity, especially in the low humidity region below 20%[21]. A TiO_2-La_2O_3 porous glass-ceramic sensor has been reported[22]; however, the pore size control of the sensor was not enough and the sensitivity below 20% was unclear. Since TPGC had relatively uniform penetrating pores between 8 and 20 nm[21], the TPGC sensor could be employed in humidity sensing in the low humidity region advantageously. The impedance of 10^9 Ω at 0% is, however, not high enough to be used practically. It should be on the order of 10^6 Ω at 0% for practical use. We are now investigating to improve the conductivity with a TPGC which contains crystals of TiO or Ti_nO_{2n-1} having higher conductivity than TiO_2.

In the sensor made by sintering, it is suggested that an ion dissolved into capillary condensation water from ceramics has a significant contribution to the sensitivity at a high humidity region[17, 18]. Since TPGC is made by acid leaching, it does not contain any ion which dissolves into capillary condensation water. Impregnation of an ion into the micropore, therefore, may be effective. As shown in Fig. 2, the impedance of TPGC sensor changed to a value of 10^4 Ω from a value of 10^5 Ω in 70% humidity by impregnation of KCl. This change may be attributable to an ionic conduction with dissolving KCl into capillary condensation water. Therefore, impregnation of an electrolyte is essential for the TPGC sensor used in a high humidity region.

Electrical and Electrochemical Properties

The TPGC electrode worked successfully in cyclic voltammetry. The cyclic voltammogram of the Pd-TPGC electrode shown in Fig. 3 is comparable to that of TPGC electrode[23]. In Fig. 4, both currents of hydrogen adsorption (region E) and desorption (region A) increased gradually with increasing of scan cycles. The gradual increase of the currents may be attributable to activation of surface hydroxyls on the microporous wall of the TPGC electrode, since hydrogen ion can be adsorbed via surface hydroxyls on this wall[21]. This implies that the current increase is attributable to an increase of surface area of electrode to the direction inside of the pore. The surface area of the TPGC electrode had been evaluated by cyclic voltammetry of hexacyanoferrate(II) ion. It was about 6.5 times as large as that of a Pt electrode having the same opening, in initial stage of scanning [23]. The surface area of the electrode may increase further with scan cycles, in regard to hydrogen adsorption. For other molecules, however, the diffusion to the inside of the electrode will be different depending on the size of the molecule. The effective surface area, therefore, should be evaluated with a standard molecule like hexacyanoferrate.

In the application of TiO_2 to photocatalysts, precious metals like Pt, Ru, Pd which had a charge separation efficiency were generally supported on the TiO_2 crystal[11-13]. The cyclic voltammogram of the Pd-TPGC electrode under irradiation shown in Fig. 4 had a remarkable difference in photo-induced current from the voltammogram of the TPGC without Pd [23]. The cyclic voltammogram of the TPGC electrode showed a quite unstable photo-induced current[23], which may be related to an irregular electron transfer or a re-oxidation of electrode by photo-generated holes. The photo-induced current in the Pd-TPGC was, however, relatively stable and showed a current which was higher by about 120 mA than the current under dark. The charge transfer and the re-oxidation of electrode might be smoothened with a supported Pd which has charge separation efficiency.

The photo-induced hole on the TPGC electrode is expected to have an activity to decompose an organic molecule, just as in the case of TiO_2 crystal[12,13]. In this work, photo-decomposition of phenol was examined. In Fig. 5, a quite unstable current attributable to a response of phenol on a Pd-TPGC electrode is observed. The photo-decomposition was then put into practice holding the applied voltage at 0 V, 1.0 V and 1.7 V. The gas chromatogram of the sample held at 0 V was the same as that of the phenol reagent employed in and nothing occurred at the voltage. The photo-decomposition occurred in the sample held at 1.0 V and 1.7 V, where a photo-induced current was clearly observed in the cyclic voltammogram as it is shown in Fig. 4. As shown in gas chromatograms in Fig. 6, the peak of phenol completely disappeared in the sample held at 1.7 V, and some peaks which may be attributable to the products of partial decomposition of phenol were observed in the sample held at 1.0 V.

CONCLUSION

The electric conductivity of the TPGC near room temperature was about 10^{-7} S m^{-1}. The TPGC humidity sensor worked well sensing the humidity. The sensor may be employed in humidity sensing in the low humidity region below 20% advantageously. The Pd-TPGC electrode worked successfully in cyclic voltammetry and showed a stable photo-induced current in the cyclic voltammogram. The Pd-TPGC electrode showed a photo-response to phenol added in electrolyte solution and worked successfully as a photo-electrochemical catalyst in the decomposition of phenol. It was clear that the Pd-TPGC electrode may be used as a photo-catalyst under an applied voltage like particles of TiO_2 crystal[12-13]. Details for the photo-electrochemical decomposition of organic molecules employing Pd-, Pt- or Ru-TPGC are now under investigation.

REFERENCES

[1] T. Kokubu and M. Yamane, "Preparation of porous glass-ceramics of the TiO_2-SiO_2 system", J. Mater. Sci., 20, p 4309-4316 (1985).
[2] T. Kokubu and M. Yamane, " The stability of mother glass for porous glass-ceramics of the TiO_2-SiO_2 system", J. Mater. Sci., 23, p 2449-2456 (1988).
[3] T. Kokubu and M. Yamane, " Incorporation of transition metal in porous glass-ceramics of TiO_2-SiO_2 system", J. Mater. Sci., 25, p 2929-2933, (1990).
[4] H. P. Hood and M. E. Nordberg, " Method of treating bolosilicate glasses", U. S. Patent 2215039, (1940).
[5] R. Maddison and P. W. MacMillan, " The production and characterization of materials having high specific surface areas from E glass", Glass Technol., 21, p 297-301 (1980).
[6] T. Kokubu and M. Yamane, " Thermal and chemical properties of TiO_2-SiO_2 porous glass-ceramics", J. Mater. Sci., 22, p 2583-2588 (1987).
[7] T. Kokubu, T. Matsuyama and M.Yamane, "Uranium adsorption capacity of TiO_2-SiO_2 porous glass-ceramics", J. Ceram. Soc. Japan, 99 [9] p 763-767 (1991).
[8] I. Aso, M. Nakao, N. Yamazoe and T. Seiyama, " Study of metal oxide catalysts in the olefin oxidation from their reduction behavior",

J. Catalysis, 57, p 287-295 (1979).
[9] F. Bozon-Verduraz, A. Omar, J. Escard and B. Pontvianne, " Chemical state and reactivity of supported palladium", J. Catalysis, 53, p 126 - 134 (1978).
[10] G.B.Raup and J. A. Dumesic, " Adsorption of CO, CO_2, H_2 and H_2O on titania surfaces with different oxidation states", J. Phys. Chem., 89 p 5240-5246 (1985).
[11] A. Fujishima and K. Honda, " Electrochemical photolysis of water at a semiconductor electrode", Nature , Vol. 238, 7, p 37-38 (1972).
[12] T. Kawai, T. Sakata, K. Hashimoto and M. Kawai, " The structure and the reactivity of particulate semiconductor photocatalyst", Nippon Kagaku Kai-shi (J. Chemical Soc. Japan) No.2, p 277-282 (1984).
[13] M. Gratzel, " Artificial photosynthesis: Water cleavage into hydrogen and oxygen by visible light", Acc. Chem. Res., 14 , p 376 -384 (1981).
[14] T. Y. Tien, H. L. Stadler, E. F. Gibbons and P. J. Zacmanidis, "TiO_2 as an air-to-fuel ratio sensor for automobile exhausts", Ceramic Bulletin, Vol. 54, No., p 285-288 (1975).
[15] N. Yamamoto, S. Tonomura, T. Matsuoka and H. Tsubomura, " A study on a palladium-titanium oxide schottky diode as a detector for gaseous components" Surface Science, 92, p 400- 406 (1980).
[16] T. Nitta, Z. Terada and T. Kanazawa, " Ceramics humidity sensor *humiceram* " National Technical Report (National Electric Co.) Vol. 24, No. 3, p 422-435 (1978).
[17] Y. Shimizu, H. Ichinose, H. Arai and T. Seiyama, " Ceramics humidity sensor" Nippon Kagaku Kai-shi(J. Chemical Soc. Japan) 6, p 1270-1277 (1985).
[18] Y. Shimizu, H. Arai and T. Seiyama, " Ceramic humidity sensors 1. Micro structure and humidity sensitive characteristics of Spinel type Oxides", Denki Kagaku (J. Electro-Chemical Soc. Japan) Vol. 50 No.10 , p 831-834 (1982).
[19] H. Arai, " Semiconductive humidity sensor of Perovskite-type oxide", Surface Science (J. in Japanese), Vol. 5, No.2, p 165-166 (1984).
[20] F. Fukushima, R. Makimoto, J. Terada and T. Nitta, " Humidity sensor *neohumiseramu* " National Technical Report (National Electric Co.) Vol. 29, No.3, p 459-465 (1983).
[21] T. Kokubu, Y. Nakahara, M. Yamane and M. Aizawa, " Electrical and electrochemical properties of TiO_2-SiO_2 porous glass- ceramics and the application", J. Electro-analytical Chem., accepted at August 28 (1992).
[22] Y. Shimizu, H. Okada and H. Arai, " Application of La_2O_3-TiO_2 porous glass-ceramics system to humidity sensor", Yogyo- Kyokai-Shi (J. Ceram. Soc. Japan) 95 (7) p 726- 729 (1987).
[23] T. Kokubu, " Cyclic Voltammetry of TiO_2-SiO_2 porous glass-ceramics", J. Technology and Education (Electro-Chemical Soc. Japan) Vol. 1, No.1, p 65-70 (1992).

CORROSION RESISTANCE OF PERMEABLE REFRACTORIES

Akihiro TSUCHINARI, Toshiyuki HOKII and Chikao KANAOKA*
Harima Ceramic Co., Ltd., Shaped Refractory Department,
1-3-1, Shinhama, Arai-cho, Takasago-shi, Hyogo 676,
Japan
*Department of Civil Engineering, Faculty Technology,
Kanazawa University, 2-40-20 Kodatsuno, Kanazawa-shi,
Ishikawa 920, Japan

The effects of the physical properties and chemical composition of the raw materials on the corrosion resistance of Al_2O_3, $MgO \cdot Al_2O_3$ and MgO permeable refractory materials for molten steel were investigated.
The corrosion resistance of a permeable refractory depends on the low melting point material produced by the reaction between the molten steel and the permeable refractory. The corrosion resistance improves with decreasing apparent porosity and mean pore size, and with increasing compressive strength. The corrosion resistance of the MgO system is superior to that of the Al_2O_3 and the $MgO \cdot Al_2O_3$ systems.

INTRODUCTION

In the steel-making process, blowing gas through a porous plug into the molten steel has been widely adopted in ladles and tundishes to homogenize the temperature and the composition of molten steel, and/or to float nonmetalic inclusions. Al_2O_3 porous plugs are mainly used for this purpose in Japan. Accompanying the current needs for high quality steel, the corrosion resistance of the porous plugs needs to be improved. Despite the importance of such an improvement, no precise study has been done.

This paper discusses the effects of the physical properties and chemical composition of the porous plug on the corrosion resistance.

EXPERIMENTAL PROCEDURE

Raw Materials and Preparation of Specimens

Tables 1 and 2 show the physical properties and chemical composition of the sintering materials. Al_2O_3, $MgO \cdot Al_2O_3$ and MgO are used as aggregates and the matrix powder. Twenty seven specimens were prepared with different amounts of aggregates and matrix, and with a different number of pressings with a friction press in order to change the property of the specimens.

Table 3 shows the weight percentages of aggregate and matrix. They are sintered at 1730℃ for 6 hours.

The A15, S15 and M15 in Table 3 denote a matrix content of 15 wt% in the system Al_2O_3, $MgO \cdot Al_2O_3$ and MgO, respectively.

Table 1. Physical Properties of Aggregates and Matrix.

System		Aggregate			Matrix		
		Al_2O_3	$MgO \cdot Al_2O_3$	MgO	Al_2O_3	$MgO \cdot Al_2O_3$	MgO
Apparent density(g/cm³)		3.81	3.29	3.35	-	-	-
Bulk density(g/cm³)		3.62	3.22	3.25	-	-	-
Apparent porosity(%)		5.0	2.1	3.0	-	-	-
Size (μm)	Max.	1000	1000	1000	40	150	150
	Mean	740	670	680	5	13	11

Table 2. Chemical Composition of Aggregates and Matrix.

System		Aggregate			Matrix		
		Al_2O_3	$MgO \cdot Al_2O_3$	MgO	Al_2O_3	$MgO \cdot Al_2O_3$	MgO
Chemical composition (wt%)	SiO_2	0.03	0.23	0.22	0.18	0.15	0.22
	Al_2O_3	99.72	72.98	0.09	99.07	72.76	0.09
	Fe_2O_3	0.01	0.11	0.05	0.02	0.16	0.05
	CaO	0.05	0.25	0.39	0.14	0.50	0.39
	MgO	0.02	26.36	99.21	0.02	26.56	99.21
	Na_2O	0.16	0.07	-	0.54	-	-
	K_2O	0.01	-	-	0.03	-	-
	B_2O_3	-	-	0.04	-	-	0.04

Table 3. Weight Percentages of Aggregates and Matrix.

	System	Al_2O_3			$MgO \cdot Al_2O_3$			MgO		
		A15	A10	A5	S15	S10	S5	M15	M10	M5
Agg.	Al_2O_3	85	90	95	-	-	-	-	-	-
	$MgO \cdot Al_2O_3$	-	-	-	85	90	95	-	-	-
	MgO							85	90	95
Matrix	Al_2O_3	15	10	5	-	-	-	-	-	-
	$MgO \cdot Al_2O_3$	-	-	-	15	10	5	-	-	-
	MgO	-	-	-	-	-	-	15	10	5

Porous Materials

Method of Evaluation

The pore distribution was measured by a mercury pressurizing porosimeter, and the apparent porosity and compressive strength were measured based on the Japanese standards JIS R2205 and R2206, respectively.

The corrosion resistance was evaluated by the rotary corrosion method. As shown in Fig. 1, test specimens were attached inside a cylinder rotating at 8 r.p.m. with molten steel (650 g) at 1650°C. The molten steel was replaced every 30 min. This procedure was repeated five times. Then the corroded thickness was measured at ten different points for each specimen, and the average corroded thickness was calculated.

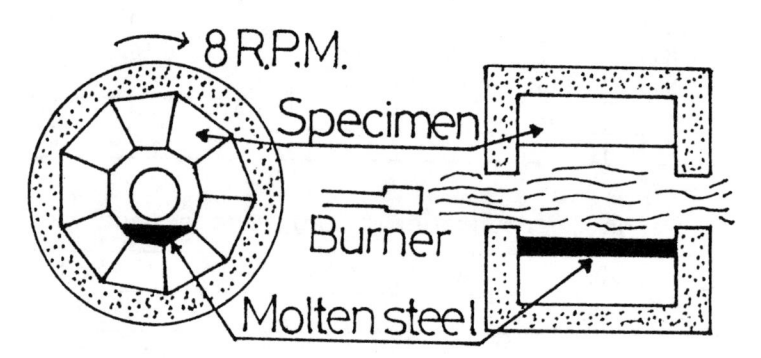

Figure 1 Apparatus for rotary corrosion test.

RESULTS AND DISCUSSION

Effects of Chemical Composition on Corrosion Resistance

Figure 2 shows the photos of Al_2O_3, $MgO \cdot Al_2O_3$ and MgO permeable refractory after the rotary corrosion test. Table 4 shows the results of the X-ray diffraction analysis of the original and the penetration layers of specimens A15-15, S15-15 and M15-15 in Fig. 2. $MgO \cdot Al_2O_3$ s. s. and MgOs. s. in Table 4 were the materials which were produced by occluding a small amount of Fe or FeO into $MgO \cdot Al_2O_3$ and MgO, respectively.

As seen in Fig. 2, the MgO system was not corroded as much as the Al_2O_3 and $MgO \cdot Al_2O_3$ systems. The material loss of the $MgO \cdot Al_2O_3$ system was very large. According to the X-ray diffraction analysis, a relatively large amount of low melting point materials, such as $FeO \cdot Al_2O_3$, FeO and $MgO \cdot Fe_2O_3$, are detected at the penetration layers of the Al_2O_3, $MgO \cdot Al_2O_3$ and MgO systems, respectively. The melting points of $FeO \cdot Al_2O_3$, FeO and $MgO \cdot Fe_2O_3$ are 1450℃, 1355℃ and 1713℃, respectively. Therefore, the corrosion resistance is considered to be closely related to the melting point of these products.

Influences of Physical Properties on Corrosion Resistance

Figures 3~5 show the correlation of the corrosion rate with the apparent porosity(Po), the mean pore size($\bar{d}$) and the compressive strength(Cs). Although the data are

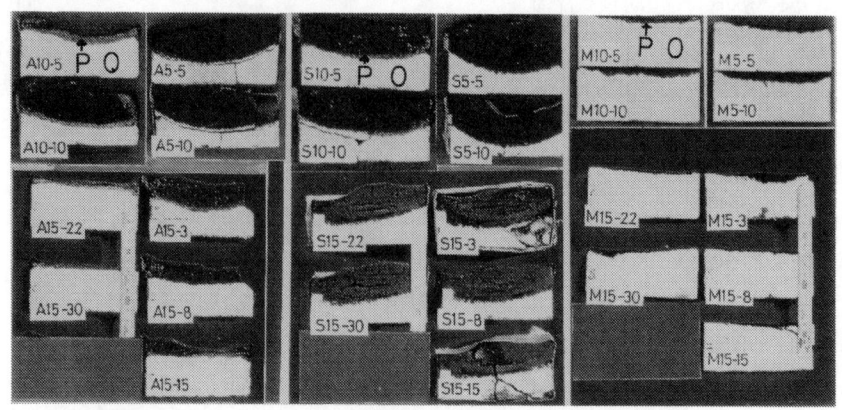

(Al₂O₃ system)   (MgO·Al₂O₃ system)   (MgO system)

Figure 2 Photos of specimens after rotary corrosion test.
         Material loss: MgO < Al₂O₃ < MgO·Al₂O₃
         Penetration thickness: MgO < MgO·Al₂O₃ < Al₂O₃
         Key: A5-10 Matrix content (Wt%)
              └── Number of pressings
         O: Original layer, P: Penetration layer

Table 4. Result of the X-ray Diffraction after Rotary Corrosion Test.

System	Al₂O₃ O	Al₂O₃ P	MgO·Al₂O₃ O	MgO·Al₂O₃ P	MgO O	MgO P
Corundum (Al₂O₃)	5+	2	–	–	–	–
Periclase (MgO)	–	–	–	–	5+	5+
β-alumina (Al₂O₃)	1–	–	–	–	–	–
Spinel s.s. (FeO·Al₂O₃)	–	1	–	–	–	–
Spinel s.s. (MgO·Al₂O₃)	–	–	–	2	–	–
Magnesia s.s. (MgO)	–	–	–	1–	–	–
Wustite (FeO)	–	–	–	1–	–	–
Spinel (MgO·Al₂O₃)	–	–	5+	–	–	–
Magnesioferrite (MgO·Fe₂O₃)	–	–	–	–	–	1
Sodium magnesium aluminum oxide (Na₂O·4MgO·15Al₂O₃)	–	–	1–	–	–	–

Numeral shows the peak intensity of the X-ray;
5⁺ > 5 > 4 > 3 > 2 > 1 > 1⁻
K denotes Kilo-count per second of the X-ray.
0.5K<1<1.0K, 1.0K<2<1.5K, 1.5K<3<2.0K, 2.0K<4<2.5K,
2.5K<5<3.0K, 3.0K<5⁺
O: Original layer, P: Penetration layer

scattered, we can conclude that the corrosion rate increases with decreasing Po and $\bar{d}$, and with increasing Cs.

According to a previous published theory (1), the penetration thickness is proportional to the root of the apparent porosity (Po) and the pore size ($\bar{d}$). Further, the dissolution of the permeable refractory is inversely proportional to Cs. Therefore the experimental corrosion rate is plotted against $\sqrt{Po\bar{d}}/Cs$ in Fig. 6. As shown in Fig. 6, the corrosion

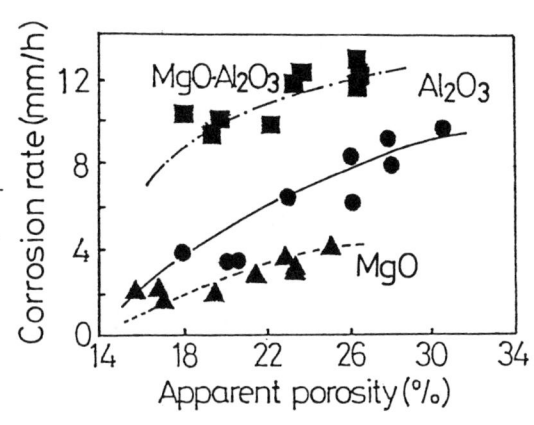

Figure 3 Influence of apparent porosity on corrosion resistance.

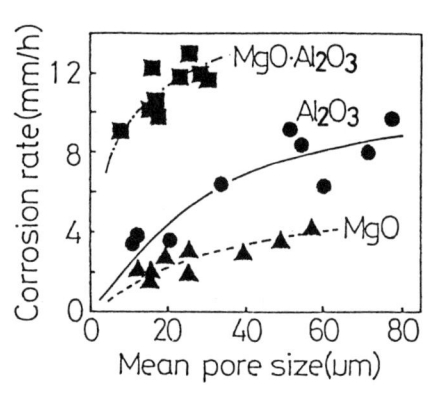

Figure 4 Influence of mean pore size on corrosion resistance.

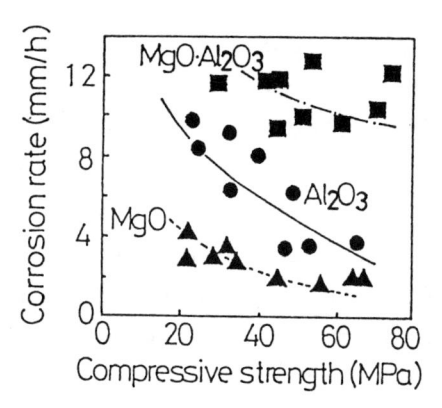

Figure 5 Influence of compressive strength on corrosion resistance.

Porous Materials

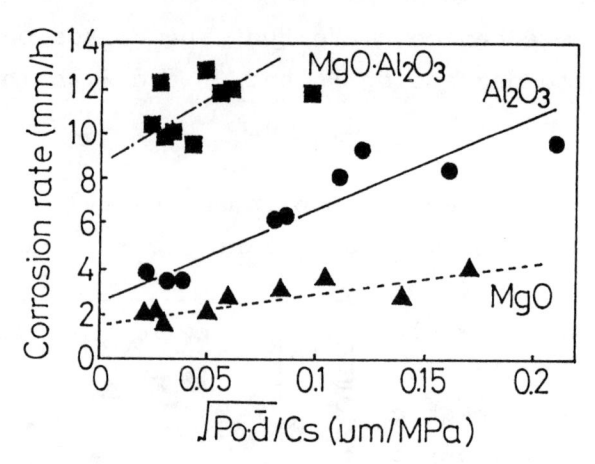

Figure 6 Corrosion rate as function of $\sqrt{Po \cdot \bar{d}}/Cs$ with Po the apparent porosity, $\bar{d}$ the mean pore size and Cs compressive strength.

rate has a linear relation with $\sqrt{Po\bar{d}}/Cs$. The MgO system shows the lowest corrosion rate for a given $\sqrt{Po\bar{d}}/Cs$.

CONCLUSIONS

The effects of the physical properties and chemical composition of the permeable refractory on the corrosion resistance for molten steel were investigated. The major results are as follows:
1) The corrosion resistance of a permeable refractory depends on the low melting point material produced by the reaction between the molten steel and the permeable refractory.
2) The corrosion resistance improves with decreasing

apparent porosity and mean pore size, and with increasing compressive strength.
3) The corrosion resistance of the MgO system was superior to that of the Al_2O_3 and the $MgO \cdot Al_2O_3$ systems.

REFERENCE

1) N. Nameishi, A. Inoue, K. Furukawa, "A study on mechanism of slag penetration into refractories (1st Report)", Taikabutsu, 22, 70-74 (1970).

HEXAALUMINATE-RELATED COMPOUNDS AS THERMALLY STABLE CATALYST MATERIALS

Masato Machida[a], Toru Shiomitsu, Hiroshi Inoue, Koichi Eguchi, and Hiromichi Arai
Department of Materials Science and Technology, Graduate School of Engineering Sciences, Kyushu University, Kasuga, Fukuoka 816 Japan
[a] Present address ; Faculty of Engineering, Miyazaki University Miyazaki 889-21 Japan

Hexaaluminate compounds with layer crystal structures can retain large specific surface area at high temperatures when the powders are prepared by hydrolysis of metal alkoxides. From high-resolution electron microscopic observation, the large surface area is related to the thin planar morphology of which basal plane is parallel to the (001) plane. Such morphology is caused by anisotropic grain growth which was confirmed by the change of the crystal morphology during heat treatment at 1300°C for ca. 10^4 h. The SIMS study showed the anisotropic oxygen self-diffusion in a hexaaluminate single crystal. Monatomic layers between oxygen close packed spinel blocks in the layer structure appear to be a preferential diffusion route of oxide ions. The anisotropic grain growth of hexaaluminate is consistent with the anisotropic diffusion.

INTRODUCTION

Development of thermally stable catalysts is required for high temperature application of catalytic processes. Alumina, being one of the conventional catalyst supports, can not be used above 1100°C because of significant sintering. This results in a drastic decrease of surface area to ca. 1 m^2/g. From a systematic study on the effect of additives on thermal stabilization of alumina, we have revealed that the loss of surface area due to sintering can be successfully suppressed when the additives lead to the formation of a hexaaluminate phase[1]. In particular, hexaaluminate fine particles prepared by hydrolysis of corresponding alkoxides are useful as a thermally stable catalyst support [1-4]. The prominent feature of this powder is the resistance to sintering, i.e., a large surface area above 20

m^2/g and resultant pore structure necessary for catalytic reactions can be retained at high operating temperatures above 1200 °C.

As reported previously, the large surface area retention was obtained for various hexaaluminates with different types of cations of mono-, di-, and trivalence and the thermal stability must be a structure-dependent property of these compounds. To clarify the relation between the thermal stability and the crystal structure, information on the grain growth is required. In this study, therefore, we have studied the grain growth behavior of the hexaaluminate fine particles during an isothermal heat treatment. The change in surface area, grain size, and particle morphology were measured as a function of heating period. We have also discussed the relation between the grain growth and crystal structure from a view point of the solid-state diffusion process.

EXPERIMENTAL

Sample Preparation

The sample employed for an isothermal heating was the manganese-substituted hexaaluminate, $Sr_{0.8}La_{0.2}MnAl_{11}O_{19}$, in which the Mn ions partially replace the Al site as a catalytically active component. The powder sample was prepared by hydrolysis of metal alkoxides. Calculated amounts of Sr metal and $Al(OC_3H_7)_3$ were stirred in guaranteed-grade 2-propanol until dissolution was complete. Hydrolysis was performed by introducing an aqueous solution of nitrates of manganese and lanthanum dropwise into the alcoholic solution. The resulting gel was finally calcined at 1300 °C for 5 h to produce the hexaaluminate phase.

Heat Treatment and Characterization

The powder sample of $Sr_{0.8}La_{0.2}MnAl_{11}O_{19}$ pressed into a disk was submitted to isothermal heating at 1300 °C for 9800 h in air. In order to evaluate the heat resistance of the hexaaluminate catalyst, the changes in crystal structure, surface area and microstructure were measured in the heating course. The specific surface area was measured by the BET method using nitrogen adsorption. A SEM/EDX system was used for microstructural observation and for quantitative analysis of the cationic composition. The crystal size distribution of primary particles was obtained by TEM observation.

SIMS Analysis of Oxygen Self Diffusion

Oxygen self-diffusion in a single crystal of barium hexaaluminate was analyzed by SIMS after diffusion annealing in $^{18}O_2$ at 1200-1600 °C. Oxygen self-diffusion coefficients along different crystal axes were calculated from depth profiles of ^{18}O concentration in the single crystal.

RESULTS AND DISCUSSION

Crystal Structure and Particle Morphology of Hexaaluminate

The hexaaluminate compounds can be divided into two structural types, i.e., β-alumina and magnetoplumbite. Basically, these structures consist of spinel blocks and mirror planes, which are stacked alternatively along c axis to form a pseudo-layer structure as shown in Fig. 1. A spinel block is composed of Al^{3+} and O^{2-} ions, having the same rigid close-packed structure as spinel oxides. Each spinel blocks are weakly held together by a rather open layer (mirror plane), in which various large cations such as alkali, alkaline earth or rare earth metals can be accommodated. The major difference between β-alumina and magnetoplumbite lies in the content and arrangement of the ions within the mirror plane layer. One of the interesting features concerning this layered crystal structure can be observed from the crystal morphology of sol-gel derived fine particles as follows.

Crystal morphology of fresh $Sr_{0.8}La_{0.2}MnAl_{11}O_{19}$ sample was studied by HREM observation after calcination at 1300°C for 5 h (Fig. 2). The sample crystallized as a thin planar particle with a thickness of about 20 nm which is about one tenth of the diameter. Selected area diffraction patterns and high-resolution images were taken when the electron beam is perpendicular or parallel to the facet. An incident beam normal to

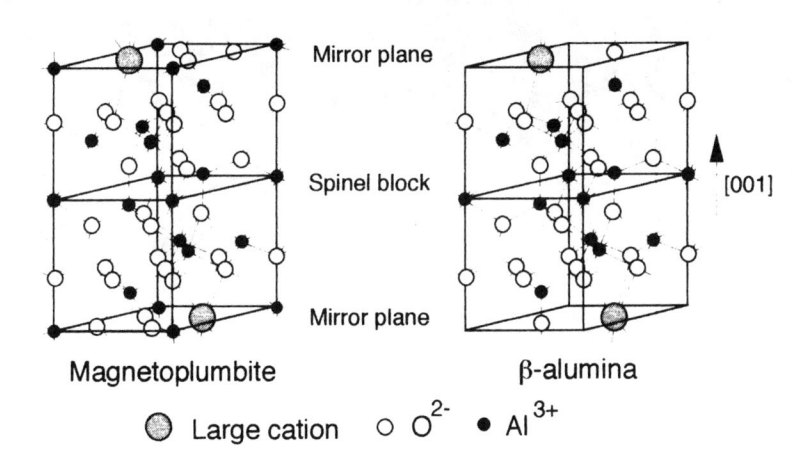

Figure 1 Crystal structure of hexaaluminate compounds

the facet showed the hexagonal diffraction patterns due to the [001] zone axis, whereas an incident beam parallel to the hexagonal facet showed the diffraction pattern due to the [110] zone axis. Thus, the facet is parallel to the (001) plane of hexaaluminate structure. The crystal structure of the $Sr_{0.8}La_{0.2}MnAl_{11}O_{19}$ in this study seems to be based on magnetoplumbite type because the sample was produced by the partial replacement of $SrAl_{12}O_{19}$ with magnetoplumbite structure. However, such an anisotropic crystal shape, being observed for other β-alumina and their derivatives, appears to be the structurally dependent property of the hexaaluminate family which consists of the layer structure. The structural image with an incident beam normal to the (110) plane showed the evidence of an array of spinel layers with ca. 1 nm thick along the c axis. On the contrary, the image of flat basal plane symmetrically oriented with [001] parallel to the incident beam is characterized by ca. 0.6 nm spaced white dots in a hexagonal array. From these images, it can be clearly observed that the basal surface of facets is precisely parallel to mirror planes at the lattice level.

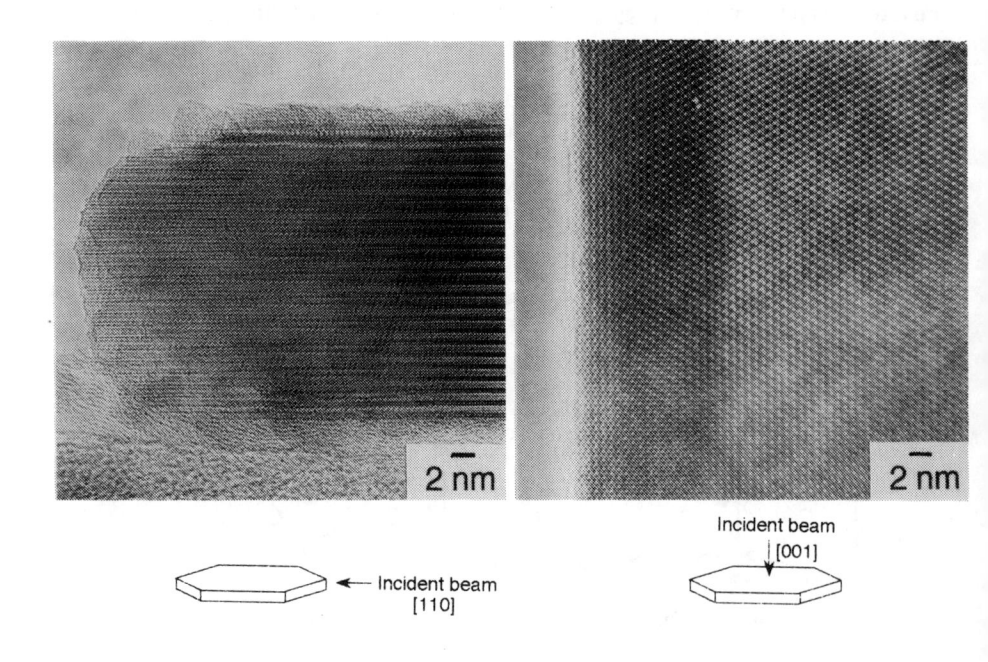

Figure 2 High-resolution transmission electron micrographs of alkoxide-derived $Sr_{0.8}La_{0.2}MnAl_{11}O_{19}$.

Change in Surface Area and Microstructure

An isothermal heat treatment of hexaaluminate powders ($Sr_{0.8}La_{0.2}MnAl_{11}O_{19}$) was carried out at 1300°C for 9800 h in air. X-ray diffraction showed that phase separations or precipitation of other phases rarely occurred by the heat treatment. The deviation of the cation composition during heating was also negligible from EDX analysis. The active species for the catalytic reaction, Mn, was retained constantly in the heating of 9800 h without volatilization. This means that the Mn species remain stable in the lattice and therefore the Mn-substituted hexaaluminate catalysts can be used at high operating temperatures for long periods.

The BET surface area of $Sr_{0.8}La_{0.2}MnAl_{11}O_{19}$ was measured as a function of heating period as shown in Fig. 3. Although the heat treatment gradually reduced the surface area, 4200 h of heating was necessary to reduce it to half (ca. 10 m^2/g) of the initial surface area. This is enough to demonstrate the substantial potential of thermal stability for long periods at high temperatures. SEM observation showed that the particle size of the hexaaluminate increased with the heating period as evident from the appearance of large hexagonal facets, whereas a local agglomeration was not observed even after heating for 9800 h (Fig. 4). Since the sample showed no crystallographic change during the heating, the loss in surface area apparently results from the grain growth of the planar particles.

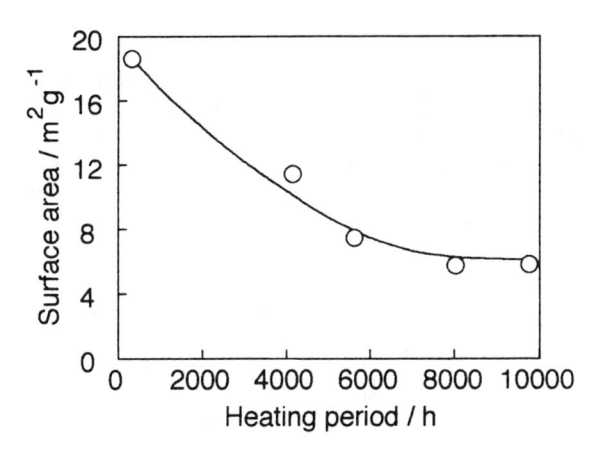

Figure 3 Change of surface area of $Sr_{0.8}La_{0.2}MnAl_{11}O_{19}$ during heat treatment at 1300°C.

5 h   5700 h   7500 h

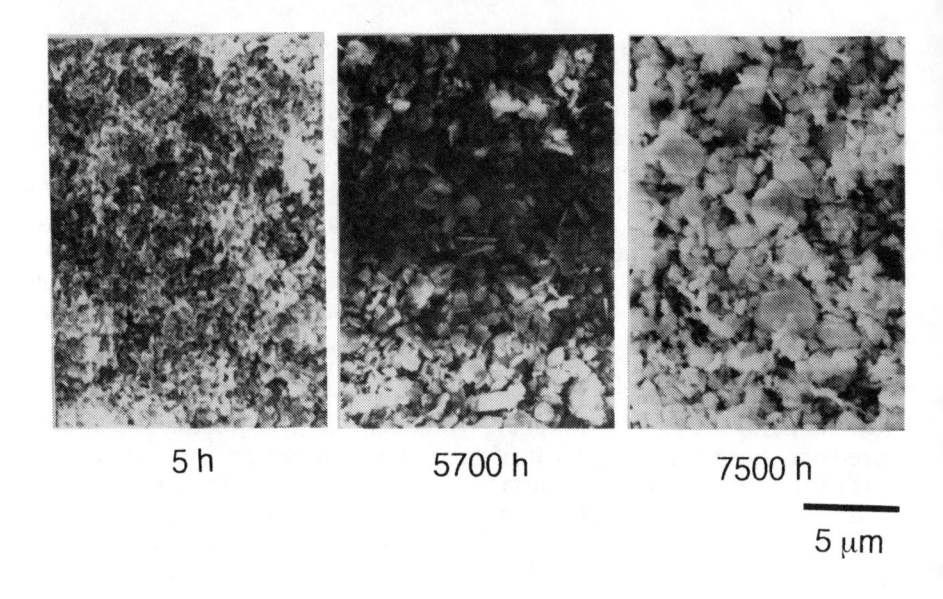

5 μm

Figure 4   Microstructures of of $Sr_{0.8}La_{0.2}MnAl_{11}O_{19}$ after heat treatment at 1300°C.

Change in Crystal Size and Crystal Morphology

Change in crystal size distribution and crystal morphology of the sol-gel derived hexaaluminate fine particles were studied by TEM. Crystal size was measured for one hundred particles when the incident electron beam is normal or parallel to the (001) plane. These values correspond to the crystal diameter and the crystal thickness of the planar facet, respectively. Figure 5 shows the change in average values of crystal diameter and crystal thickness as a function of heating period. These values are statistically meaningful because the specific surface area calculation based on the assumption that the sample is composed of columnar particles with these dimensions agreed with the apparent BET surface area (Fig. 3). It is noted that the crystal diameter increased faster than the crystal thickness in the initial heating process because of the anisotropic crystal growth normal to [001]. However, the crystal growth along both directions is saturated after 4000 h of heating. This also corresponds to the saturation of surface area as shown in Fig. 3. The relative increase of the thickness leads to the decrease in surface area because the specific surface area of

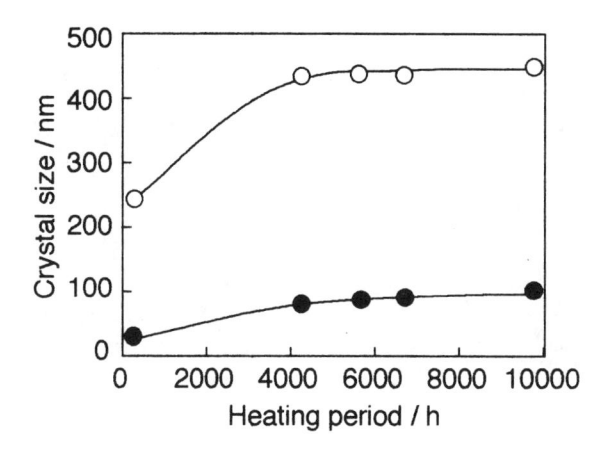

Figure 5 Change of average particle size of $Sr_{0.8}La_{0.2}MnAl_{11}O_{19}$ during heat treatment at 1300°C.
○ diameter of hexagonal facet   ● thickness of hexagonal facet

planar particles depends on the thickness rather than on the diameter. In the case of hexaaluminate, however, an increase in the thickness is significantly suppressed by the anisotropic grain growth along the (001) plane. This is the reason why the large surface area can be retained for a long period.

Surface Microstructure of Hexaaluminate Fine Particles

The surface microstructure of the powder sample after 9800 h of heating is well reflected by an anisotropic grain growth as observed with a high-resolution SEM (Fig. 6). The surface of basal plane parallel to (001) is flat and smooth except for several parallel growth steps. The thickness of these unit steps is less than 10 nm which corresponds to the stacking of a few spinel blocks. On the contrary, a curved side plane of the facet contains a lot of streaks on the surface. This surface roughness undoubtedly indicates that the grain growth proceeded significantly on the side plane of facets rather than on the basal plane as speculated below. Hexaaluminate structure consists of the stacking of oxygen close-packed spinel blocks as shown in Fig. 1. Continuous spinel formation along the (001) plane is expected to easily proceed because of the absence of possible barriers to the lattice construction. Stacking of each spinel block, however, will proceed more slowly due to discontinuous

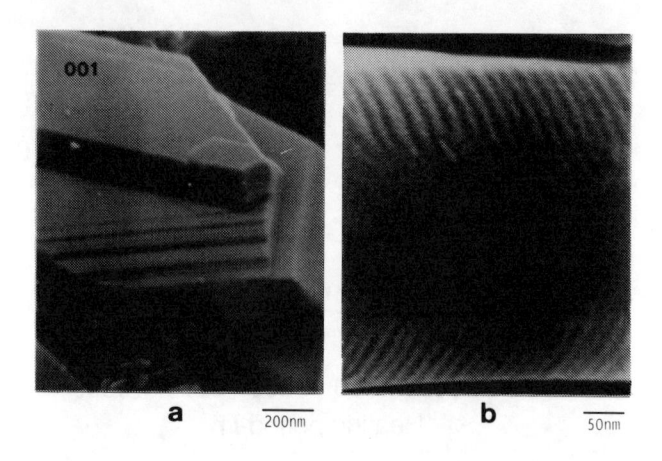

Figure 6 Surface microstructures of hexagonal facets of $Sr_{0.8}La_{0.2}MnAl_{11}O_{19}$ after heat treatment at 1300°C for 9800 h. a) The basal plane with parallel steps, b) The side plane with streaks.

insertion of mirror plane layers. Therefore, on the basal plane, the grain growth appears to proceed through the formation of one dimensional steps and a subsequent two-dimensional step growth on a smooth surface. While such a layer growth on the basal plane brings about the increase in thickness of a hexagonal facet, its growth rate will be lower than that parallel to the facet. Consequently, the anisotropic grain growth appears to be closely related to the difference of the growth mechanism between basal and side planes of the facet.

Relation between Grain Growth and Solid-State Diffusion

The basic interpretation of the relationship between the pseudo-layer crystal structure, sintering and grain growth requires information on oxygen self-diffusion. The SIMS analysis was employed to determine the oxygen self-diffusion in the different crystallographic directions. Figure 7 shows the diffusion coefficient as a function of reciprocal diffusion temperature. Diffusivities of oxygen normal to the c axis ($\perp c$) was an order of magnitude larger than those along the c axis ($\parallel c$). The diffusion coefficient in the $\parallel c$ direction was close to that in $MgAl_2O_4$ spinel oxide. These results indicate that the pseudo-layer structure of hexaaluminate influences the diffusion of oxide ions, i.e., loosely packed intermediate monolayers between closely packed spinel blocks, which spread in parallel to the (001) plane, are likely to be preferential diffusion route of oxide ions. One of the structure-dependent properties which can be speculated from the results of this study

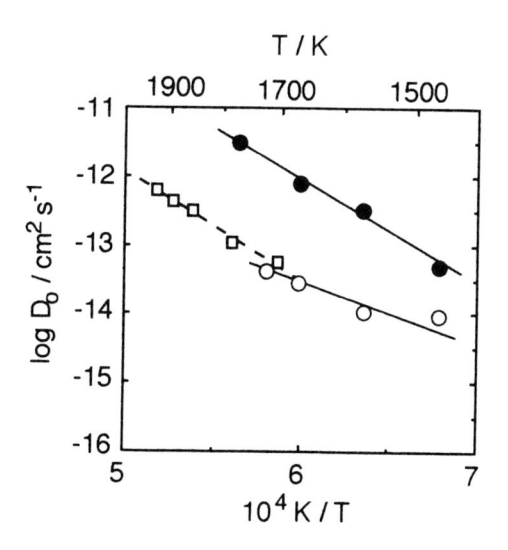

Figure 7 Temperature dependence of oxygen self-diffusion coefficients.
● $Ba_{0.75}Al_{11.0}O_{17.25}$, ⊥ c
○ $Ba_{0.75}Al_{11.0}O_{17.25}$, ∥ c
□ $MgAl_2O_4$ (reported by Ando and Oishi [7])

is the relation to the anisotropic growth of hexaaluminate fine particles. When the rate of the solid-state sintering process is determined by solid state oxygen diffusion as reported for $MgAl_2O_4$, diffusion in the mirror plane, which plays a key role in the transfer of constituent ions from bulk to surface, will lead to preferential grain growth along the ⊥c direction. This is in contrast to $MgAl_2O_4$ powders, i.e., isotropic oxygen diffusion results in spherical morphology. The anisotropic grain growth of hexaaluminate fine particles is consistent with the anisotropic diffusion in the pseudo-layer structure.

CONCLUSION

This study revealed that anisotropic grain growth, which results in planar morphology with a small aspect ratio, is the reason for the large surface area retention of hexaaluminate powders. From electron microscopic observation, hexaaluminate crystallized as thin planar particles of which basal planes are exactly parallel to the (001) plane. Anisotropic grain growth was confirmed by a morphological change during heat treatment using a long period (9800 h) and is also obviously seen in the difference in surface microstructures. The anisotropic grain growth reflects the pseudo-layer structure, i.e., the rate of spinel block formation is higher than that of stacking of spinel blocks along c axis due to insertion of mirror plane layers. SIMS analysis on oxygen self-diffusion in a single crystal hexaaluminate shows that loosely packed intermediate layers (mirror planes) between closely-packed spinel blocks is likely a preferential diffusion route of oxide ions. The anisotropic grain growth of hexaaluminate fine particles appears to be accelerated by this preferential mass transfer process in a sintering process.

ACKNOWLEDGEMENT

The authors are greatly indebted to Dr. H. Haneda, National Institute for Research in Inorganic Materials, for SIMS measurement of this study .

REFERENCES

1. M. Machida, K. Eguchi, and H. Arai, "Effect of Additives on the Surface Area of Oxide Supports for Catalytic Combustion," J. Catal., 103 [2] 385-393 (1987).
2. M. Machida, K. Eguchi, and H. Arai, "Analytical Electron Microscope Analysis of the Formation of $BaO6Al_2O_3$," J. Am. Ceram. Soc., 71 [12] 1142-47 (1988).
3. M. Machida, K. Eguchi, and H. Arai, "Catalytic Properties of $BaMAl_{11}O_{19}$ (M=Cr,Mn,Fe,Co, and Ni) for High-temperature Catalytic Combustion," J. Catal., 120 [2] 377-86 (1989).
4. M. Machida, K. Eguchi, and H. Arai, "Effect of Structural Modification on the Catalytic Property of Mn-substituted Hexaaluminates," J. Catal., 123 [2] 477-85 (1990).
5. N. Iyi, S. Takekawa, and S. Kimura, "Crystal Chemistry of Hexaaluminates; β-alumina and Mgnetoplumbite Structures," J. Solid State Chem., 83 [1] 8-19 (1989).
6. M. Machida, T. Shiomitsu, K. Eguchi, H. Haneda, and H. Arai, "Secondary Ion Mass Spectrometric Analysis of Anisotropic Oxygen Self-diffusion in Barium Hexaaluminates Single Crystals," J. Mater. Chem., 2 [4] 455-458 (1992).
7. K. Ando and Y. Oishi, "Oxygen Self-diffusion in $MgAl_2O_4$ Single Crystal" J. Chem. Phys.,61, 625 (1974).

Section IV. Applications

APPLICATION OF CERAMIC FOAM

T. Masuda, K. Tomita, and T. Iwata
Yahata Works, Krosaki Corporation
1-1 Higashihama-machi, Yahatanisi-ku, Kitakyushu City, Japan

Ceramic foam has a unique structure consisting of a three-dimensional skeleton, and various applications in many fields are now under study. In this study, the oil trapping ratios and oil recovery ratios of ceramic foam, which are now being studied for practical use as filters for commercial kitchens (grease filters), were measured. It was found that an oil trapping ratio as high as 78% can be obtained and that the oil recovery ratio is lower than that of the conventional baffle type filters. The low recovery ratio is mainly due to the network structure: clogging of pores as well as stagnant oil in pores, cracks, and voids in the skeleton. Consequently, it was revealed that the reduction of the pore clogging by glass coating the skeleton surface, significantly improved the recovery ratio.

INTRODUCTION

In the kitchens of hotels and restaurants, grease filters are attached to the inlets of exhaust ducts in order to trap oil mist generated during cooking. Metal filters are currently used for these filters, for such types as mesh, baffle, and fiber. However, the use of ceramic foam is being considered because of its high trapping ratio[1]. Figure 1 shows the practical use of a filter. The filter is attached to the inlet of the exhaust duct over a heating and cooking table. The inlet of

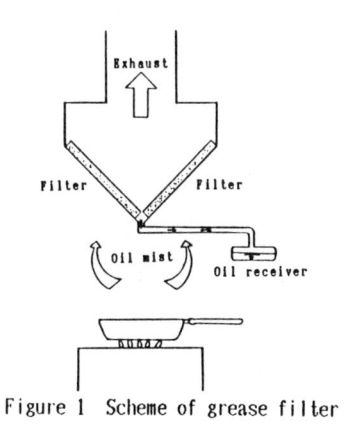

Figure 1 Scheme of grease filter

the exhaust duct has an angle of about 45°. The oil trapped by the filter drops from the filter and is collected by a receiver under the filter. The purpose of the filters is to prevent contaminated air in kitchens from going outside and to avoid staining the insides of exhaust ducts with oil. In particular, in recent years exhaust ducts in large buildings and underground markets have became very long, so that excessive oil contamination leads to very expensive cleaning and poses a potential fire hazard. Requirements for grease filters are as follows:
(1) high oil removal ratio
(2) high oil recovery ratio, and
(3) low pressure loss.
The oil trapping ratio is the most important property, and it is believed that the higher the ratio, the better. The standard authorized by the Japan Food Service Equipment Association recommends an oil trapping ratio of 60% or higher. The recovery ratio is the ratio of oil returning to a receiver under the filter to the trapped oil in the filter. In the case of a low recovery ratio, a large amount of oil is accumulated in the filter and there is a high risk of fire. The standard authorized by the Japan Food Service Equipment Association recommends a recovery ratio of 80% or higher. With respect to pressure loss, a lower loss is desirable, because it contributes to the improvement of the capacity of exhaust blowers and the amenity of kitchens. In this study, basic filter tests were conducted to examine thses three requirements, in order to further improve the performance.

EXPERIMENTAL PROCEDURE

Two types of ceramic foam were used in this study. Physical properties of the ceramic foam are listed in Table 1.

Table 1. Properties of Ceramic Foam

	A	B
Porosity(%)	86	83
Density (kg/m^3)	390	410
Modulus of Rupture (MPa)	1.03	1.18
Puressure loss (mmAq.)	4.5	3.5
Chemical Composition(%)	Al_2O_3:80, SiO_2:15	

The pore size of the ceramic foam was denoted as the number of cells. The relation between the number of cells and the pore diameter ware shown in Table 2.

Table 2. The Relation Between the Number of Cells and Pore Diameter

No. of Cells	Pore Diameter (mm)
6	1.5 ~ 2.5
13	0.8 ~ 1.2
20	0.4 ~ 0.6

The outline of a testing system is shown in Fig. 2[2]. The testing system is composed of an oil generator, an exhaust duct, and a filter attached to its air inlet. Oil mist is generated by dropping oil and water simultaneously on a frying pan at 390°C. The oil and water are supplied at a rate of 150 g/h and 480 g/h, respectively. The generated oil mist is drawn by the exhaust duct with a wind velocity of about 1 m/s, and is trapped by the filter surface. The filter is attached at an inclination of 45° so that the trapped oil flows down along the ceramic foam and is recovered by an oil receiver. In this test, the trapping ratio (T) and recovery ratio (R) were calculated by

$$T = 100 \times (W_r + W_t)/W_s \qquad (1)$$

$$R = 100 \times W_r/(W_r + W_t) \qquad (2)$$

where W_r is the weight of the recovered oil, W_t is the weight of the trapped oil and W_s is the weight of the used oil. The pressure loss was measured using a duct equipped with a manometer.

Figure 2 Schematic representation of the testing system

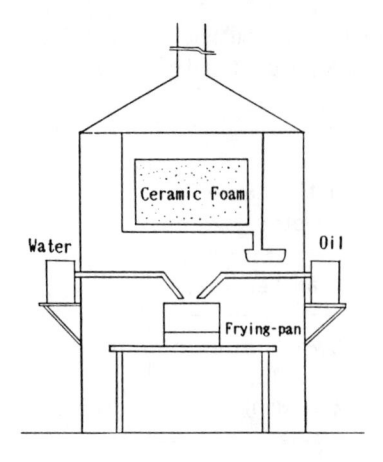

Experimental Conditions

Wind velocity : 1.1 ～ 1.2 m/s

Rate of supplying oil : 150 g/h

Rate of supplying water : 450 g/h

Heating temperature : 390°C

Testing time : 3 ～ 12 h

RESULTS AND DISCUSSION

Trapping Ratio and Pressure Loss

Figure 3 shows the relation between the number of cells and the trapping ratio for filters having a thickness of 15, 25 and 35 mm. Testing was performed for 3 hours using the testing system explained earlier. The removal ratio of all samples except the one with 6 cells, which had a thickness of 15 mm, exceeded the target value, 60%. As the number of cells increased, the oil trapping ratio is improved and the one with 20 cells is close to 80%. This shows that the material is quite excellent compared with the stainless steel baffle type. With respect to thickness, almost no difference is found in the case of 13 cells and 20 cells. This suggests that when the filter mesh is sufficiently small, oil mist trapping occurs near the filter surface.

The relationship between the number of cells and the pressure loss for each thickness is shown in Fig. 4. The ceramic foam filters have low pressure losses compared with the conventional stainless steel baffle filter. The greater the number of cells, the higher the pressure loss. Considering that the pressure loss of the conventional stainless steel baffle type is about 10 mmAq., a thickness of 15 mm for 20 cells is the limit for practical applications.

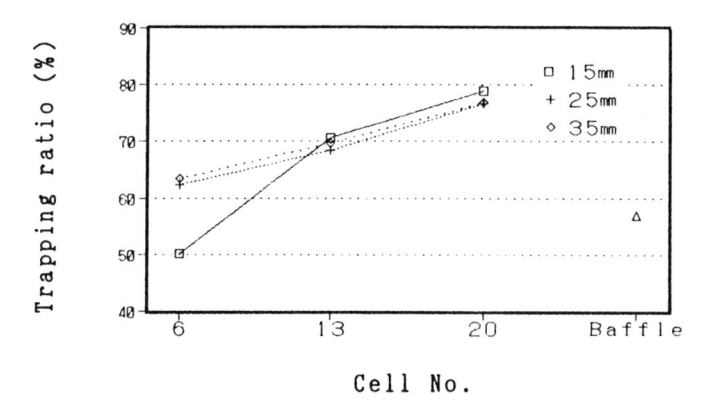

Figure 3 Relation between the number of cells and the trapping ratio

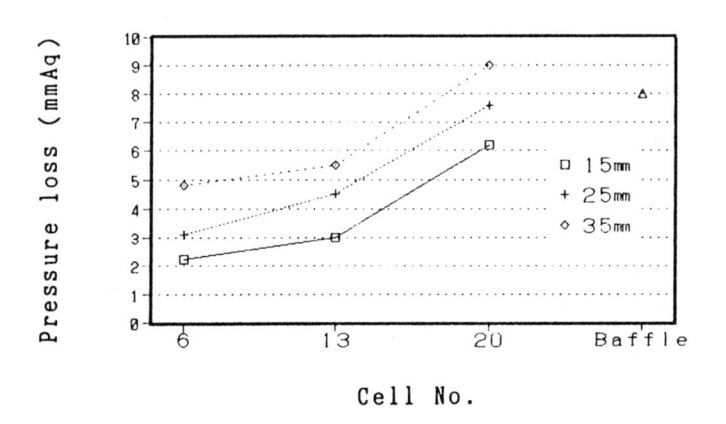

Figure 4 Relation between the number of cells and the pressure loss

Recovery Ratio

The relationship between the number of cells and the recovery ratio is shown in Fig. 5. Testing was performed continuously for 12 hours using the testing system explained earlier. During the testing the filter weight and the amount of recovered oil was measured in order to calculate recovery ratios after 3,6, and 9 hours. With a decrease in the number of cells, the recovery ratio was improved. However, it did not reach the target value of 80%. A low recovery ratio means that oil tends to stay in the filter and drops slowly. Table 2 lists the measured result of changes in the amount of retention of the conventional stainless steel filter and the ceramic foam filter. This table illustrates that the ceramic foam is more liable to retain oil than the stainless steel filter.

Table 3. Chenges in the amount of the oil retention

	Ceramic foam			Baffle filter
	#6	#13	#20	
After 6h	286g	306g	330g	34g
After 12h	345g	375g	403g	54g

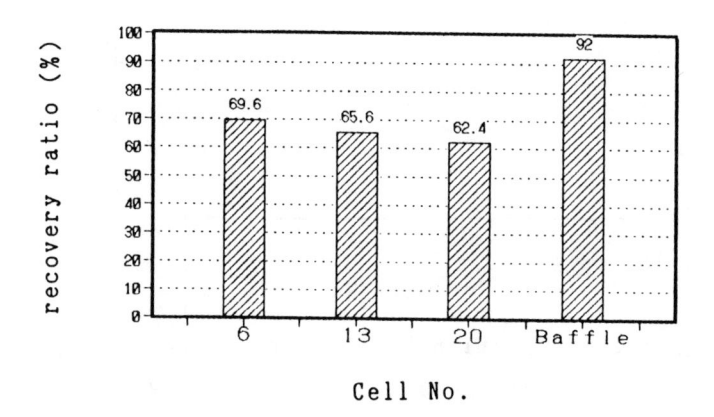

Figure 5 Relation between the number of cells and the recovery ratio

Improvement of the Ceramic Foam

Probable causes of greater oil retention are:
(1) Shape (three-dimensional skeleton structure)
(2) Clogging
(3) Oil penetration into cavities
(4) Oil penetration into ceramic pore

During testing, the oil formed drops and stayed in the cells. Figure 6 shows the clogging of the filter, this might trap oil and prevent it from dropping. Figure 7 shows cavities in the skeleton of the ceramic foam. If there are cracks or large pores in the surface of the skeleton, the oil trapped in these cavities will be retained.

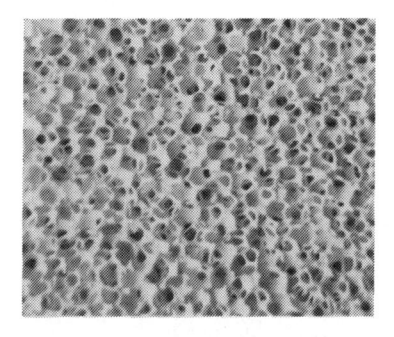

Figure 6　The clogging of the filter

Figure 7　Cavities in the skeleton

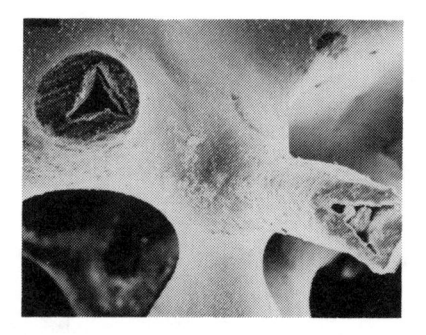

Consequently, in order to improve the recovery ratio, the following measures were taken:
(1) reduction of clogging, due to the improved forming method
(2) glass coating of skeleton surface.

As a result, a high-performance oil mist filter has been developed. Figure 8 shows the surface of the new filter. As shown in this photograph the new product has almost no clogging, due to the improved forming method. Figure 9 shows the glass coating. The skeleton surface is completely covered with glass, which prevents oil from penetrating into pores or cavities in the skeleton. The recovery ratios of the new oil mist filter are shown in Fig. 10. The recovery ratio after 12 hours reached 80.5% and has been considerably improved. Figure 11 shows the progress of the amount of oil retention. The oil retention decreased compared to that before improvement.

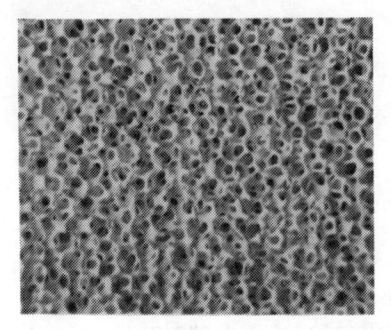

Figure 8 The surface of the improved filter

Figure 9 Glass coating of skeleton surface

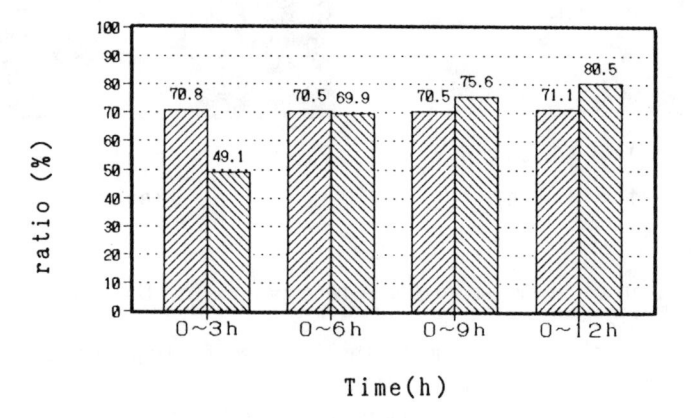

Figure 10 Measurements of trapping raio and recovery ratio

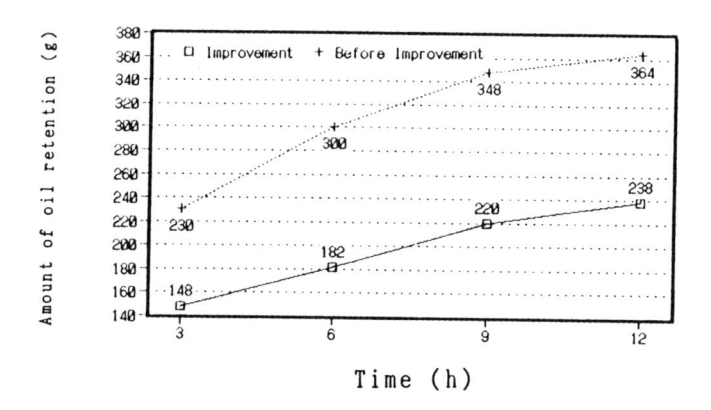

Figure 11 The progress of the amount of oil retention

CONCLUSION

In order to utilize ceramic foam as a grease filter, the filter performance was tested. In the case of the oil trapping ratio, the influence of the number of cells and filter thickness were tested.
It was found that the trapping ratio of a filter with 13 cells is 70%, and that of a filter with 20 cells is 78%. Furthermore, with respect to filter thickness, it was revealed that a sample with more than 13 cells shows no difference within the range of 15-35 mm of filter thickness. It was found that the recovery ratio is much lower than that of the conventional stainless steel baffle filter. The low recovery ratio is mainly due to the skeleton form which causes clogging of the pores as well as stagnant oil in pores, cracks, and voids in the skelton. Consequently, it was revealed that the reduction of the pore clogging and the glass coating on the skeleton surface substantially improved the recovery ratio.

REFERENCE

1. M. Ezaki and H. Nagai, Ceramic Filter For Range Hood, 26010-1992, Japan Pat.

2. Tokyo Shobouchyo, Hi wo shiyousuru setubitou no gijutukijun, 364-365, Yobouzimusinsa Kansakijun, 3th ed., Tokyo Bousai Sidoukyoukai, 1992.

MICROPOROUS GELS AS DESICCANTS

Sridhar Komarneni and Prakash B. Malla, Materials Research Laboratory, The Pennsylvania State University, University Park, PA, 16802, USA

Porous solids, such as silica gels, zeolites and many others have been used as desiccants. These readily available desiccant materials, however, have not been optimized for the specific purpose of desiccation in gas-fired cooling and dehumidification equipment. The objective of this research is to design and develop silica gels which are microporous and which exhibit moderate Type I Brunauer adsorption isotherms having separation factors between 0.07 and 0.1. A series of silica gels have been synthesized by the hydrolysis and polymerization of tetramethylorthosilicate (TMOS) by varying different factors, such as amount of water for hydrolysis, presence or absence of solvent, pH and amount of chemical additive. The gels were characterized by water and nitrogen sorption measurements. The pore size and its distribution and the polarity of gels have been varied at will by the appropriate chemical processing. Silica gels with moderate Type I adsorption isotherms have been developed for gas-fired cooling and dehumidification equipment by a proprietary process. The ability to design and develop silica gels of different isotherm shapes may also be useful in the areas of catalysis, waste disposal, sensors, etc.

INTRODUCTION

Microporous solids, such as silica gels, zeolites, clays and many others have been used as sorbents, catalysts and molecular sieves. The effectiveness or the limitation of these materials for a specific purpose in the fields of application mentioned above depends on the pore size and its distribution, and the nature (polar or nonpolar) of pore surfaces.

The present paper deals with the optimization of the pore size and the polarity in silica gels to achieve ideal desiccants for use in gas-fired cooling and dehumidification equipment. It has been determined (1) that the ideal shape of the desiccant isotherm for use in gas-fired cooling/dehumidification systems has a

To the extent authorized under the laws of the United States of America, all copyright interests in this publication are the property of The American Ceramic Society. Any duplication, reproduction, or republication of this publication or any part thereof, without the express written consent of The American Ceramic Society or fee paid to the Copyright Clearance Center, is prohibited.

separation factor ranging from 0.07 to 0.1 in accordance with the isotherm equation:

$$X = \frac{P}{R + P - RP}$$

wherein X is the normalized loading fraction of water, P is the relative pressure of water and R is the separation factor. Additional properties of the ideal desiccant are low heat of adsorption, high water adsorption capacity, high diffusivity of water, and high chemical and physical stability.

Most of the commercially available desiccant materials have not been developed for the specific purpose of providing space cooling. Therefore, the object of this research was to synthesize microporous silica gels exhibiting water sorption isotherms having separation factor between 0.07 and 0.1 for use as desiccants in gas-fired open space air-conditioning and dehumidification equipment.

Silica gels can be synthesized by the hydrolysis and polymerization of various starting materials such as alkoxides, colloidal silica alkali, silicate solution, and silica fumes. The advent of solution-sol-gel (SSG) processing has led to the design and synthesis of microporous gels by controlled hydrolysis, polymerization and condensation of especially metal alkoxides. The hydrolysis and condensation reactions of alkoxide can be represented by the following equations (2):

$$\equiv Si - OR + H_2O \underset{\text{esterification}}{\overset{\text{hydrolysis}}{\leftrightarrow}} \equiv Si - OH + ROH$$

$$\equiv Si - OR + HO - Si\equiv \underset{\text{esterification}}{\overset{\text{alcohol condensation}}{\leftrightarrow}} \equiv Si - O - Si\equiv + ROH$$

$$\equiv Si - OH + HO - Si\equiv \underset{\text{hydrolysis}}{\overset{\text{water condensation}}{\leftrightarrow}} \equiv Si - O - Si\equiv + H_2O$$

where R is an alkyl group such as C_2H_5, CH_3, etc.

The above hydrolysis and condensation reactions are affected by the starting precursors, amounts of water, presence or absence of solvent, chemical additives, type of catalyst such as acid or base or their nature, temperature, aging and drying conditions.

EXPERIMENTAL

Preparation of Gels

Silica gels were prepared in a closed system by mixing tetramethoxysilane (TMOS), $Si(OCH_3)_4$ with deionized water in different molar ratios with or without methanol as solvent. The H_2O to TMOS molar ratios (X) used were 4.1, 12.3 and 24.6. When methanol was used as a solvent, a methanol to TMOS molar ratio of 5.22 was used in all cases but the H_2O to TMOS ratios varied as above. Another group of gels were made with HCl as the catalyst and these gels were made from 10 ml of TMOS and 5 ml of 10^{-1} or 10^{-2} or 10^{-3} or 10^{-4} $\underline{M}$ HCl (H_2O/TMOS molar ratio = 4.1). All the chemicals were mixed in plastic bottles at room temperature (~27°C). When methanol was used as a solvent, TMOS was first mixed with methanol and then water was added. Gelation was allowed to take place at room temperature. Another group of gels were prepared with different amounts of a chemical additive. The details are, however, proprietary. The resulting gels were calcined at 200°C or at 260°C for 30 minutes in a furnace to dry and remove organics. The xerogels obtained by the above calcination were ground and heated again in an oven at 200°C before characterization by water and N_2 adsorption.

Water and Nitrogen Isotherms

Water sorption isotherms were measured by a volumetric method using a computer-interfaced sorption device (3). The above xerogels were degassed at 200°C $\geq$ 5 hours before the water sorption measurements. The constant volume in which the sample was exposed with a high accuracy pressure transducer (MKS Instrument, Inc., Model 390H). The apparatus is precise and sensitive, inasmuch as the pressure transducer could detect a pressure difference of + 3 µg of water. Nitrogen adsorption isotherms of selected xerogels degassed at 200°C were measured with a Quantachrome Autosorb-1 Automated Gas Adsorption system at liquid nitrogen temperature.

RESULTS AND DISCUSSION

Water Adsorption Isotherms

Figure 1 shows the water adsorption isotherms of three different gels which were made using three different water-to-TMOS molar ratios. The gels with the lowest and highest ratios exhibited adsorption isotherms which are close to Type I indicating the presence of micropores whereas the gel with an intermediate X showed mesoporous nature. Nitrogen adsorption and desorption isotherms showed that the gel sample with the lowest X showed Type I adsorption isotherm and micropores (Figure 2A), whereas the gel sample with intermediate X (= 12.3) showed a Type IV isotherm and mesopores (Figure 2B). The N_2 and H_2O adsorption studies give consistent results. The presence of methanol as a solvent and diluent led to higher water adsorption capacities (Table 1) with higher separation factors (Figure 3) which indicate an increase in mesopores compared to the gels prepared in the absence of methanol as solvent (Figure 1). Water sorption isotherms of gels made with different concentrations of acid are given in Figure 4. The best isotherm shape (close to a separation factor, R of 0.1) was achieved with the gel prepared using 10^{-1} M HCl. Some gel properties derived from the water adsorption isotherms are given in Table 2. The pore-size, surface area and sorption capacity went through a minimum in the 10^{-3} to 10^{-1} M HCl concentration. This is apparently related to the isoelectric point of silica which lies between the pH values of 1 and 2. In an acid system, hydrolysis of the alkoxide groups takes place rapidly

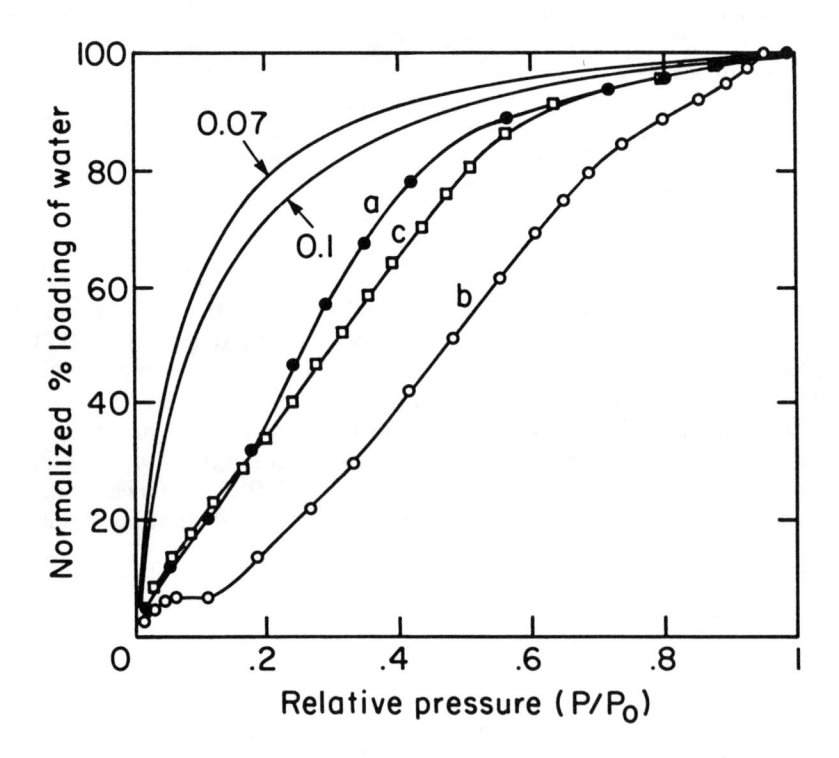

Figure 1   Water adsorption isotherms of silica gels made by using different water to TMOS molar ratios without methanol as solvent: (a) 4.1, (b) 12.3, and (c) 24.6.

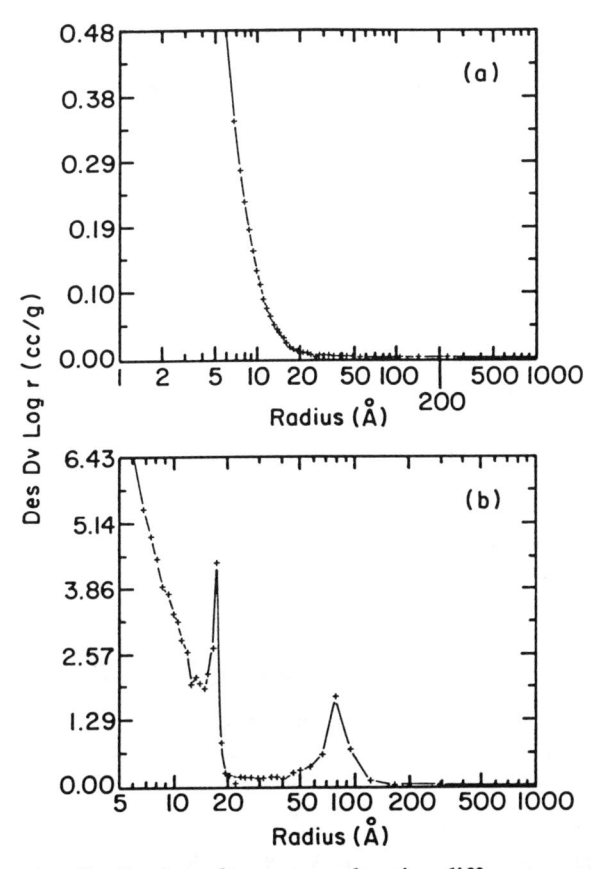

Figure 2   Pore size distribution of two xerogels using different water to TMOS ratios (X) in the absence of methanol solvent: (a) X = 4.1, and (b) X = 12.3.

Table 1.   Water Adsorption Capacities and Surface Areas of Gels Made Using Different Water-to-TMOS Ratios.

H_2O to TMOS Molar Ratio (X)	Water Adsorption, Capacity, g/g — No Methanol Solvent	Water Adsorption, Capacity, g/g — Methanol Solvent	Surface Area, m^2/g — No Methanol Solvent	Surface Area, m^2/g — Methanol Solvent
4.1	0.2458	0.3026	662	492
12.3	0.4160	0.2962	558	624
24.6	0.3782	0.2946	690	700

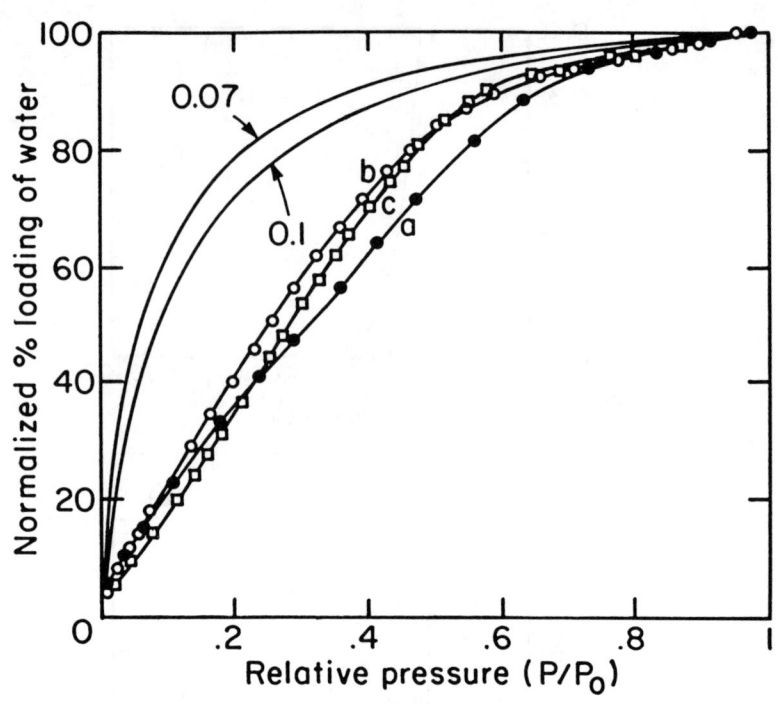

Figure 3   Water adsorption isotherms of silica gels made by using different water to TMOS molar ratios in the presence of methanol as solvent: (a) 4.1, (b) 12.3 and (c) 24.6.

Table 2.   The Pore Sizes, Surface Areas and Sorption Capacities of Silica Gels Made by Using Different HCl Concentrations.

HCl Concentration, M	Average Slit Width[1] (Å)	Surface Area, m^2/g	Sorption Capacity g/g (P/P_o=0.95)
0.0	9.3	714	0.279
10^{-4}	8.4	583	0.247
10^{-3}	7.2	382	0.171
10^{-2}	7.2	392	0.173
10^{-1}	7.2	372	0.180

[1] Average slit width (pore-size) estimated from t-plot of water

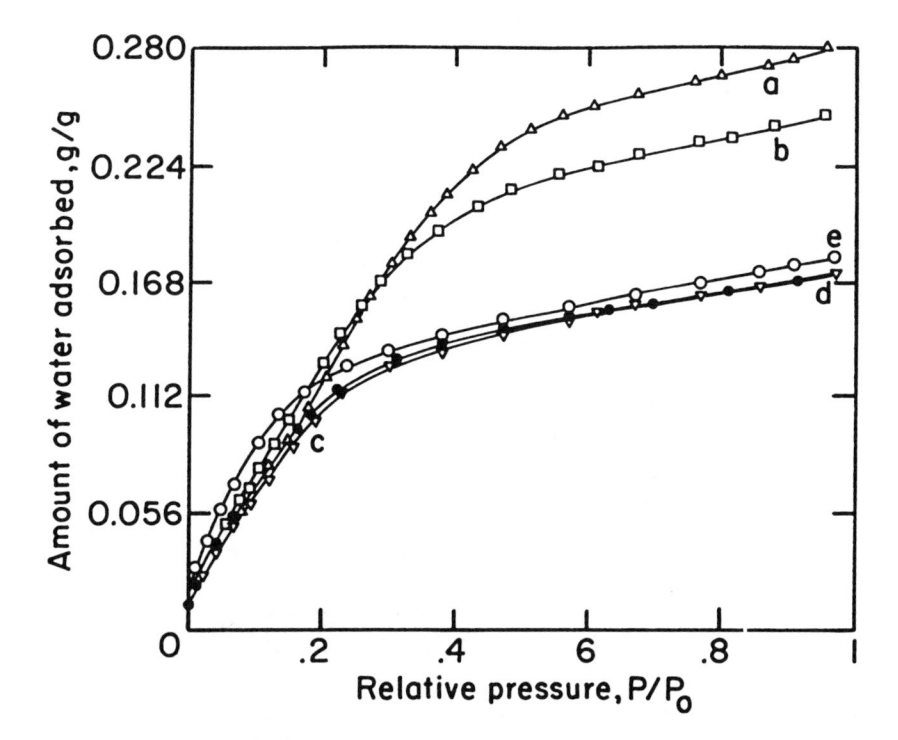

Figure 4 Water adsorption isotherms of silica gels made by using different concentrations of HCl: (a) 0, (b) 10^{-4}, (c) 10^{-3}, (d) 10^{-2}, and (e) 10^{-1}.

while the condensation reaction is slow (4). This type of reaction sequence leads to weak cross-linking of the condensed species which can interpenetrate and shrink greatly as the solvent is removed during gelation and this, in turn, leads to dense microporous gels (5). At the isoelectric point, the H^+ and OH^- ions on the surface are at a minimum and the above sequence of reactions are apparently favored. These results are consistent with our earlier results where a different 'X' ratio was used (6).

Adsorption isotherms of gels prepared with a chemical additive are shown in Figure 5. Increasing the amount of chemical additive X increased the separation factor and type V adsorption isotherms resulted indicating weak interaction of water with the surfaces. The lowest amount of additive (A) with an [A/A + Si] ratio of 0.02 led to Type I adsorption isotherm with a separation factor close to 0.1. Figure 6 shows the adsorption-desorption isotherms of this gel and this result clearly shows that a desired microporous gel (proprietary) has been prepared with a separation factor of 0.1 which may be useful as a desiccant in gas-fired cooling and dehumidification equipment.

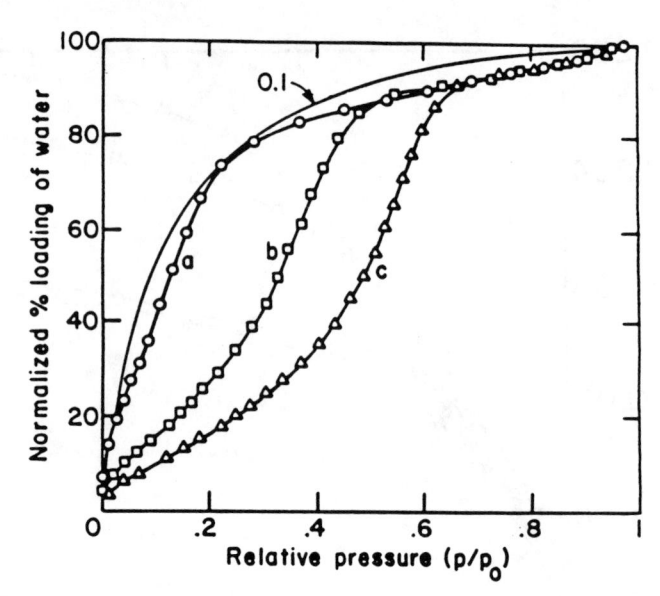

Figure 5 Water adsorption isotherms of silica gels prepared with a chemical additive (A) using different [A/A + Si] molar ratios: (a) 0.02, (b) 0.1, and (c) 0.2

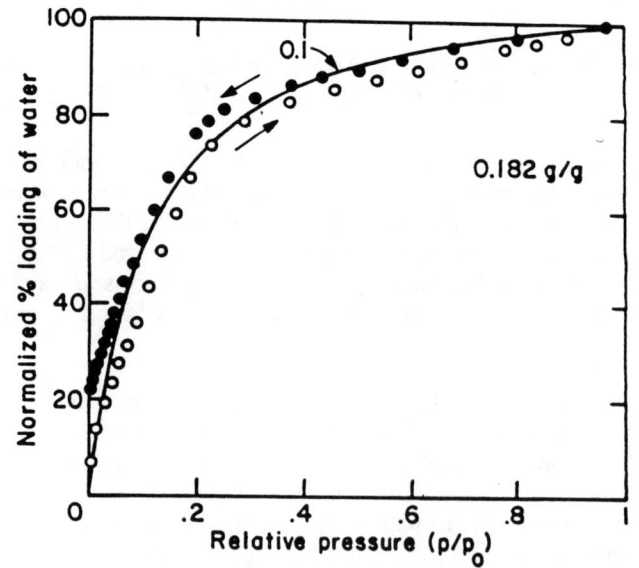

Figure 6 Water adsorption (O) and desorption (O) isotherms of silica gel prepared with a chemical additive (A) using an [A/A + Si] molar ratio of 0.02.

CONCLUSIONS

Microporous silica gels with moderate Type I Brunauer adsorption isotherms have been prepared from tetramethoxysilane by using acid as catalyst and adding minor amounts of other chemicals. These gels may be useful as desiccants in gas-fired cooling and dehumidification systems.

ACKNOWLEDGEMENTS

This research was supported by Gas Research Institute under Contract No. 5087-260-1473. Experimental assistance by C. L. Liu and S. Fakhouri is appreciated.

REFERENCES

1. R. K. Collier, T. S. Cole, and Z. Lavan, "Advanced Desiccation Materials Assessment" Final Report under Contract No. 5084-243-1089, Gas Research Institute, Chicago, Illinois, (1986).
2. C. J. Brinker, "Hydrolysis and Condensation of Silicates: Effects on Structure", J. Non-Cryst. Solids, 100 [1] 31- 50, (1988).
3. S. Yamanaka and S. Komarneni, "Apparatus for Measuring Liquid Vapor Adsorption and Desorption Characteristics of a Sample", United States Patent No. 5,058,442, October 22, (1991).
4. R. A. Assink and B. D. Kay, "Sol-Gel Kinetics I, Functional Group Kinetics", J. Non Cryst. Solids, 99, 359-70 (1988).
5. C. J. Brinker and G. W. Scherer, Sol-Gel Science, The Physics and Chemistry of Sol-Gel Processing. Academic Press, New York, (1990).
6. P.B. Malla, S. Komarneni, H. Taguchi and H. Kido, "Probing the Nature and the Structure of Pores in Silica Xerogels by Water Sorption: The Tetramethyl Orthosilicate-Hydrogen Chloride/Fluoride System", J. Am. Ceram. Soc. 74 [12] 2988-95 (1991).

ELECTROSTATIC FORMATION OF CERAMIC MEMBRANE

Hideo Yamamoto and Senichi Masuda
Dept. of Bioengineering, Soka University
Tangi-cho, Hachioji-city, Tokyo 192 (JAPAN)

A new method for forming a ceramic membrane was devised. Ultra fine particles of silicon nitride synthesized by thermally activated chemical vapor deposition (CVD) were deposited on an outer wall surface of a porous ceramic tube (substrate) by electrostatic force and sintered in an inert gas atmosphere. The ceramic-made electrode assembly using surface discharge was used for charging ultra fine particles at a high temperature. Unique ceramic membranes with a three-dimensional network for use as fiber filters were obtained. The effective pore size was around 0.04–0.8 μm in diameter, and the porosity was extremely large(%).

INTRODUCTION

This paper describes about a new method of forming ceramic membranes using ultrafine particles prepared by chemical vapor deposition(CVD). Recently, ultrafine particles prepared by the CVD process have generated considerable interest as advanced industrial materials because of their superior chemical and physical properties, purity, and size uniformity. For the present, however, their production cost is very high, and there are many handling difficulties to use them as industrial materials in large quantities. For instance, it is necessary to develop new handling techniques, such as separation, classification, mixing, dispersion, formation, etc. to prevent contamination of the particle surface. From these viewpoints, we have proposed and investigated a new process for forming a fine porous ceramic membrane, as one of the advantageous uses of ultrafine particles. The principle of the method is as follows:

The first step is to prepare ultrafine particles by a thermally activated CVD method. The second step is to charge the particles electrostatically, directly after production, and deposit them on a surface of a porous ceramic substrate using electrostatic force in the DC-field. The electrostatic deposition produces an unusual structure of a particle layer, as shown in Fig. 1. This structure is the formation of the field-induced pearl-chains of particles. The next step is to sinter the particle layer very carefully. We expect that the field-induced pearl-chains are likely to form a fiber matrix, in such a way that each contact point between particles is sintered together to form fibers, and fibers are then intertwined with each other to form the membrane. The processes from particle production to sintering are carried out continuously in an inert gas atmosphere. This can prevent contamination and agglomeration of the particles in the handling processes. We call this method "Electrostatic Formation of Ceramic Membrane"(EFCM).

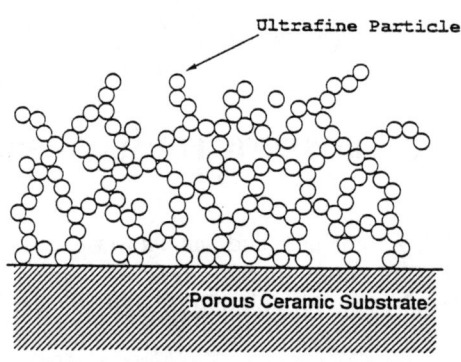

Figure 1 Structure of a particle layer

The purpose of this paper is to illustrate the principle of the EFCM, the structure of the EFCM-produced membrane, and the results of basic performance tests of the membrane prepared from silicon nitride particles.

EXPERIMENTAL APPARATUS

Figure 2 shows a schematic diagram of experimental apparatus for producing a membrane from ultrafine particles of silicon-nitride. It consists of a reactor for preparing particles and a membrane-forming section. These two sections are located in series in a common quartz tube(**1** in Fig. 2). The reactor consists of four coaxial cylinders including the common quartz tube(**1**). In the membrane-forming section, a cylindrical ion source using high-frequency surface discharge(**3**) and a wire electrode(**4**) are installed in order to charge the particles delivered from the reactor. The ion source(**3**) is supplied with a high-frequency high-voltage by a power supply(**5**). Frequency is 10kHz and voltage is 10-20 kVpp. The wire electrode(**4**) is supplied with a high DC-voltage which is below 5kV by a power supply(**6**). A porous ceramic substrate(**7**) is set around the wire electrode. The substrate utilized in this experiment is a sintered alumina tube. Its outer and inner diameters are 10 and 7 mm, respectively, and its length is 140 mm. Its mean pore size is 10 μm according to catalog data.

The reaction system using a gaseous process of metal-chloride and ammonia is one of the useful methods to produce ultrafine particles of metalnitrides such as Si_3N_4, BN, AlN, TiN, etc. For preparing silicon nitride particles in this experi-

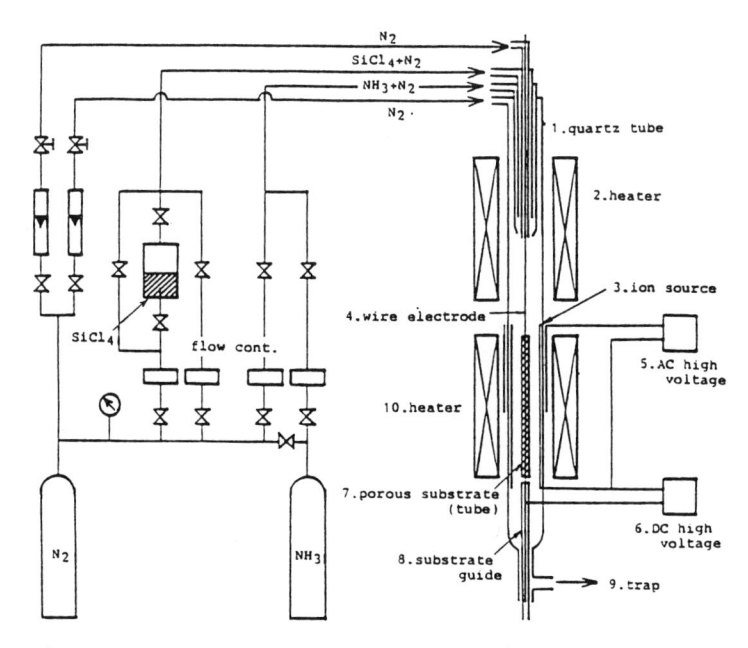

Figure 2 Schematic diagram of experimental apparatus

ment, silicon tetrachloride and ammonia gas reaction system was used. This reaction process is usually carried out under an excess ammonia gas condition so that ammonium chloride deposits are formed as the by-product. This deposition onto the ceramic substrate can be prevented by heating the forming section, above around 340°C, which is the sublimation temperature of ammonium chloride. Therefore, we heated the membrane-forming section above 500°C.

Figure 3 shows the details of reactor tubes. Silicon tetrachloride and ammonia carried with N_2 gas are introduced into the middle tubes separately, and mixed at the reaction zone to form ultrafine particles. Particles

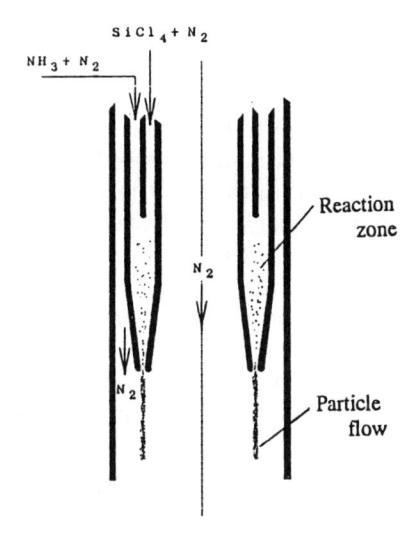

Figure 3 Reaction tubes

Porous Materials

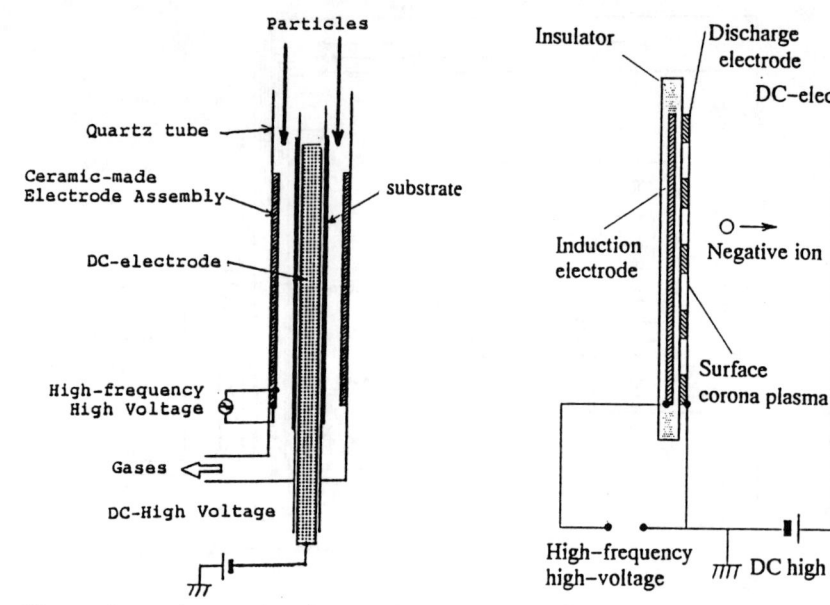

Figure 4 Membrane-forming section

Figure 5 Electrode assembly

prepared flow down in a form of a thin cylindrical cloud that is sandwiched between the inner and outer sheath N_2 gas streams.

Figure 4 gives the details of the membrane-forming section. Particles delivered from the reactor are charged and travel across the DC-field to be deposited on the outer wall of the substrate. A ceramic-made cylindrical electrode assembly is used to form a cylindrical ion source. Figure 5 is a cross sectional view of the ceramic-made electrode assembly. It consists of discharge and induction electrodes. They are insulated from each other with a thin alumina layer. When a high-frequency high-voltage is applied between them, a stable surface corona discharge takes place uniformly on its surface. This can form a stable plasma ion source even under high temperature and complex gas compositions. But, the plasma consists of both negative and positive ions. Setting a DC-electrode at the center of the ceramic-made electrode assembly and supplying a DC high-voltage between them, monopolar ions are emitted from the plasma source and travel across the charging zone. Particles passing through here are charged monopolar and travel across the charging zone to be deposited on the substrate. The substrate can be moved up and down, as well as can be rotated, so that a uniform deposit can be obtained. The particle layer deposited on the substrate is sintered to form a membrane in an inert gas atmosphere without being taken out from the system.

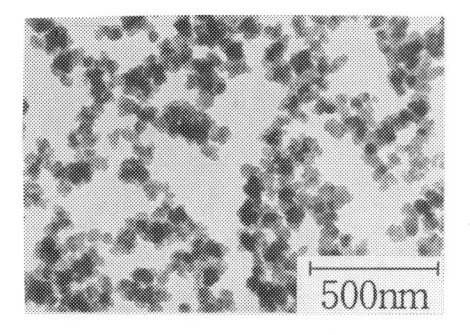

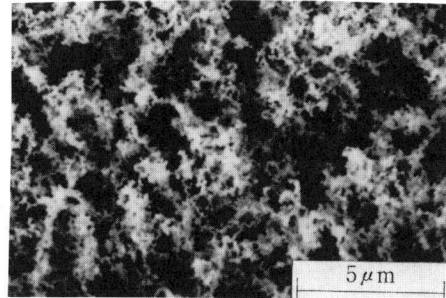

Figure 6 TEM photo of particles   **Figure 7** Surface of deposited particle layer

MEMBRANE FORMATION

The pore size of the membrane formed by sintering of the deposit is expected to depend on the size of the original particles. Figure 6 is TEM photograph of the particles prepared in this experiment. The reactive conditions are, the concentration of silicon tetrachloride ($SiCl_4$); 1.4 vol%, the ratio of ammonia(NH_3) to $SiCl_4$; 6, and reaction temperature; 1200°C. As indicated in the photograph, the particles are of uniform size of several tens of nanometers. Figure 7 is a photograph of the surface of the particle layer deposited on the substrate before sintering. Original particles are shown in Fig. 6. The structure is unique to electrostatic deposition. It shows the formation of the field-induced pearl-chains of particles. This particular structure is likely to be the essential factor for generating the very unique membrane structure after sintering. The deposit structure could be controlled by changing one or both of the DC- and AC-voltages.

THE STRUCTURE OF THE SINTERED MEMBRANE

To form a better membrane, it is necessary to sinter the particle layer to provide as porous a layer as possible, with large voids. In the sintering process, careful control of the parameters such as temperature, time, gas atmosphere etc. is essential. In this experiment, we determined these parameters in an empirical way by observing the resultant membrane structure using SEM analysis.

The silicon nitride particles prepared in this experiment could be sintered at about 1200°C in N_2 gas atmosphere without sintering aid. This is much lower than that normally associated with production of the structural silicon-nitride materials. For this reason, these particles are considered very fine and amorphous, or the deposit layer is very thin. As this is a very interesting phenomenon, we plan to study it in more detail in the future.

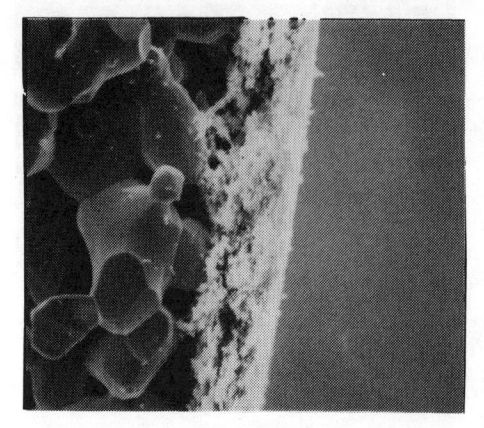

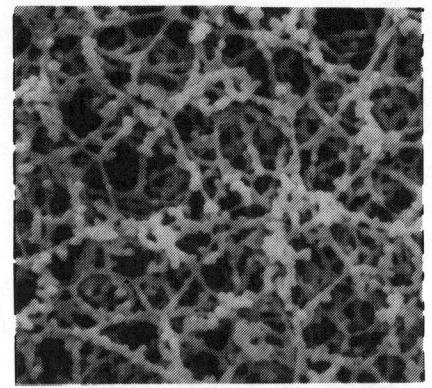

Figure 8 Cross-section of the membrane **Figure 9** Surface of the membrane

Figure 8 is a cross-sectional SEM photograph of the membrane. It shows an approximately 20 μm-thick tight membrane on the porous alumina substrate. Figure 9 shows the surface structure of the membrane. A unique membrane matrix with a three-dimensional network is observed. Such structures have resulted from the structure of the electrostatic particle deposition. A membrane with this structure has a large porosity just like a porous fiber filter. The effective pore size of the membrane is several micrometers, as observed.

BASIC CHARACTERISTICS OF THE MEMBRANE.

Figure 10 shows the gas pressure drops across the substrate and that coated with the membrane. The difference represents the pressure drop of the membrane alone, which is very small, although the pore size of the membrane is about 1/50 of that of the substrate as shown in Figs. 11 and 12. They are in the same magnification. This is because the membrane has a very porous and three-dimensional network structure, and its thickness is much thinner than that of the substrate; the latter is 20 to 1500 μm in thickness.

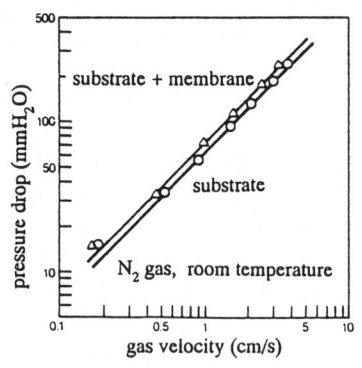

Figure 10 Gas pressure drop of the membrane as a function of gas velocity

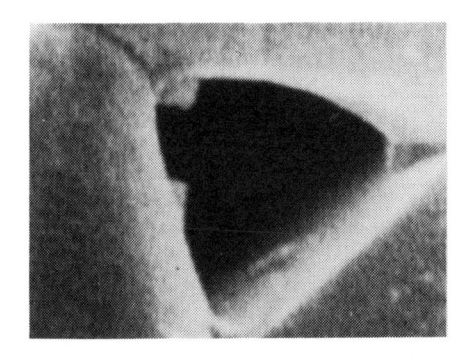

Figure 11 Surface of the substrate

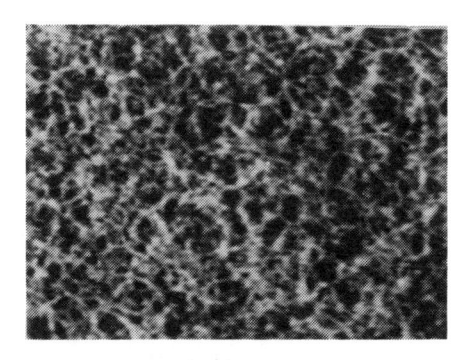

Figure 12 Surface of the membrane

Figure 13 shows the results of a separation experiment using fine silica particles dispersed in boiling water. Silica consists of mono-dispersed particles with a 1.8 μm diameter. The rejection of particles is 100% at the water temperature of 100°C and under the filtration pressure of 1.5 kPa. The flux in steady state is around 0.1m³/m²hr. These are very good separation performance results.

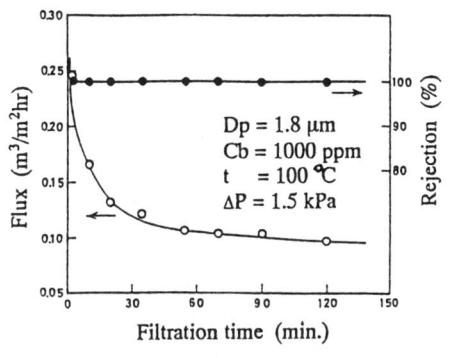

Figure 13 Removal of silica particles

CONTROL OF MEMBRANE STRUCTURE

There is a strong need for membranes with smaller pores. There are many factors which control the pore size in this forming method as shown in Fig. 14. Controlling the size of the starting particles is one of the major factors that can be used to control the pore size. When the size of starting particles is smaller, the pore size will be smaller. One of the factors to control the particles size is the reactant gas concentration.

Figure 15 shows the effects of the concentration of silicon tetrachloride on particle diameter at the reaction temperature of 1200°C. The median size was measured by the centrifugal sedimentation method. The ratio of ammonia to silicon tetrachloride was fixed at 6. The particle size decreased as the concentration of silicon tetrachloride decreased. Three membranes were made by using three kinds of particle sizes, **A, B, C** as shown in Fig. 15.

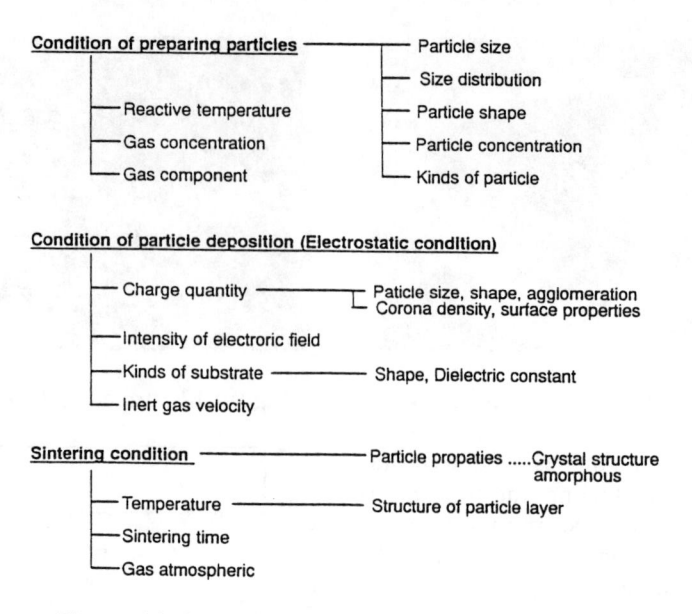

```
Condition of preparing particles ──── Particle size
                         │            ├ Size distribution
                         ├ Reactive temperature
                         ├ Gas concentration      ├ Particle shape
                         └ Gas component          ├ Particle concentration
                                                  └ Kinds of particle

Condition of particle deposition (Electrostatic condition)
         ├ Charge quantity ─────┬ Paticle size, shape, agglomeration
         │                      └ Corona density, surface properties
         ├ Intensity of electroric field
         ├ Kinds of substrate ──── Shape, Dielectric constant
         └ Inert gas velocity

Sintering condition ──────── Particle propaties .....Crystal structure
         │                                              amorphous
         ├ Temperature ──────── Structure of particle layer
         ├ Sintering time
         └ Gas atmospheric
```

Figure 14 Control factors of membrane pore size

Figure 16 shows the membrane formed by using the particle size **A**, and the particles were the largest particles. They resulted in a very large pore-size (Fig. 16).

Figure 17 shows the membrane formed by using the particle size **B**, which resulted in a smaller pore size than that formed with **A**.

Figure 18 indicates the membrane formed by using the smallest particle size **C**. The pore size is very small, i.e., about 1μm or less. It is about one tenth that of membrane B. These results indicate that the pore size of the membrane can be controlled to some extent by controlling the prepared or deposited particles as membrane materials.

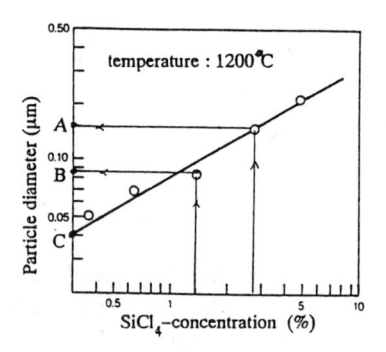

Figure 15 Change in particle as a function of SiCl$_4$ concentration

Figure 19 shows the results of separation tests of the membranes B and C, using the mono-dispersed polystyrene latex system of various particle sizes. The latex

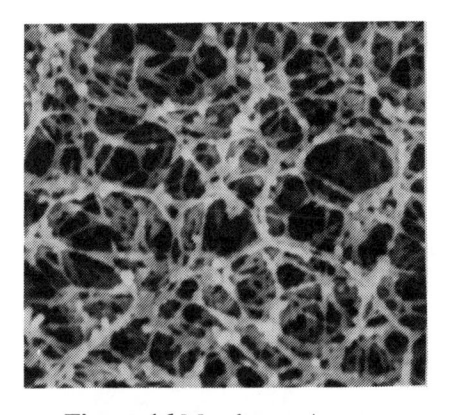

Figure 16 Membrane A

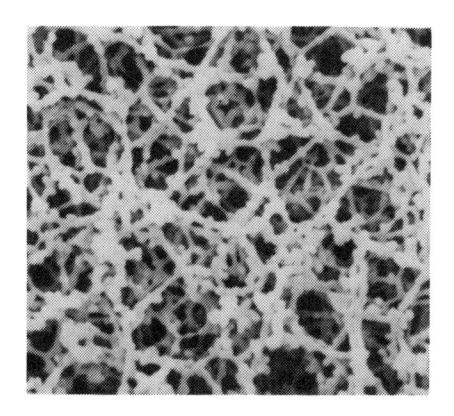

Figure 17 Membrane B

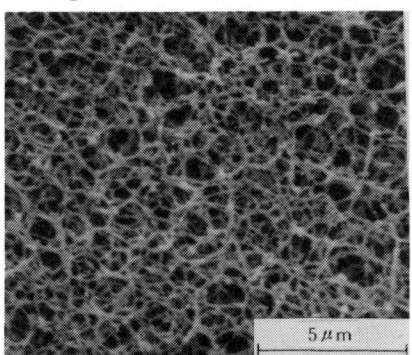

Figure 18 Membrane C

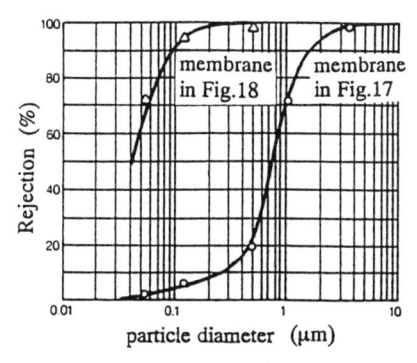

Figure 19 Rejection of the membrane as a function of particle diameter

concentrations were 100 ppm or less. The separation efficiency corresponds to the pore size distribution of each membrane and agrees well with the results obtained by SEM photos. Both membranes have a very sharp separation efficiency. The 50%-rejection size of the membrane B is about 0.8 μm. On the other hand, that of the membrane C is 0.04 μm, which is about one-tenth that of membrane B. This pore size is useful for ultrafiltration.

CONCLUSIONS

This paper has reported a new method for ceramic membrane preparation using CVD ultra-fine particles and it is called "Electrostatic Formation of Ceramic Membrane (EFCM)". The membranes produced by the EFCM method with silicon nitride have a three-dimensional network structure and very high porosity. This structure has a special advantage as it allows various fluids to pass through it with a very low pressure drop. The unusual membrane structure appears to be generated from the unique structure of the electrostatic deposition of particles in a form of the field–induced pearl–chains. The pore size of the membrane can be controlled to some extent by controlling the size of the particles prepared and used as the membrane materials. One of the membranes produced in this study has micropores of 0.04 μm on the average. The separation efficiency of this membrane is fairly sharp. This membrane is suitable for various applications including ultrafiltration(UF). We are exploring other application areas for these membranes. We are also investigating the effectiveness of the EFCM for ultrafine particles of other materials such as TiO_2, Al_2O_3, SiO_2, and teflon.

APPLICATION OF HIPED POROUS METALS FOR AN ELECTROCHEMICAL DETECTOR

Yoshihiro Okumoto, Kozo Ishizaki and Akifumi Yamada
Nagaoka University of Technology
Nagaoka, Niigata 940-21, Japan

Porous platinum with high open porosity and many penetrating pores is produced by a modified HIP method. The amount of penetrating pores increases by passing pressurized gas through a platinum green body during sintering. By decreasing the HIPing pressure after sintering, the permeability of the porous platinum sample improved 20% compared with the normal HIPing process. The porous platinum was applied as a new filter-type electrode in an electrochemical detector. A detection response with high stability was observed and the peak height was proportional to the concentration of the pilot sample.

INTRODUCTION

Open porous metals have many applications, for example, as electrodes [1]. High specific surface area and open porosity are required for industrial applications, especially as electrodes. However, it is difficult to produce porous metals by normal sintering, because metal powders have a narrow sintering temperature range.

Recently, Ishizaki et al. applied capsule-free hot isostatic pressing (HIPing) for the production of porous materials [2]. Metals with high open porosity can be easily produced by this method. To obtain a porous metal with high permeability, the authors have modified the capsule-free HIP method. In the new HIP method, penetrating pores are produced by forcing pressurized gas to travel through samples by alternating the HIPing pressure during sintering [3,4]. By using this porous metal as a working electrode in an electrochemical detector, the structure of the electrochemical detection cell becomes simpler and more compact than the reported detection cells [5] using a non-porous working electrode.

In the present study, the production of porous platinum and its application to a working electrode in an electrochemical detector are investigated.

EXPERIMENTAL

Sintering

The platinum powder used was smaller than 150 mesh. The powder was uniaxially pressed in a stainless steel mold at a pressure of 3 MPa for 60 s. The obtained pellets, 20 mm in diameter and about 0.5 mm thick, were cold isostatically pressed (CIPed) at a pressure of 10 MPa for 60 s.

The CIPed platinum samples were capsule-free HIPed (Kobe-Steel System 3, Professor-HIP) for 60 s in argon gas at a pressure of 100 MPa and a temperature of 1 300 °C or 1 400 °C. Four different HIP treatments were carried out. Sample **A** was sintered by the normal capsule-free HIPing at 1 300 °C, and **C** at 1 400 °C. Sample **B** was sintered by applying the new HIP sintering method at 1 300 °C, and **D** at 1 400 °C. The HIPing pressure decreased abruptly after sintering in the new HIP method. The details of the new HIP sintering method were reported previously [4]. A special crucible was used to force a flow of argon gas through the platinum samples by decreasing the HIPing pressure.

The permeability of the samples was measured by using nitrogen gas [4]. The gas flow and pressure drop through the sample were measured by a flowmeter and a mercury manometer. The relative density and open porosity were calculated by the Archimedean method [6].

Working Electrode

Fig. 1 shows the configuration of the produced electrochemical detector. The porous platinum working electrode, B in Fig. 1, was produced by the following method. The electrode with a diameter of about 4 mm was cut from the sintered sample. The platinum wire was soldered to a spot on the porous platinum electrode, which was sealed in the Teflon holder by applying epoxy resin to the gap between the electrode and the holder.

The stainless steel block, E in Fig. 1, and the saturated calomel electrode (SCE) were used as counter electrode (CE) and reference electrode (RE), respectively. The three electrodes were assembled as shown in Fig. 1.

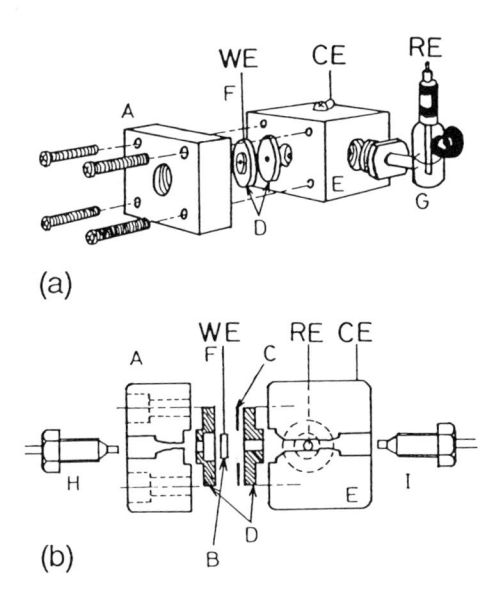

Figure 1 Configuration of electrochemical detection cell.
a) Three dimensional diagram of the cell. b) Cross section of the cell. A: Polyflon block, B: working electrode (porous platinum), C: Teflon spacer, D: Teflon holder, E: counter electrode (stainless steel block), F: platinum lead wire, G: reference electrode (SCE), H: inlet fitting and I: outlet fitting.

Electrochemical Detection with a Porous Platinum Working Electrode

The block diagram of the detection system is shown in Fig. 2. The carrier is led by the pump (B in Fig. 2, JASCO, Intelligent HPLC Pump 880-PU). When the carrier from the pump, streaming in a Teflon tube with a diameter of 0.65 mm, flowed to the injection valve (C, RHEODYNE, Model 7125), the pilot sample is injected. The electric potential of the working electrode versus SCE was controlled by the potentiostat (E, Fuso, model 312). The electrolytic current generated from the oxidation-reduction reaction between the pilot sample and the working electrode is detected by a recorder (F, SEKONIC, SS.250F).

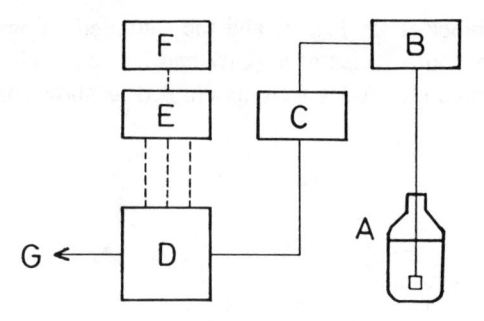

Figure 2 Block diagram of flow system and electronics.
A: carrier reservoir, B: pump, C: injection valve and sample injector, D: electrochemical detection cell, E: potentiostat, F: recorder and G: drain.

Potassium chloride was used as an carrier, and potassium ferrocyanide as a pilot sample. The oxidation reaction for this detection is:

$$[Fe(CN)_6]^{4-} \longrightarrow [Fe(CN)_6]^{3-} + e \qquad (1)$$

The principle of this electrochemical detection is based on Faraday's law of electrolysis as

$$Q = \int i \cdot dt \propto n \cdot F \cdot N \qquad (2)$$

where Q is the total electric charge, i the electric current, t the detecting time, n the number of electrons generated from the electrochemical reaction, F the Faraday's constant and N the amount of pilot sample. In this electrochemical detection, the amount of pilot sample could be calculated by measuring the electric charge.

The properties of the porous platinum working electrode were evaluated as following order:

1) The voltammogram (current versus electric potential of working electrode curve) is measured by scanning the electric potential of the working electrode to determine the applied electric potential E_{app}.
2) The electric current is measured by applying E_{app} to the working electrode.
3) The relation between concentrations of pilot samples and the peak height of the chart are measured.

RESULTS AND DISCUSSION

Effects of HIPing Pressure on Porous Platinum

Fig. 3 shows the effect of alternating HIPing pressure on the permeability of porous platinum. The flow of nitrogen gas through the sample was assumed to be a laminar flow which is expressed by Equation (3)(Darcy's low).

$$\frac{Q_f}{A} = \frac{D_s}{\eta} \cdot \frac{\Delta P}{L} \qquad (3)$$

where Q_f is the quantity of gas flow, ΔP the pressure drop through the sample, A the measured area, L the sample thickness, D_s the permeability coefficient and η the viscosity of nitrogen gas. By alternations of HIPing pressure, the permeability of the porous platinum increases. The values of D_s calculated by Equation 3 are listed in Table 1.

Fig. 4 shows the influence of the alternating HIPing pressure on the relative density and the open porosity. It is seen that the relative density and the open porosity did not change significantly between the different HIPing methods. It can be considered that the weak bridges of platinum sample are broken and the pore-tunnel shape becomes less curved by the effect of forcing the gas through platinum samples during sintering. Therefore, the permeability of porous platinum was improved by the new HIP method.

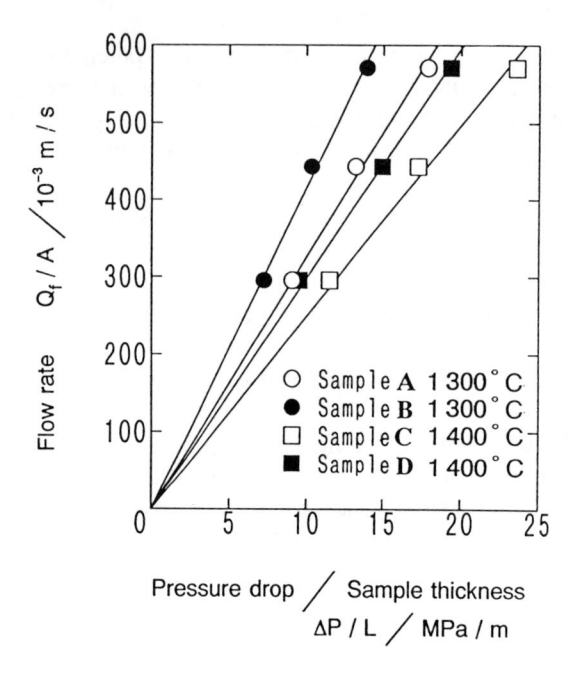

Figure 3 Effects of alternating HIPing pressure on a permeability of the porous platinum. Permeability of porous platinum increases by alternation of HIPing pressure. The square and circle marks mean sintering temperature of 1 300 °C and 1 400 °C, respectively. The open and solid marks indicate normal capsule-free HIPing and one cycle alternation HIPing, respectively.

Table 1. Permeability Coefficient, D_s / m^2

Type of HIP treatment	Sintering Temperature 1 300 °C	1 400 °C
Normal HIPing	5.6 ×10^{-13}	4.3 ×10^{-13}
1 Cycle Alternation	7.2 ×10^{-13}	5.2 ×10^{-13}

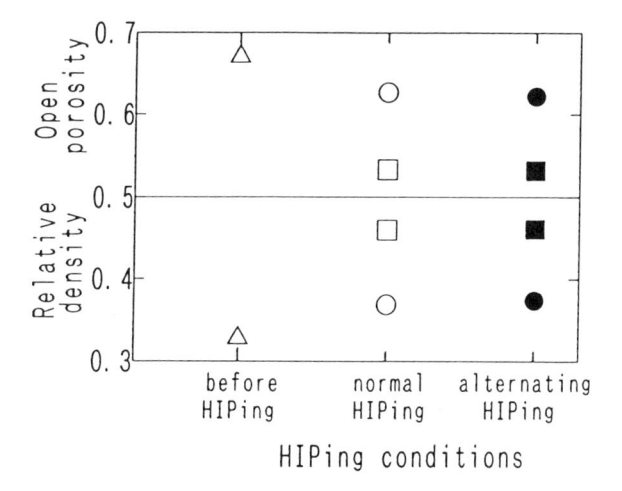

Figure 4 The influence of HIPing condition on the relative density and the open porosity. There is no significant change of the relative density and open porosity between the sintered porous platinum produced by alternating HIPing pressure and by the normal capsule-free HIP method. The open triangles correspond to CIPed pellets, other marks have the same significance as Fig. 3.

Porous Platinum as a Working Electrode

The properties of the porous platinum electrode were investigated by using platinum sample **D**. A SEM micrograph of the surface of sample **D** is shown in Fig. 5.

Figure 5 SEM photograph of the surface of porous platinum sintered at 1 400°C using the new HIP method.

Porous Materials 321

The voltammogram of carrier and pilot sample with carrier is shown in Fig. 6. The dashed line and solid lines indicate the curve of the carrier and the curve of the pilot sample with the carrier, respectively. The $i-E$ curve clearly indicates the limiting current at the potential of + 0.85 V versus SCE. Therefore, the electrolysis was carried out at this potential E_{app} = + 0.85 V versus SCE).

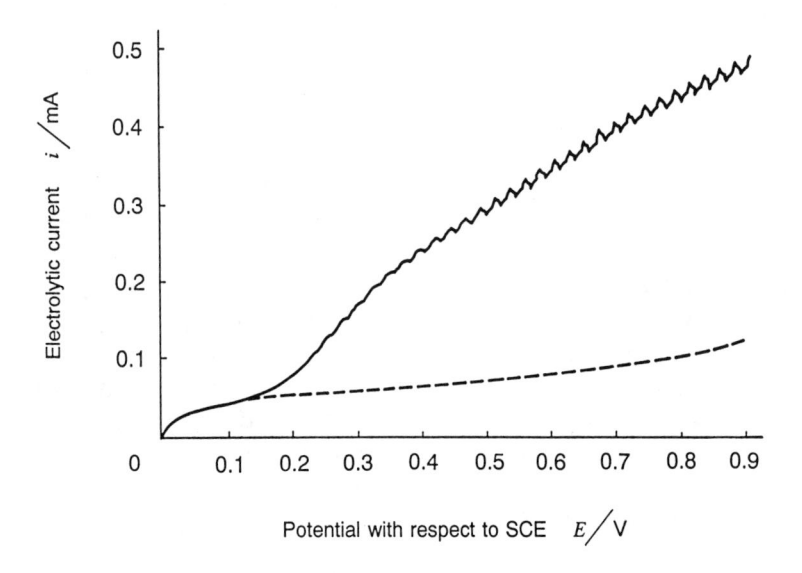

Figure 6 Voltammogram of carrier and pilot sample with carrier. The dotted line indicates the curve of the carrier and the solid line indicates the curve of the pilot sample with the carrier.

Fig. 7 shows the $i-t$ curves obtained by detection of several concentrations of the pilot samples. The flow rate of carrier was controlled at 1 ml/min. Sharp peaks are obtained and the response is stable enough to use this porous platinum as a working electrode. Assuming the area of peaks being proportional to the peak height, the relationship of the peak height to the concentration of pilot sample was investigated, see Fig. 8. The peak heights were proportional to the concentrations of pilot sample as shown in Fig. 8. This relationship is a very important factor to determine that the porous platinum can be used as a working electrode. Therefore, this porous platinum sample D can be used as a new filter-type working electrode for an electrochemical detector.

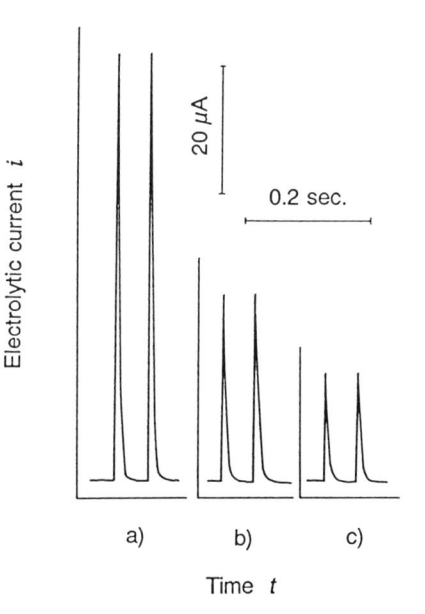

Figure 7 Current vs. time curves obtained by detection of several concentrations of the pilot samples (potassium ferrocyanide).
a)   1 mmol/l
b)   0.5 mmol/l
c)   0.25 mmol/l

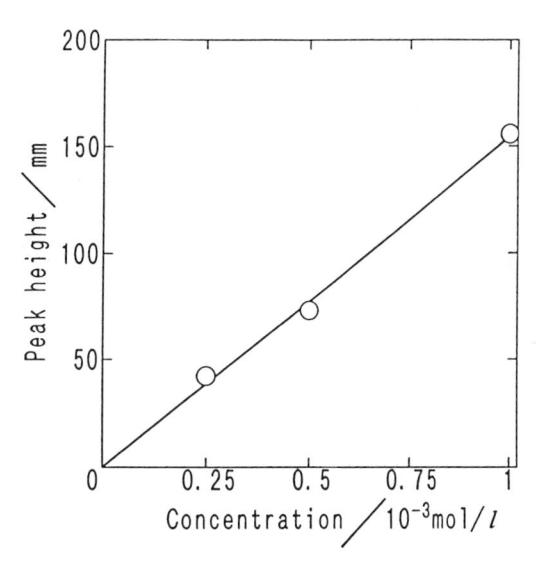

Figure 8 The relationship between the peak height in the i–t chart and concentration of the pilot sample. The peak height is linearly proportional to the concentration of the pilot sample.

Porous Materials 323

CONCLUSIONS

The amount of penetrating pores in porous platinum increases by alternating the HIPing pressure during sintering. The obtained platinum sample was applied as a working electrode in an electrochemical detector. The following conclusions are revealed from the present results:
1) The permeability of the sintered porous platinum increases about 20% by alternating the HIPing pressure compared to normal capsule-free HIPing.
2) A high stability of detection response is achieved by using the porous platinum working electrode in an electrochemical detector. The peak height is linearly proportional to the concentration of the pilot sample.

REFERENCES

1) G.E. Freud and B. Chinaglia, " Sintered Platinum for Cardiac Pacing", The International Journal of Artificial Organs, **4**, 5, 238-242, 1981
2) K. Ishizaki, S. Okada, T. Fujikawa and A. Takata, Germany Pat., P 40 91 346.5(1991). US Pat., No.5, 126, 103 (1992)
3) K. Ishizaki, A. Takata, and Y. Okumoto, Patent pending.
4) Y. Okumoto, A. Takata, and K. Ishizaki, "Manufacture of Penetrating-Pores Metals by High Gas Pressure", Materials Research Society Symposium Proceeding Vol.251: Pressure Effects on Materials Processing and Design, edited by K. Ishizaki, E. Hodge and M. Concannon, 115-120, 1992
5) Y. Takata and G. Muto , "Flow Coulometric Detector for Liquid Chromatography", Analytical Chemistry, **45**, 11, 1864-1868, 1973
6) K. Ishizaki, A. Takata and S. Okada, "Mechanically Enhanced Open Porous Materials by HIP Process", J. Jpn. Ceram. Soc., **98**, 533-540, 1990

APPLICATION OF POROUS ALUMINA CERAMICS FOR CASTING MOLDS

Yoshihito Kondo, Yutaka Hashizuka, Shojiro Okada
Kagawa Industrial Technology Center
587-1 Goto-cho, Takamatsu 761, Japan
and
Motoaki Shibayama
Shikoku Research Institute Incorporated
2109-8 Yashimanishi-machi, Takamatsu 761, Japan

Although slip casting using gypsum enables one to make a large complex and dense green body, the surface of the green body is contaminated with calcium and sulfur. In the case of a resin mold, it is necessary to apply external pressure because of its poor wettability with water. In this study, the wettability and casting rate of porous alumina mold was compared to a gypsum mold. Porous alumina ceramics produced by a presintering method show very high wettability value as compared with gypsum. Therefore, porous alumina ceramics provided much larger penetration pressure and higher rate of casting than gypsum.

INTRODUCTION

Advanced ceramics are steadily extending the engineering applications with the developments of submicron powder technology and new processing. Physical properties of advanced ceramics are influenced by several kinds of defects produced in the forming process, and large defects can not be removed even by the HIP(hot isostatic pressing) sintering process. Although new forming technologies, such as CIP(cold isostatic pressing) and injection molding, have been developed, slip casting, which has been used in the forming of traditional ceramics, enables one to make a large complex and dense green body.

Both wear and dissolution of gypsum molds produce contamination of calcium and sulfur on the surface layer of the green body. The porous resin mold provides wear resistance and contamination free properties but requires application of pressure to the suspension during casting because of its poor wettability with water[1]. In this study, porous alumina ceramics as casting molds were examined and compared to gypsum with regards to wettability with water and rate of cast layer.

MEASUREMENT OF WETTABILITY WITH WATER

There are several methods of measuring the wettability with a liquid. In this study, the wettability with water for both porous alumina and gypsum was measured using a water penetration method.

The penetration velocity, v, of a liquid in porous material is expressed by the following equation[2]:

$$v = \frac{dl}{dt} = \frac{r^2 \Delta P}{8\eta l} \tag{1}$$

where l is the penetration distance of the liquid in the porous material, r the average pore radius, η the viscosity of liquid, and ΔP the penetration pressure. If the penetration phenomenon is based on the external pressure, Δp, and the wettability with liquid, then the following expression is obtained.

$$\Delta P = \frac{2\gamma_L \cos\theta}{r} + \Delta p \tag{2}$$

where γ_L is the surface tension of water, and θ the contact angle between water and the porous material. From Eqs.(1) and (2), Eq.(3) is obtained

$$v = \frac{dl}{dt} = \frac{r^2}{8\eta l}\left(\frac{2\gamma_L \cos\theta}{r} + \Delta p\right) \tag{3}$$

When Δp equals 0 under the condition without external pressure, the integration of Eq.(3) under the boundary condition, $t=0$ and $l=0$, yields Eq.(4).

$$L^2 = \frac{r\gamma_L \cos\theta}{2\eta} t \tag{4}$$

The penetration weight, w, of liquid at time, t, in a cylindrical porous material with sectional area, s, is described by the following equation.

$$w = s L \varepsilon \rho_L \tag{5}$$

where, ε and ρ_L is the porosity of porous material and the density of liquid, respectively. From Eqs.(4) and (5), Eq.(6) is obtained.

$$\left(\frac{w}{s}\right)^2 = \varepsilon^2 \rho_L^2 \frac{r\gamma_L \cos\theta}{2\eta} t \tag{6}$$

Since γ_L is known for water, the contact angle, θ, may be estimated by measuring the change of water penetration weight with time.

Specimens of about 50 mm diameter and 20 mm thickness were prepared for both gypsum and porous alumina ceramics. The surrounding side wall of specimens was sealed with putty and a water penetration was made only on the circular base of the specimens, as shown in Fig. 1. Water penetration weight was measured at a temperature of 25℃.

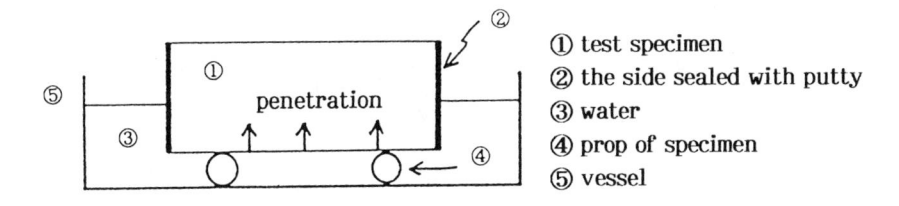

Figure 1 Schematic diagram of water penetration method.

WETTABILITY OF GYPSUM

Several kinds of gypsum molds with different water contents, as shown in Table 1, for α- and β-types (C300C and A grade by Noritake–Nittoh–Gypsum Co.) were prepared to size of 55 mm diameter and 25 mm thickness. The average pore sizes of the gypsum molds, measured by a mercury pressure porosimetry (Calro Erba Strumentazione, Macropore 120 and porosimeter 2000), were 4~5 μm for α-type and 4.5~6 μm for β-type.

Figure 2 is the change of porosity as a function of water content in the gypsum molds. Judging from the linear relation between porosity and water content, the porosity of gypsum mold depends on water content and is independent of the type of gypsum. In general, α-type is applied to the casting mold for advanced ceramics and β-type for traditional ceramics.

Table 1. Composition of Gypsum and Water.

Type of gypsum	Gypsum weight (g)	Water content (g)			
		1	2	3	4
α-type	100	40	45	50	55
β-type	100	60	65	70	75

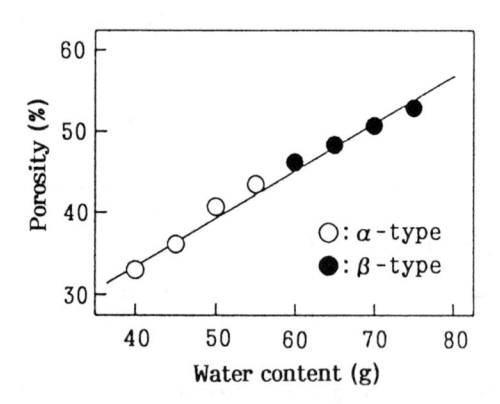

Figure 2 Change of porosity as a function of water content in gypsum molds.

In order to estimate the contact angle between water and gypsum molds, the penetration weight of water was measured at times of 10, 20, 30, and 60 sec. The relationship between the square of penetration weight of water per unit area, $(w/s)^2$, and the penetration time is shown in Fig. 3. As seen from Eq.(6), $(w/s)^2$ increased linearly with an increase in penetration time. The slope of the straight line, K, is described by the following equation from Eq.(6):

$$K = \varepsilon^2 \rho_L^2 \frac{r \gamma_L \cos \theta}{2\eta} \tag{7}$$

Judging from Eq.(7), the slope, K, becomes larger with increasing porosity and pore size. As the pore size is nearly constant in Fig. 3, the slope, K, of the gypsum mold becomes larger with increasing porosity. The free contact angles, θ, between water and gypsum molds were 89.3 degree for α- type and 88.0 degree for β-type.

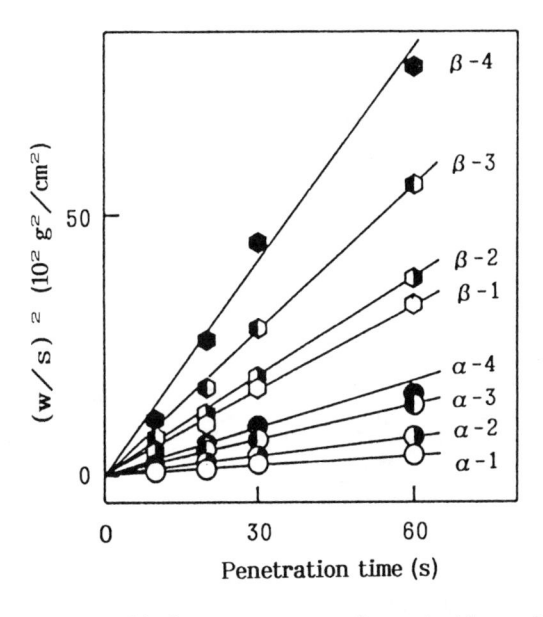

Figure 3 Relationship between square of penetration weight of water per unit area, $(w/s)^2$, and penetration time.

Porous Materials

WETTABILITY OF POROUS ALUMINA CERAMICS

As raw materials for the porous alumina ceramics, #600 white fused alumina powder (Showadenko CO.) and alumina powders with average grain sizes of 3 and 1.4 μm (Showadenko CO., A-42-3 and AL-45) were used. After these raw powders were mixed with 3 wt% binder (Chukyo-Yushi CO., D-830) by ball milling, the three mixtures were granulated using a spray-dryer and then pressed uniaxially under 20~80 MPa. The pressed green bodies were heated at a rate of 150°C/h to 1600°C for #600 white fused alumina, to 1400, 1550, 1600°C for 3 μm alumina and to 1400, 1500, 1550°C for 1.4 μm alumina.

Figure 4 A and B indicate the change of porosity with respect to forming pressure for the alumina ceramics using alumina powders of 3 and 1.4 μm, respectively. Although the decrease of porosity with the increase of forming pressure and sintering temperature is a general tendency, the porosity of porous alumina sintered at 1500 and 1550°C decreases gradually as the forming pressure increases up to 50 MPa, and then rapidly with further increasing forming pressure, as shown in Fig. 4(B).

The average pore sizes of porous alumina ceramics, which decreased slightly with the increase of forming pressure and sintering temperature, were 5~8 μm for #600 white fused alumina, 0.9~1.1 μm for grain size of 3 μm and 0.4~0.55 μm for 1.4 μm. The pore size was proportional to the grain size in all three consolidation methods[3].

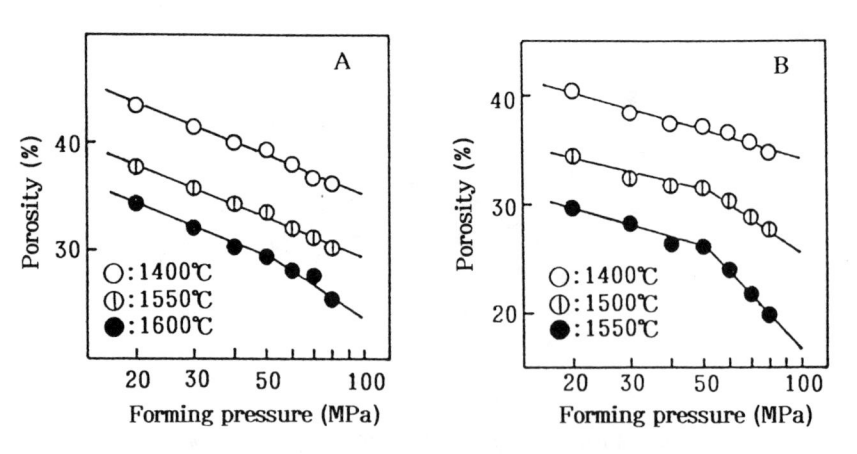

Figure 4 Change of porosity to uniaxial forming pressure in porous alumina ceramics. (A:3μm alumina powder, B:1.4μm)

The change of $(w/s)^2$ to penetration time for porous alumina ceramics with about 40% porosity and different pore sizes, and with about 0.45 μm pore size and different porosities, are shown in Figs. 5 and 6 respectively. The linear relationship between $(w/s)^2$ and penetration time of the porous alumina is observed to be similar to that of the gypsum mold. A high penetration rate is predicted for porous alumina ceramics with large porosity as well as pore size by Eq.(7).

The contact angles, θ, between water and porous alumina ceramics, calculated by Eq.(7), are listed in Table 2. Although the wettability with water of porous alumina ceramics must be essentially constant, it becomes larger at lower sintering temperatures and with raw materials of smaller grain size. These phenomena are caused by the differences in the pore shape and the surface state of alumina grain. The porous alumina ceramics have better casting properties as compared with gypsum, because of their higher wettability with water.

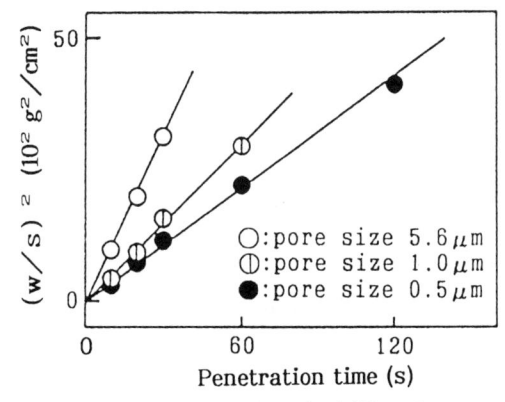

Figure 5 Relationship between $(w/s)^2$ and penetration time for porous alumina ceramics with about 40% porosity and different pore sizes.

Table 2. Contact Angle, θ, between Water and Porous Alumina Ceramics.

Raw material	#600	3 μm			1.4 μm		
Sintering temp. /℃	1650	1400	1550	1600	1400	1500	1550
θ degree	83.8	81.3	82.0	82.7	78.7	81.0	82.5

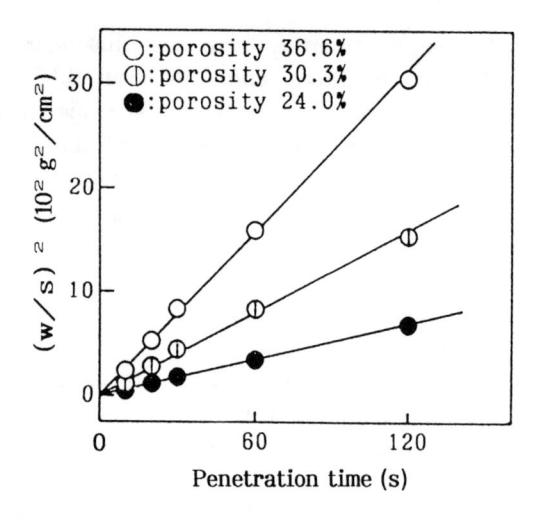

Figure 6 Relationship between $(w/s)^2$ and penetration time for porous alumina ceramics with about $0.45\,\mu$m pore size and different porosities.

CASTING TESTS OF ALUMINA

The casting tests for the porous ceramics were carried out without external pressure using the alumina suspension with average particle size $0.6\,\mu$m (Showadenko Co., AL-160SG), dispersant 0.4 wt% (Chukyo-Yushi Co., D-305) and concentration 45 vol%, and the thickness of the cast layer at a casting time of 10 min was measured.

As the penetration phenomenon is only owing to the wettability with water without external pressure, the penetration pressure, ΔP, in casting is expressed by equating Δp to zero in Eq.(2). The porous ceramics as a casting mold, with the high wettability value and the small pore size, give a high penetration pressure.

Figure 7 shows the relationship between the penetration pressure of porous ceramics mold and the thickness of the cast layer. Most of the porous alumina ceramics indicate a larger penetration pressure than the gypsum. The ceramics with small pore size give a higher penetration pressure. The thickness of the cast layer becomes larger with increasing penetration pressure, and one of porous alumina ceramics with the highest penetration pressure gives about 3 x the times thickness in comparison to the gypsum mold.

SEM photographs for surface layers of casting molds after the casting test are shown in Fig. 8. Although the surface layer of gypsum mold is shown to be clogged with alumina particles, there are no clogging in the porous alumina ceramics with 0.5 μm pore. To prevent clogging on the surface layer of mold, it is necessary to use the porous alumina ceramics mold that has pores smaller than the particle size of the alumina suspension.

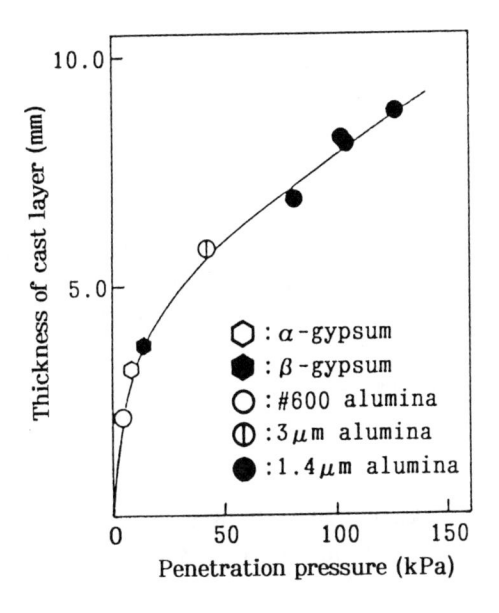

Figure 7 Relationship between the penetration pressure of porous ceramics mold and the thickness of cast layer.

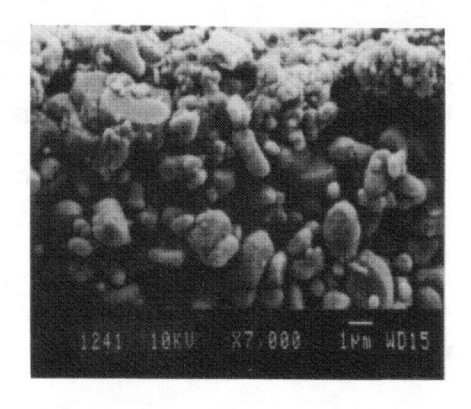

(A) α-type gypsum (B) Porous alumina with 0.45 μm pore

Figure 8 SEM photographs of surface layers after casting.

CONCLUSIONS

The porous alumina ceramics are applied as a casting mold. The wettability with water and the rate of casting are compared with a gypsum mold. The following conclusions are obtained:
(1) The porous alumina ceramics indicate high wettability as compared with gypsum.
(2) The porous alumina ceramics with fine pores provide a much larger penetration pressure and a higher rate of casting than the gypsum.
(3) There is no clogging on the surface layer of porous alumina ceramics after casting.
(4) The alumina molds have superior properties for slip casting.

REFERENCES

1. H. Mizuta, Y. Shibazaki, K. Sakai, M. Katagiri and H. Fujimoto, "Slip cast forming and sintered body of high pure alumina", J. Japan Soc. Powder and Powder Metallurgy, 35,619-624(1988)
2. M. Koishi and T. Tsunoda, Surface chemistry of powder, Nikkan-Kogyo-Shinbunsya, 1978, P.59 (in Japanese)
3. H. Suzuki, S. Takagi, H. Morimitsu and S. Hirano, "Microstructure control of porous silica glass with monodispersed spherical silica particles", J. Ceram. Soc. Japan, 100, 272-275(1992)

PROSPECTS FOR OBTAINING A SUPERCONDUCTING FILTER TO PURIFY OXYGEN FROM ARGON

Yuichi SAWAI and Kozo ISHIZAKI, Nagaoka University of Technology, Nagaoka, Niigata 940-21, Japan
Shigeki HAYASHI, Osaka Sanso Kogyo Ltd., 670 Kami, Sakai, Osaka 593, Japan,
and
Ravi JAIN, The BOC Group, 100 Mountain Ave., Murray Hill, New Jersey 07974, USA

A possible superconducting filter, through which argon can pass and oxygen can not pass, is discussed theoretically. Oxygen and argon are paramagnetic and diamagnetic materials, respectively. The mixture of oxygen and argon can be separated in a magnetic field higher than 4 T^2/m of BdB/dx, where B is the magnetic flux density and dB/dx is its gradient. Such a magnetic field can be obtained by a superconducting filter. Because magnetic flux does not pass through a superconducting body of the filter, and instead goes along the penetrating pores, B in the penetrating pores and dB/dx on the surface of the filter are very high, which allows separation of argon from oxygen.

INTRODUCTION

To produce pure oxygen, there currently is not conventional way to remove argon impurities from oxygen. It is known that oxygen and argon exhibit paramagnetic and diamagnetic behavior, respectively. A magnetic force, F, acting on a material in a magnetic field is given by

$$F = M\frac{dH}{dx} = \frac{M}{\mu_0}\frac{dB}{dx} \qquad (1)$$

where H is the magnetic field, B is the magnetic flux density, μ_0 is the relative permeability (1.26×10^{-6} in SI unit), and M is the magnetization; which is given by

$$M = \chi_m H = \frac{\chi_m}{\mu_0} B \qquad (2)$$

where χ_m is a magnetic susceptibility. A magnetic susceptibility of paramagnetic material (oxygen) is positive and of a diamagnetic material (argon) is negative. According to Equations (1) and (2), the force acting on oxygen is opposite to the force acting on argon. In other words, a mixture of oxygen and argon can be separated in a magnetic field. In the present paper, a superconducting filter to remove argon from oxygen is discussed. Such a filter can be obtained by applying the Meissner effect of superconducting materials.

DISCUSSION

According to Equations (1) and (2), the magnetic force acting on a magnetic material is given by

$$F = \frac{\chi_m}{\mu_0^2} B \frac{dB}{dx} \qquad (3)$$

The magnetic susceptibility of liquid oxygen is 4.74×10^{-9} (SI unit)[1].

If 4 T^2/m of *BdB/dx* is applied, the magnetic force acting on liquid oxygen in a unit volume is

$$F = \frac{4.74 \times 10^{-9}}{(1.26 \times 10^{-6})^2} \times 4 = 12\ 000 \quad (N/m^3) \tag{4}$$

This is the force which can lift up liquid oxygen itself. According to this method, a strong magnet with 4 T^2/m of *BdB/dx* is necessary. But it is difficult and expensive to obtain a high value of *BdB/dx* by using a standard permanent magnet or a coil.

A porous superconducting material is useful to obtain a *BdB/dx* value higher than 4 T^2/m. The Meissner effect shows that the magnetic flux goes outside of a superconducting body and does not pass through it. If there is a hole in a superconducting body as shown in Fig. 1, a magnetic flux passes through the hole.

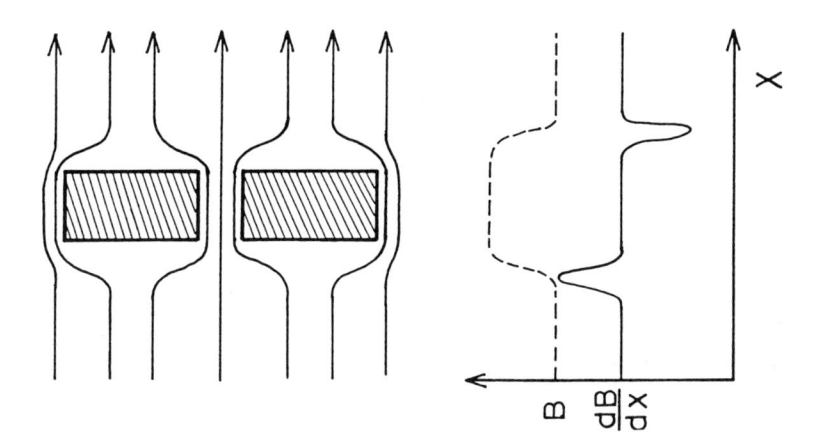

Figure 1 Superconducting material (in the superconducting state) in a magnetic field. Magnetic flux passes through a hole and outside of the superconducting body. *B* in the hole and *dB/dx* on the surface increase.

The magnitude of the magnetic flux density B in a hole is higher than that of outside, and the gradient dB/dx on the surface of the superconducting material becomes very high.

The magnetic flux around the superconducting filter in the normal state is shown in Fig. 2(a). If the filter in Fig. 2(a) is cooled down below the critical temperature T_c, the magnetic flux behaves as shown in Fig. 2(b). The magnetic flux passes through the penetrating pores. The magnitude of dB/dx increases to a maximum on the surface.

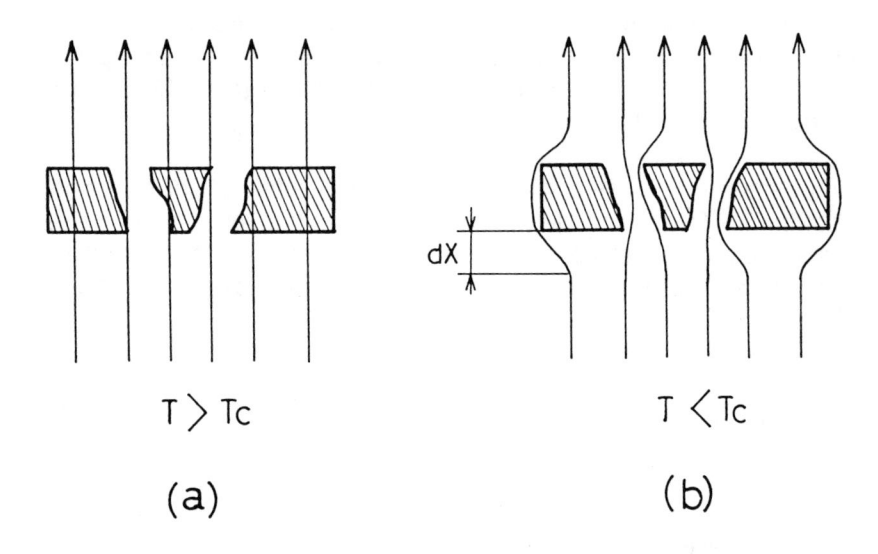

Figure 2 The superconducting filter (a) in the normal state and (b) in the superconducting state.

The magnetic field in the pores (B_{pore}) is always higher than that of outside (B_{out}). B_{pore} is given by

$$B_{pore} = K B_{out} \quad (K>1) \tag{5}$$

The value of K is a function of the configuration of pores, and it is about 4–10 when the open porosity is about 0.3 (see appendix). It means B_{pore} is at least 4 times higher than B_{out} because B is proportional to the density per unit area of the magnetic flux line shown in Fig. 2. A computer simulation of the magnetic field around superconducting materials shows that the length dx, shown in Fig. 2(b), is less than 10^{-2} m. Therefore if B_{out} is 0.1 T, B_{pore} is about 0.4–1 T and dB/dx on the surface of the filter is about 30–90 T/m. This simple calculation shows that a sufficient BdB/dx to separate oxygen and argon is possible to obtain by using the superconducting filter, and 0.1 T of B created by a standard permanent magnet or a coil.

Up to now we assumed that a magnetic flux does not pass through superconducting materials. But in a magnetic field higher than the lower critical magnetic field H_{C1}, some part of B_{out} passes through the superconducting body as shown in Fig. 3 (compare with Fig. 2(b)). It means that the magnitude of B_{pore} decreases and dB/dx also decreases. Such a state is called the mixed state. In the mixed state, about $B_{out}/3$ passes through the superconducting body, and K decreases. It can be solved by applying higher B_{out} or using a superconducting material with high H_{C1}. Therefore, the material with high critical temperature T_C, H_{C1} and H_{C2} is necessary. $YBa_2Cu_3O_7$ (123 phase) has 90 K of T_C, H_{C1} and H_{C2} are 0.1 and 90 T at 77 K respectively[2]. The 123 phase is suitable for this purpose.

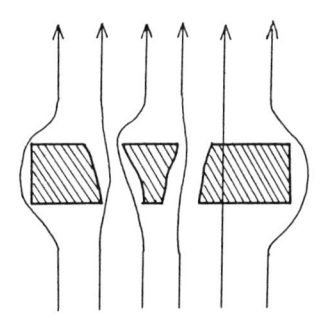

Figure 3 The superconducting filter in the mixed state (in the magnetic field $H_{C1}<H<H_{C2}$). Some magnetic flux pass through the superconducting body.

CONCLUSION

It has been theoretically discussed that to remove argon impurities from oxygen is possible by using a superconducting filter in a magnetic field. If the superconducting filter in a magnetic field is cooled down below the critical temperature, a magnetic flux passes through only penetrating pores and the gradient of the magnetic flux dB/dx on the surface of the filter becomes very high. Oxygen and argon are paramagnetic and diamagnetic materials, respectively, and it is possible that the mixture of both gases can be separated in a magnetic field higher than 4 T^2/m of BdB/dx. $YBa_2Cu_3O_7$ is one of the suitable materials for the superconducting filter.

REFERENCES

[1] S. Iida, Magnetic Field in the Materials, p. 435, Electromagnetic volume II, second edition, Maruzen, Japan, 1971
[2] K. Fueki and K. Kitazawa, Discovery of the superconducting oxide in Bi-Sr-Ca-Cu-O system, p.55, Chemistry of Superconducting Oxides, first edition, Kodansha, Japan, 1988
[3] R. Kondo, Porosity and thermalconductivity, p. 251, Porous Materials, first edition, Gihodo, Japan, 1991

APPENDIX

Calculation of the pore configuration factor K from the open porosity is as follows. Now we are assuming that the filter is thin enough, so the cross section area of a penetrating pore is almost constant through the pore. Shape of a penetrating pore can be assumed straight. First, K is given by

$$K = \frac{S_{tot}}{\Sigma S_{pore}} \qquad (1)$$

where S_{tot} is a projection area of the filter and S_{pore} is a cross section area of a penetrating pore. The open porosity of the porous material is given by

$$P_o = \frac{V_o}{V} \qquad (2)$$

where V_o is the total volume of the open pore and V is the volume of the porous material. The volume of the penetrating pores porosity V_p, which is given by $V_p = \alpha V_o$, where α is about 0.3–0.8[3], and K can be calculated as

$$K = \frac{S_{tot}}{\Sigma S_{pore}} \propto \frac{V}{V_p} = \frac{1}{\alpha P_o} \qquad (3)$$

Assuming that the open porosity is 0.3, this value is calculated about 4–10.

EFFECT OF POROUS MATERIALS ON THE GENERATION AND THE GROWTH OF BUBBLES IN AERATION

Nobuyuki UENO
Nakajima Corporation NEU Division
1135-1, Nakamori, Danshi, Takamatsu-shi, Kagawa 761, Japan

This paper demonstrates that generation and growth of bubbles are closely related to the wetting behavior of a liquid on porous materials during aeration. The wetting behavior was evaluated as the contact angle of pure water (10 μl) on the surface of each material. In fundamental tests, the relationship between the generation of bubbles and the contact angle of porous media are discussed using a silicone oil treated glass capillary tube and an untreated one as examples. The contact angles were 24.3° and 100.1° respectively. In the actual bubbling test, two typical porous media of the same structure were used. The test specimens were made according to the same method as used for grinding wheel manufacture. The abrasive grains of fused alumina #80 were bonded with phenolic resin and glass. The contact angle of glass (24.3°) widely differed from the contact angle of phenolic resin (100.7°). The relationship between the outflow and the internal pressure of the media was investigated. In order to observe the momentary bubble generation, a high-speed video camera was used. Test results show that the bubble size grows as the contact angle increases.

INTRODUCTION

Aeration is used for cleaning water and dissolving gases such as oxygen and ozone. The materials of porous media used for aeration are plastic, metal, or ceramic. They are selected according to their mechanical, thermal and chemical properties respectively. During aeration, dissolving efficiency is dependent on the bubble size. As to the relationship between the dissolving oxygen and the bubble size, it has been observed in experiments that the smaller the bubble, the higher dissolving efficiency becomes. Thus, bubble size is one of the most important factors in aeration. It is widely known that the bubble size is controlled by the pore size of the porous media, and is in proportion to the pore size. If the generation and the growth of bubbles are examined in detail, it is necessary to consider the interface of solid, liquid and vapor. The bubble is generated in the solid-liquid interfacial area, so that the bubble size depends on the solid-liquid interfacial energy ; i.e., the wetting behavior of a liquid on porous material (the contact angle). From this point of view, it is considered that materials of porous media influence the generation of bubbles considerably. However, little detailed work of the relation between the bubble size and porous material has been reported. Therefore, in the fundamental test of this paper, the generative mechanisms of bubbles and the effects of contact angle are discussed. In order to observe the relationship between the generation of bubbles and the materials of porous media, glass capillary tubes were used as the simplest for aeration. In the actual bubbling test, two porous media are made of different materials for an aeration model, and the bubbles which flow out from each porous media are observed. This paper focuses on the aeration bubbles.

EXPERIMENTAL PROCEDURE

Measurement of the Contact Angle

The contact angle can be observed from the configuration of a liquid on a solid surface. Therefore, the measurement was made by reflected mirror and micrometer equipped camera, and analyzed accurately with computer. The liquid (25°C, 10μl pure water) was quietly dropped on the surfaces of each specimen using the micro pipette. The surfaces of the specimen were cleaned by degreasing cleaner and ultrasonic washing apparatus. Figure 1 shows the process of measurement. The contact angles of porous materials of the actual bubbling test were not measured directly, so that the general borne-silicate glass plates (76mm×26mm×1.3mm) were used instead of the glass bonded materials. The specimen (ϕ30×3mm) of phenolic resin bonded materials was made by hot pressing at 180 °C in 30 min.

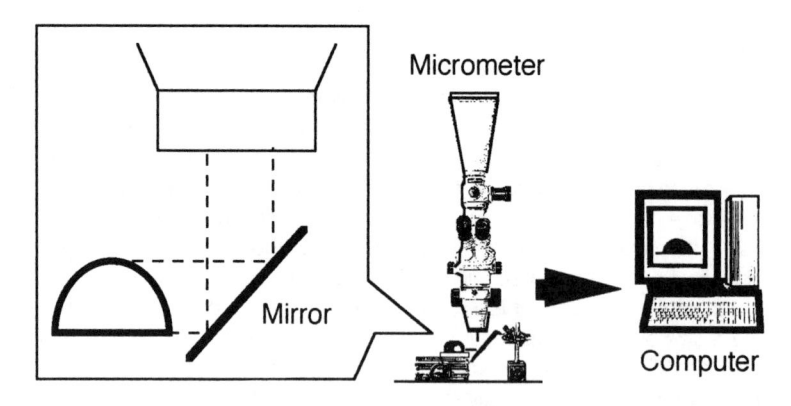

Figure 1 Measurement of contact angle

Observation of Bubble on the Glass Capillary Tube

Figure 2 shows the experimental apparatus of the fundamental aeration test using two glass capillary tubes with 0.5 mm diameter holes. The edges of each tube were polished to mirror finish with fine diamond grains. The polished edge of one tube was treated with silicone oil to have a large contact angle, and then was dried at 120 °C for 1 h. The other glass capillary tube was untreated.

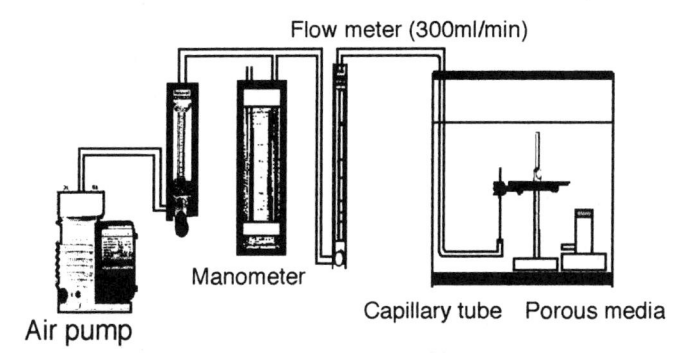

Figure 2 Apparatus of aeration test

Figure 3 shows the mechanism of the water proof effect of silicone oil. In this reaction, R and R' are in the high molecular radical and hydrophobic groups, so that the surface of the silicone treated glass capillary tube changes the surface of hydrophobic property. The photograph frames of a bubble from the glass capillary tube were taken by high-speed video camera. The bubble portraits were treated graphically by computer.

Figure 3 Mechanism of waterproof effect of silicone

Porous Materials

Actual Bubbling Test

In the actual bubbling test of porous media, two specimens ($\phi 30 \times 3$ mm) were made according to the same method used for grinding wheel manufacture. The specimen was composed of approximately 44 V% bone, 22 V% bond, and 34 V% pore. The aggregate material was fused alumina #80. The bond materials were glass and phenolic resin that had a similar contact angle to that of the silicone treated glass. The porous media were made to fix these specimens on one side of the polyvinyl chloride pip. The observation of this test was carried out by the same apparatus as for the fundamental aeration test (Fig. 2). The volume of air flowing through the porous media was controlled by the valve of the flow meter between the air pump and the differential manometer. The bubbles were observed photographically.

RESULTS AND DISCUSSIONS

Contact Angle

Figure 4 shows the liquid shapes and values of the contact angle of each specimen. Here, the values of the contact angles were calculated using analytical computer co-ordinates. In each specimen, the difference between right and left contact angle was due to the gravitational effect and the inhomogenous wetting of the liquid and the specimen. The contact angle of silicone treated glass and phenolic resin bonded were higher than that of untreated glass. If one considers the solid-liquid interfacial energy (Υ_{SL}), it shows that Υ_{SL} of silicone treated glass (or phenolic resin bonded) is higher than Υ_{SL} of untreated glass (as shown in Fig. 9).

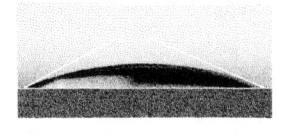

26.08°   22.44°
(a) Untreated glass (glass bonded)

97.03°   103.21°
(b) Silicone treated glass

99.29°   102.06°
(c) Phenolic resin bonded

Figure 4 Comparison of contact angle

Fundamental Test Using Glass Capillary Tube

Figure 5 is a 0.03-second interval photo of the variation in bubble shapes which flow out from the glass capillary tube at air flowing above 30 ml / min. In this continuous test, four patterns of the photo are repeated. It should be noted that there is significant difference between the contacting surface on glass capillary tube of untreated glass and silicone treated glass. The generation and the growth of a specific bubble were observed by the regulation of the air flowing valve.

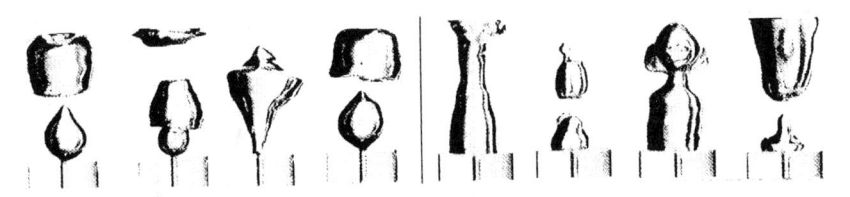

(a) Untreated glass   (b) Silicone treated glass
Figure 5 Variation in bubble shape at continuous bubbling test

Figures 6 and 7 show the cycle of these bubbles each which was generated and then flew out from each glass capillary tube. The bubble on silicone treated glass grew remarkably in comparison with the bubble on the untreated one. In order to express the variation in Fig. 6 and 7, the individual bubble areas of the momentary photo was calculated by computer. Figure 8 plots the variation in the bubble area. The bubble cycle times of untreated glass and silicone treated glass were 0.15 s and 1.08 s, respectively. The last bubble area of silicone treated glass was 4.13 times as large as the untreated glass. Assuming that these bubbles are a perfect sphere, the comparison of the bubbles are shown in Table 1.

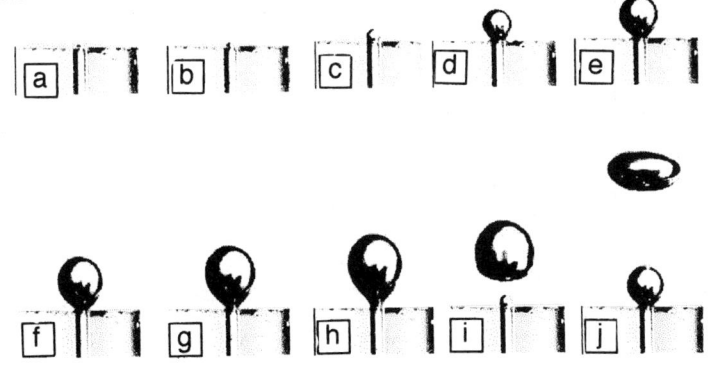

Figure 6 Variation in bubble on untreated glass

Porous Materials   347

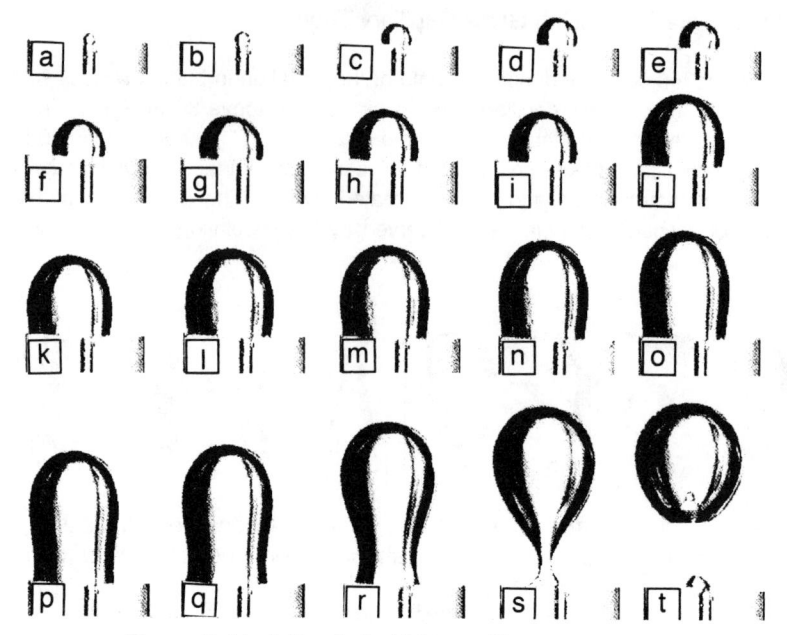

Figure 7 Variation in bubble on silicone treated glass

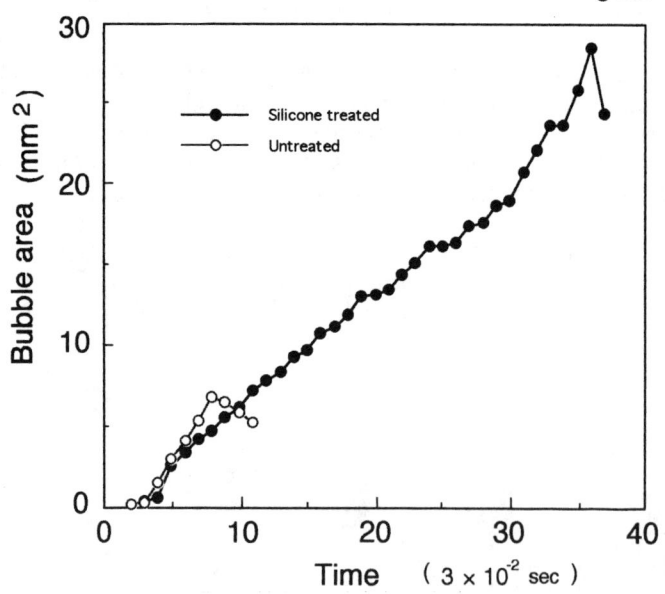

Figure 8 Variation in cross-sectional area of each bubble

Table 1. Comparison of each Bubble

	Untreated	Silicone treated
Bubble size (Diameter mm)	2.96	6.02
Volume (mm^3)	13.63	114.46
Surface area (mm^2)	27.60	114.00
Bubble number ratio	8.4	1
Total surface area (mm2)	231.84	114.00

Table 1 shows that the total surface area of untreated glass is about 2 times as much as that of the silicone treated glass. If one considers that the dissolving efficiency is dependent on the total surface area of bubbles, the dissolving efficiency of the untreated glass doubles.

This large difference between the untreated glass and silicone treated one can be explained in figure 9 and the relation

$$\gamma_{SV} = \gamma_{SL} + \gamma_{LV} \cos\theta$$
$$\gamma`_{SL} = \gamma`_{SV} - \gamma`_{LV} \cos\theta$$

as follows: the solid-liquid interfacial energy ($\gamma`_{SL}$) is increased by silicone treatment, therefore, a force is required to extend the contacting surface on the glass capillary tube. Consequently, the bubble grows largely as if it has grown out of a large pore size tube.[1],[2]

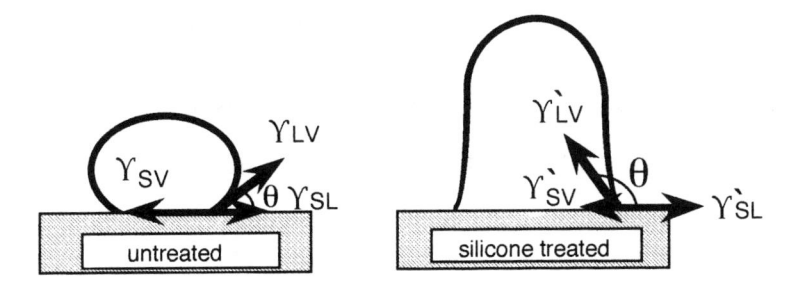

Figure 9 Comparison of interfacial energy

Porous Materials

Actual Bubbling Test

Figure 10 shows the comparison of microstructures of glass bonded and phenolic resin bonded media. The average grain size and pore size were 250 μm and 150 μm, respectively. As shown in this figure, both media were similar. Figure 11 plots the relationship between the outflow and the internal pressure with incoming air of 100, 200 and 300 ml / min. It can be seen from this figure that the phenolic resin bonded media is superior to the glass bonded in air permeability. This can be explained as follows : the solid-vapor interfacial energy (γvs) of the glass bonded media is higher than phenolic resin bonded one, and the growth of the bubble is retarded by this energy (Fig.9). Therefore, the internal pressure of glass bonded media is higher than that of phenolic resin bonded. Figure 12 shows the comparison of bubbles of each media.

(a) Glass bonded   (b) Phenolic resin bonded

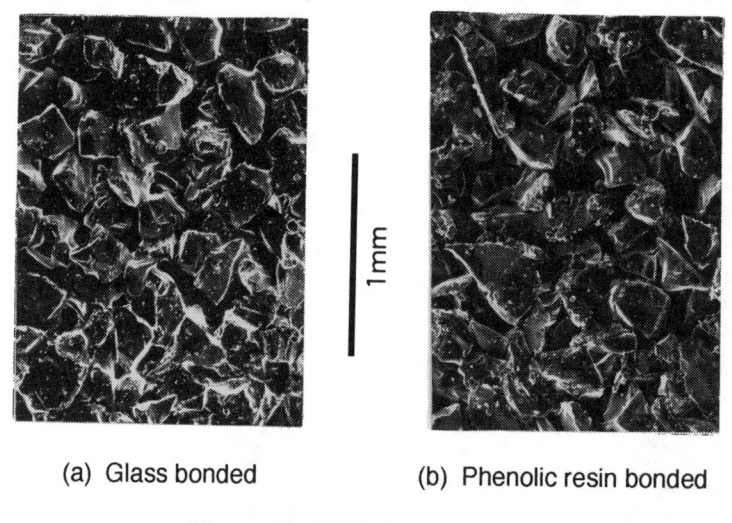

Figure 10 SEM photographs of each media

Figure 11 Comparison of internal pressure and outflow of each media

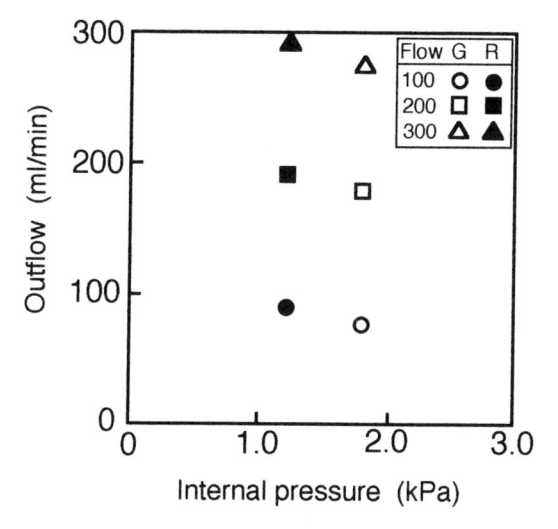

G : Glass bonded
R : Phenolic resin bonded

(a) Glass bonded   (b) Phenolic resin bonded

Figure 12 Comparison of the bubbles of porous media

CONCLUSIONS

From the fundamental test using a glass capillary tube, it was clear that the generation and the growth of the bubble were influenced significantly by surface chemical properties of the porous materials.
The results of this study are summarized :

(1) The bubble sizes from the high-contact angle materials were larger than those from the low-contact angle materials.

(2) The internal pressure of the porous material was not related to flowing air, and was determined by the structure and the material.

REFERENCES

1. W. D. Kingery,"Role of Surface Energies and Wetting in Metal-Ceramic Sealing," Bull. Am. Ceram. Soc., 35, 108 (1956).
2. W. D. Kingery,"Plausible Concepts Necessary and Sufficient for the Interpretation of Ceramic Grain Boundary Phenomena," J. Am. Ceram. Soc., 57, 83 (1974).

DEVELOPMENT OF POROUS CERAMICS FOR A NEGATIVE-PRESSURE DIFFERENCE IRRIGATION SYSTEM

Minoru KUBOTA and Takayuki KOJIMA*
Kyuemon Co., Ltd. 1050-1 Magarikawa, Nishiarita-cho, Nishimatsuura-gun, Saga-ken 849-41, JAPAN
*Saga University, Faculty of Agriculture,
1 Honjo-machi, Saga-shi, Saga-ken 840, JAPAN

This study was carried out to utilize porous ceramic materials in agricultural fields. The negative pressure difference system of porous ceramic pipes was introduced. An economical porous ceramic pipe of uniform open pores was fabricated by the sintering of low grade silica powder. This system could be available for all plants without exception. For example, the harvest period of lettuce was reduced by 30 days and the growth of fine roots on porous materials was promoted compared with conventional plantations.

INTRODUCTION

The principle of negative pressure difference irrigation system was first introduced by B.E.Livingston in 1930. Since then, this method has been studied for practical purposes by many researchers[1], but it did not come into wide use even at present. As compared with conventional irrigation, it may certainly be said that this method is very economical because of its minimum water consumption.

In this study, the obstacles to success were revealed in details and the methods of wide application were proposed.

OUTLINE OF NEGATIVE PRESSURE IRRIGATION SYSTEM

As the water in the soil is partly absorbed by the plant and partly evaporates on the surface of land, a water pressure (h_s) in the soil increases. When a porous pipe in the neighborhood of root zone in the soil is saturated with water and the water pressure in the porous pipe reaches a constant negative pressure (h_p), a negative pressure difference ($\Delta h = |h_s - h_p|$) is generated between the soil and the porous pipe.

If the negative pressure of water in the soil is higher than that in the porous pipe, water transfers from the porous pipe to the soil. The schematic is shown in Fig. 1.

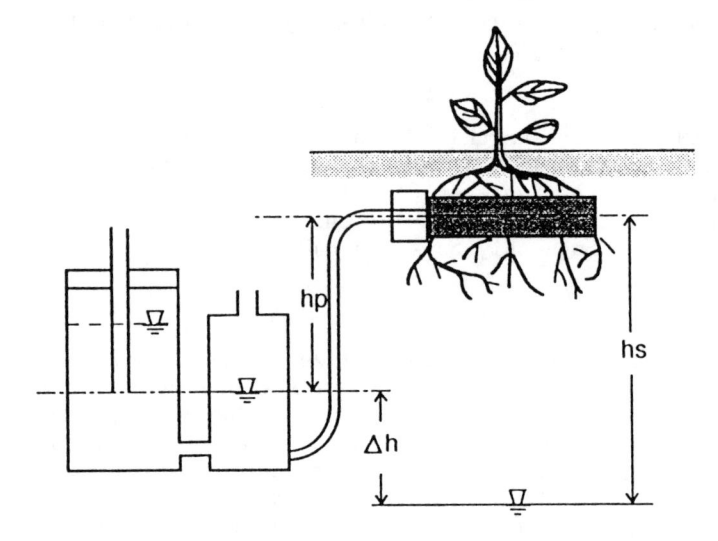

Figure 1 Schematic of negative pressure irrigation system.

PRELIMINARY TEST ACCORDING TO CONVENTIONAL SYSTEM

A preliminary investigation was carried out to determine any limitations of the negative-pressure system.

Negative Effect of Dissolved Gas

A negative pressure is maintained by the surface tension of water in the open pore. A dissolved gas in the water in the ceramic porous pipe and in the piping tube becomes separated due to a temperature change and remains in the pipe. Consequently, a state of negative pressure no longer exists and the water transfer stops.

Economic Fabrication of the Porous Pipe

Porous ceramic pipes are usually made of earthen ware, unglazed pottery, or airstones made of abrasive grains. Authors concluded that one of the most important thing to

develop the negative preesure difference irrigation system was the manufacture of an appropriate porous pipe.
Necessary characteristics for the porous pipe are:

a) Uniform open pore and no gas absorption by the negative pressure difference.
b) Higher porosity enough to supply water for the plant.
c) Cheaper cost.

Silica of low grade was sintered with 40% porosity. This porous ceramic pipe is enough to supply water in this irrigation system and lower cost has the advantage of agricultural use. Pore distribution of this newly developed porous materials is shown in Fig. 2.

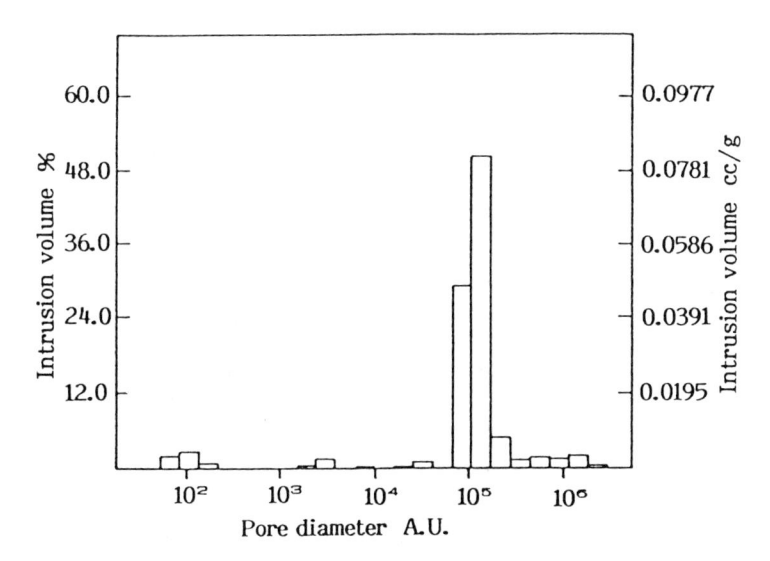

Figure 2 Pore distribution of newly developed porous material.

IMPROVEMENT OF NEGATIVE PRESSURE DIFFERENCE IRRIGATION SYSTEM

Concerning the irrigation system itself, further tests were done in order to remove the gas more rapidly. The porous ceramic pipes were linked in series by using vinyl chloride tubes. By means of siphonage, a state of negative pressure was generated and it was possible that the water circulated easily and the developed gas was removed rapidly.
The schematic of this system is shown in Fig. 3. The relation

among individual negative pressure difference is expessed as follows:

$$h_p(1) + h_p(2) + h_p(3) > H_p(4a) + H_p(4b) + H_p(4c) + H_p(4d)$$

In case of a conventional system, it is necessary to set the system on a flat plateau with few undulations, because the negative pressure difference h_p must be strictly controlled. In case of this improved system, however, the selection of the farm place becomes very simple.

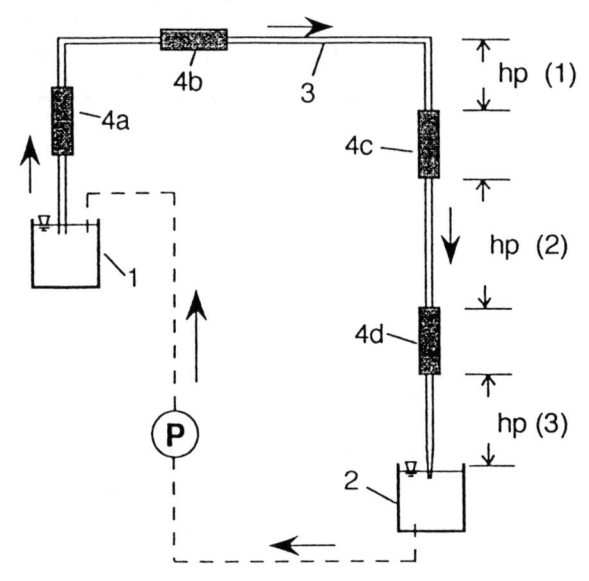

Figure 3 Improved negative pressure irrigation system.

$$h_p(1) + h_p(2) + h_p(3) > H_p(4a) + H_p(4b) + H_p(4c) + H_p(4d)$$

Both construction and running costs are very economical. For instance, porous ceramic pipes are connected by vinyl chloride tubes of 8 mm diameter, and the electric power is only 10 watt for the experimental farm of 270 m^2.
There are two disadvantages for this system. It is necessary to set a pump to transfer the water from one tank to another tank. However, this is not always expensive. Another important disadvantage is a nonuniformity in negative pressure

for each porous ceramic pipe, resulting in a difference in water supply. This problem was worked out by selecting the soil material.

In present agriculture, an addition of compost or of soil modifier and plowing labor are indispensable to prevent the soil solidification and the air shortage of the soil. When this system was used, it was possible to adopt to remove the fine soil, which has been thought to be unfavarable for agriculture, instead of the soil of lager particle size. Fine soil has capillary action and forms the water holding zone when contacting with the porous ceramic pipe.

PRACTICAL CULTIVATION TEST

The test cultivation using negative pressure difference irrigation system was done in order to judge whether this method is proper or not concerning such various plants as rice, corn, soybean, tomato, cucumber, eggplant, melon, radish, carrot, cabbage, lettuce, grape perry, orange, cherry, dahlia, bamboo, and pine. It became clear that this method is suitable for all plants without exception.

Furthermore, tests in the experiment station of Saga prefecture were intended for practical use concerning lettuce, cabbage, tomato, and strawberry. For the harvesting of lettuce, it takes usually 60 days. When this system was used, it became possible to cut the cultivation time by 30 days. It was made clear that the soil of 400cc enabled the cultivation of lettuce, cabbage and strawberry without hard labor.

EFFECT OF WATER ON PLANT GROWTH

It became clear that the type of water exerted influence on plant growth. Recently, the water treated by magnetism is a matter of interest for plant growth. The water transmitted through the porous ceramic pipe seemed to change the quality of water, even though its mechanism is not yet clarified. Figure 4 shows the effect of type of water on the growth of lettuce in the irrigation system.
Tests on the water absorption by the root group were performed for tomatoes. The schematic of plant growth is shown in Fig. 5 and results of the tests are shown in Table 1. The root group, whose diameter is 3 mm and over, is said to be a degenerated organ and it dose not have any suction function but only serves as the passing organ of water. A root hair is a very important organ to the growth of plant.

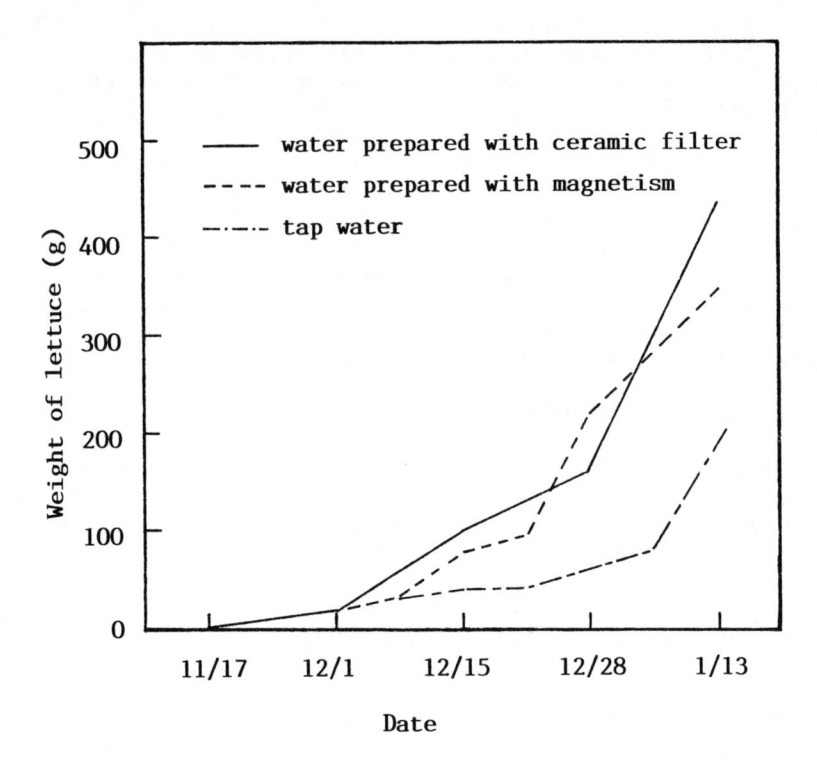

Figure 4 Growth of lettuce using several kinds of water.

Table 1. Distribution of Root Diameter (mm) and Weight Per Stump (g).

Farm land		Number per stump				Weight per stump	
	Thickness	over 5	5-3	3-1	under 1	fresh	died
Improved	Number	0.8	0.3	39.0	538.5	578.0	38.5
Conventional		0.8	4.0	34.8	67.4	107.0	8.8

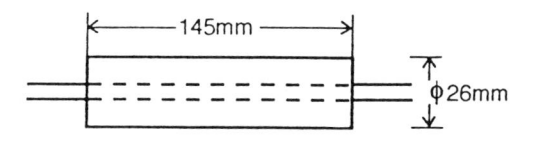

(a) Shape of porous ceramic pipe.

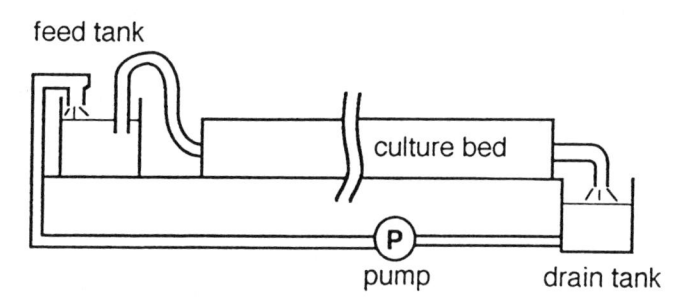

(b) Schematic diagram of newly developed irrigation system.

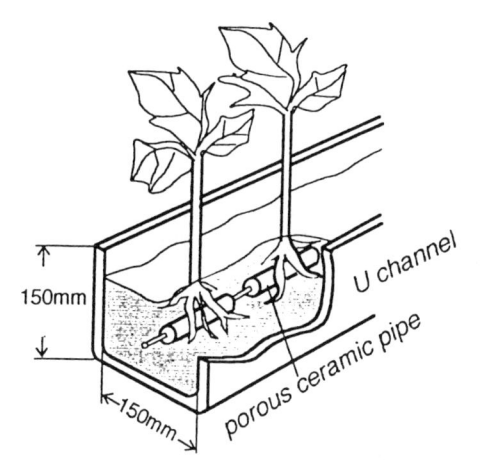

(c) General view of root zone irrigation system using porous pipe.

Figure 5 General views of improved negative pressure difference setup for cultivation of tomato.

CONCLUSION

In order to put "the negative pressure difference irrigation system" to practical use, improvements in the porous ceramic pipe and its arrangement were made. An economical porous ceramic pipe of uniform open pores was made by the sintering of low grade silica. The uniform open pores prevented the absorption of gas and kept a constant negative pressure. The pipes were connected in series and enabled the smooth passage of water regardless of the unevenness of farm land. Tests, expanded to many kinds of plants, showed the effectiveness of this improved system. This method can be applied to "afforestation" or "converting in green tract" of deserts.

REFERENCE

1) Z.Kato and S.Tejima, "Theory and Fundamental Studies on Subsurface Irrigation Method by Use of Negative Pressure", Nougyodoboku-ronbunshyu, Oct., 46-50(1982).

MERCURIC ION SENSOR WITH FIA SYSTEM USING IMMOBLIZED MERCURIC REDUCTASE ON POROUS GLASS

Motohiro Uo, Masahiko Numata*, Isao Karube** and Akio Makishima
Department of Materials Science, Faculty of Engineering, University of Tokyo
* Central Research Laboratory, Dowa Mining Co.,Ltd.
** Research Center for Advanced Science and Technology, University of Tokyo.

The mercuric ion sensor was prepared with immobilized mercuric reductase covalently coupled to porous glass. The mercuric reductase was immobilized on an arylamino derivative of porous glass with azo coupling. The mercuric ion sensor system was assembled by a flow injection analysis (FIA) system. With an enzymatic cycling reaction constructed by mercuric reductase and catalase, concentrations of mercuric ion higher than 0.025 µM (5 ppb) were detectable by this mercuric ion sensor with 0.2 ml of sample solution.
Thus, this mercuric ion sensor is suitable for mercuric detection in low concentrations using small sample volumes without any pre-treatment of the samples.

INTRODUCTION

Immobilized enzymes are playing important roles in biosensors and bioreactors. As the carrier materials for immobilized enzymes, many kinds of organic and inorganic materials have been studied. Porous silica glass is one of the typical inorganic carriers for enzymes and many kinds of enzymes were immobilized and their properties are reported in Table 1. In these cases, enzymes were immobilized with covalent bonding and typical processes are shown in Fig.1. Compared with other carrier materials, a porous glass carrier has many advantages such as its thermal and chemical durability, mechanical strength and controllability of its microstructure.

On the other hand, mercury and most of its derivatives are very toxic. Thus the development of very sensitive and rapid measurement of Hg in the environment is important. Currently, the cold-vapor atomic absorption method is widely used to detect Hg. This method is based on the chemical reduction of Hg^{2+} ions to elemental Hg with a reducing reagent as SnCl$_2$. The method is very sensitive for Hg ion; however, a complex pretreatment of samples, sophisticated equipment, and highly skilled personnel are necessary. In addition, continuous measurements are not easy.

Recently, the authors have reported about the immobilization of mercuric reductase onto a porous glass carrier and the application of the immobilized enzyme for the mercuric ion sensor[1]. This sensor makes it possible to analyze Hg^{2+} ion without pre-treatment of the sample. In the next section, the details of the mercuric reductase and the mercuric ion sensor are described.

Table 1. Enzymes Immobilized onto Porous Glass Carriers.

1. OXIDOREDUCTASE
lactoperoxidase	sulfhydryl oxidase
alcohol dehydrogenase	glucose oxidase
galactose oxidase	xanthine oxidase
L-amino acid oxidase	lipoamide dehydrogenase
urate oxidase	catalase
hydrogenase	mercuric reductase

2. TRANSFERASE
1,4-α-glucan phosphorylase

3. HYRDOLASE
nuclease	steroid esterase
lipase	alkaline phosphatase
amylase	β-amylase
glucoamylase	cellulase
β-galactosidase	invertase
β-glucuronidase	amyloglucosidase
naringinase	leucise aminopeptidase
aminopeptidase	chymotrypsin
protease	papain
urease	aminoacylase

4. LYASE
carbonic anhydrase   poly(methoxygalacturonide)lyase

5. ISOMERASE
glucose isomerase

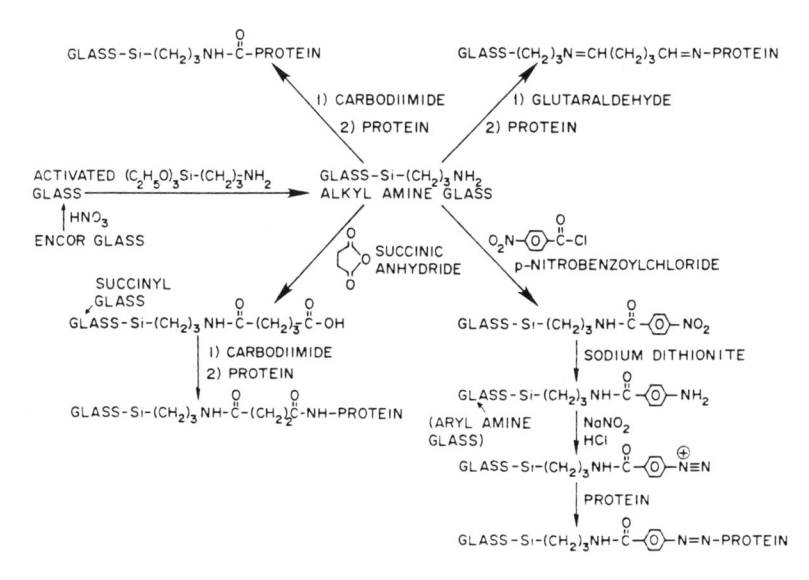

Figure 1 Methods of immobilization of enzyme on porous glass carrier[3]

MERCURIC ION SENSOR WITH IMMOBILIZED MERCURIC REDUCTASE

The mercuric reductase (NADPH:mercuric ion oxidoreductase : EC1.16.1.1) plays an important role for detoxification of mercuric compounds in some bacteria and it was purified and characterized by Fox et.al.[2]. The overall reaction catalyzed is as follows:

$$NADPH + RS\text{-}Hg\text{-}SR + H^+ = NADP^+ + Hg^0 + 2RSH \tag{1}$$

In this reaction, dimercaptide of mercuric ion (RS-Hg-SR) is reduced with NADPH and NADPH itself is oxidized to NADP$^+$. Thus, the concentration of Hg^{2+} ion can be estimated by measuring the concentration change of NADPH with various methods. In this study, mercuric reductase was immobilized onto porous glasses using various methods and their properties were studied. The immobilized mercuric reductase with azo-coupling showed a high relative activity and stability compared with the soluble enzyme. Using this immobilized mercuric reductase, the Hg^{2+} ion sensor with flow system was assembled as shown in Fig.2. The sample solution and buffer solution containing NADPH were carried into the immobilized enzyme column by peristaltic pumps and the change in NADPH concentration caused by the enzymatic reaction (as Eq.1) was detected by fluorometry.

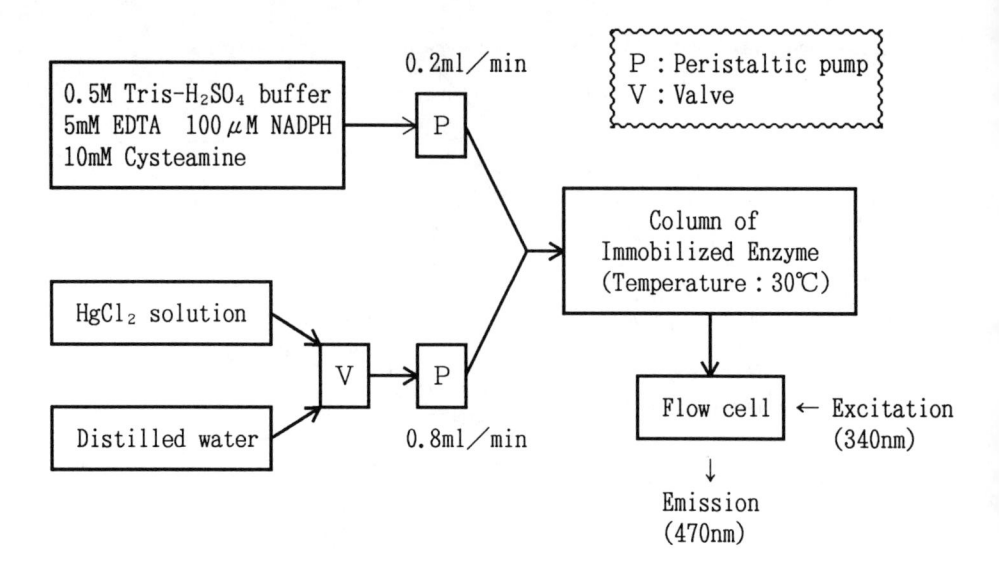

```
0.5M Tris-H₂SO₄ buffer        0.2ml/min         P : Peristaltic pump
5mM EDTA   100 μM NADPH  →  P                   V : Valve
10mM Cysteamine

HgCl₂ solution
                       → V → P
Distilled water            0.8ml/min

Column of Immobilized Enzyme (Temperature : 30℃)

Flow cell ← Excitation (340nm)
          ↓
       Emission (470nm)
```

Figure 2 Schematic representation of mercuric ion sensor with flow system.

Figure 3 shows one of the results and concentrations of Hg^{2+} ion higher than 0.5 μM (0.1 ppm Hg) could be detected. Thus, analysis of Hg^{2+} ion can be carried out by this sensor without any pretreatment of the sample or use of a complicated process.

In order to detect Hg^{2+} ion concentrations lower than 0.5 μM, an enzymatic cycling reaction with mercuric reductase and catalase was applied in this sensor. Enzymatic cycling reaction is an extremely sensitive assay using two kinds of enzymes[4]. Catalase catalyzes oxidation of elemental Hg to Hg^{2+} ion[5], thus, the enzymatic cycling reaction was described with mercuric reductase and catalase as follows:

$$\begin{matrix} NADPH \\ NADP^+ \end{matrix} \Bigg) mercuric\ reductase \Bigg(\begin{matrix} Hg^{2+} \\ Hg^0 \end{matrix} \Bigg) catalase \quad (2)$$

In this case, mercuric reductase catalyzes a reaction between the substrate (Hg^{2+}) and the indicator substance (NADPH), and catalase regenerates the substrate (Hg^{2+}). Thus, the substrate (Hg^{2+}) reacts to indicator substance (NADPH) repeatedly and the sensitivity of the assay increases compared with in the absence of catalase.

Figure 4 shows the result of this sensor with the enzymatic cycling reaction and concentrations higher than 0.03 μM (6 ppb Hg) of Hg^{2+} ion could be detected. By the cycling reaction, an excess amount of NADPH would be oxidized compared with the reaction in the absence of catalase, and the sensitivity of the sensor was increased compared with the sensor without the enzymatic cycling reactions.

Using this mercuric ion sensor, continuous measurements of mercury can be easily carried out without any pretreatment of sample. However, this system requires a relatively large sample volume (ca. 10 ml).

In this paper, in order to apply for mercury analysis in a small sample, the mercuric ion sensor was assembled with the flow injection analysis (FIA) method.

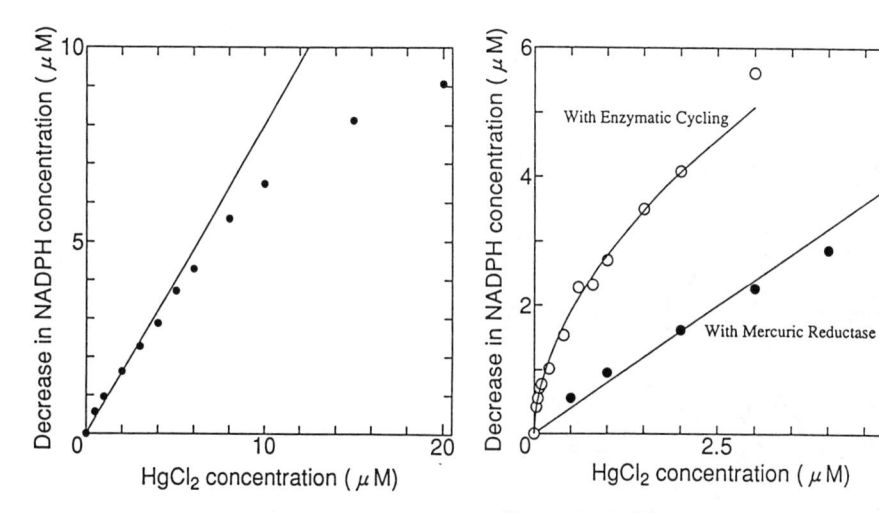

Figure 3 Calibration curve of mercuric ion sensor.

Figure 4 Calibration curve of mercuric ion sensor with enzymatic cycling and compared with the sensor with mercuic reductase

MATERIALS AND METHODS

Mercuric Reductase and Catalase

The mercuric reductase was purified by Fox's[4] method from *Escherichia coli* that was transformed by pM609 (containing mercuric reductase gene from transposon Tn501). Analytical grade of catalase (bovine lever origin), Tokyo Chemical Industries, was used without further purification.

Preparation of Porous Glass Carrier

Porous glass was prepared from sodium borosilicate glass. $10Na_2O \cdot 50B_2O_3 \cdot 40SiO_2$ (wt%) glass was treated at 580°C for 48 hours to promote the phase separation and leached in hot water for 24 hours and dried at 140°C[6]. The porosity, pore diameter and specific surface area are 55 %, 140 nm and 29 m^2/g respectively. A SEM image is shown in Fig.5 and the pore size distribution curve is shown in Fig.6. These glasses were crushed to a particle size below 0.1 mm and refluxed overnight in a 10% solution of 3-aminopropyltriethoxysilane in toluene. This alkylamine glass was refluxed overnight in chloroform containing 5% p-nitrobenzoylchloride and 10% triethyamine and washed with chloroform. The product was reacted in an aqueous solution containing 0.5 M sodium hydrogencarbonate and 0.1 M sodium dithionite at 37°C for 1 hour and boiled in an aqueous solution containing 5% sodium dithionite for 20 minutes to reduce the nitro group. The arylamino glass was washed with distilled water and dried in vacuum.

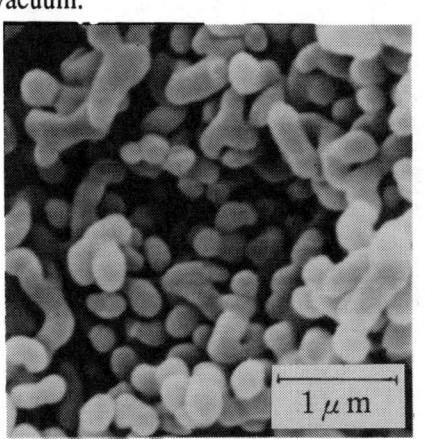

Figure 5 SEM image of porous glass carrier.

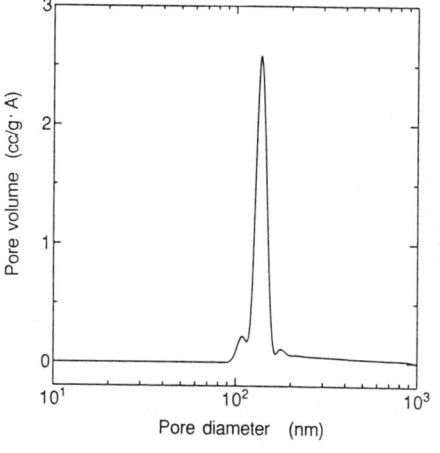

Figure 6 Pore size distribution curve of porous glass carrier.

Immobilization of Enzymes

The arylamino derivative of the porous glass was diazotized by adding 5 ml of 4 N hydro chloric acid and 50 mg of sodium nitrite and reacted at 4°C for 1 hour. The product was washed with distilled water and 0.1 M sodium phosphate buffer (pH 7.0). The mercuric reductase and catalase dissolved in 0.1 M sodium phosphate buffer (pH 7.0) was added to the glass. The amounts of immobilized enzymes were determined with photometry.

Preparation of Mercuric Ion Sensor

The mercuric ion sensor was assembled with a flow injection analysis (FIA) system as shown in Fig.7. 0.08 g of immobilized enzyme was put in the column (2 mm in diameter and 50 mm in length). The temperature of the column was maintained at 30°C. The buffer solution containing NADPH and carrier (H_2O) were carried using a plunger pump at a flow rate of 0.24 ml/min. For the measurement, 0.2 ml of the sample solution containing various contents of $HgCl_2$ was injected into the carrier flow. These solutions were mixed and carried into the column of immobilized enzyme. The NADPH concentration after enzymatic reaction was measured with optical absorption at 365 nm. The 0.2 M Glycine-NaOH buffer solution contained 10 mM cysteamine and 0.04 mM NADPH.

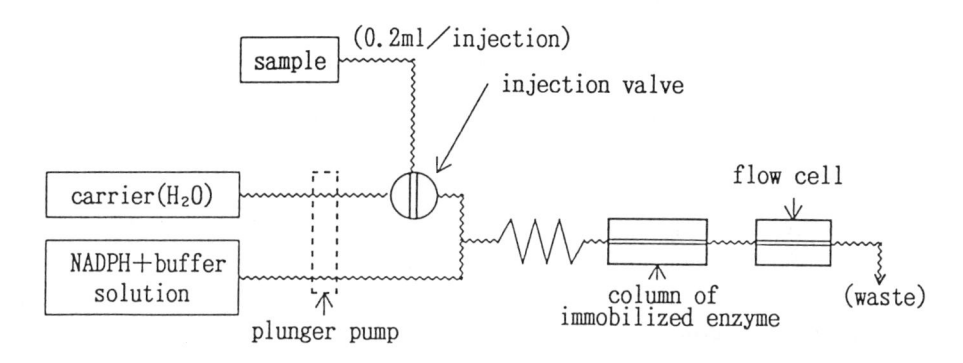

Figure 7 Schematic representation of mercuric ion sensor with FIA system.

RESULTS AND DISCUSION

The result of the Hg analysis by the FIA method is shown in Fig.8. Various concentrations of $HgCl_2$ solution (5 to 200 ppb Hg) were supplied to the FIA system and the change in absorbance (365 nm) was measured. After injection of the sample ($HgCl_2$) solution (marked in Fig.8), a decrease in absorbance at 365 nm was observed. This change in absorbance is caused by the oxidization of NADPH to NADP by Hg^{2+} ions in the presence of mercuric reductase, and it indicated the Hg^{2+} ion concentration. The response in absorbance was observed in less than 5 minutes after the injection of the sample solution and the absorbance recovered to the original value in 10 minutes. Thus, an analysis could be completed in about 15 minutes.

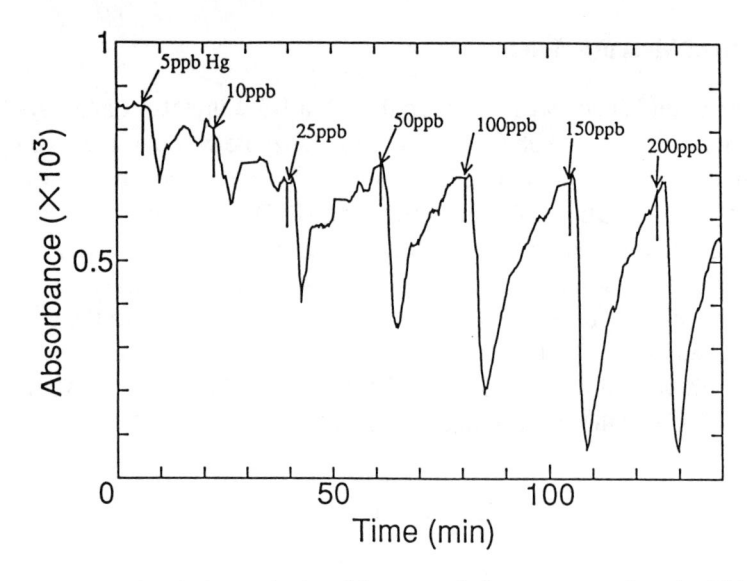

Figure 8 The result of Hg analysis with mercuric ion sensor using the FIA system.

Figure 9 shows the calibration curve for the Hg analysis. The concentration change of NADPH was converted from the change in absorbance (Fig.8) by the absorption coefficient of NADPH (3.45/mM NADPH at 365 nm). The dependence of the decrease in NADPH concentration on the $HgCl_2$ concentration was observed in the lower $HgCl_2$ range (<0.25 µM) and the minimum detectable response was obtained at 0.025 µM (5ppb).

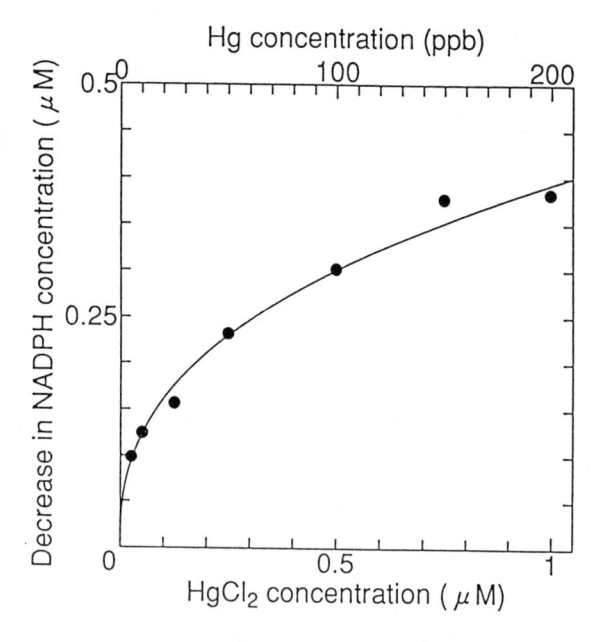

Figure 9 Calibration curve of mercuric ion sensor using the FIA system.

Table 2 shows the effect of the injected sample volume and sensitivity. The sensitivity was decreased with a decrease in the sample volume; however, half of the sensitivity was obtained with 0.05 ml of sample solution compared to the response with 0.2 ml of sample.

Table 2. Dependence of Sensitivity on Sample Volume

sample volume	0.05ml	0.10ml	0.20ml
relative sensitivity	49	73	100

As an analysis method for Hg^{2+} ion in low concentration, cold-vapor atomic absorption method is generally used. However, this method requires a large amount of sample (ca.10 ml). The mercuric ion sensor with flow system (described in section 2) also requires about 10 ml of sample solution.
Using the flow injection system, the mercuric ion sensor required a smaller sample volume. The sensitivity of the mercuric ion sensor with the FIA method was comparable to that of the atomic absorption method. The required sample volume is much smaller than that of the mercuric ion sensor with flow system and the atomic absorption method.
Wylie et.al.[7,8] proposed the application of monoclonal antibodies for the analysis of Hg^{2+} ions. This method is effective for 0.5 to 10 ppb of mercury and it is performed with only 0.1 ml. However, the stability of the antibody is a problem.

Thus, this mercuric ion sensor is suitable for mercuric detection of low concentration and small sample volume without any pre-treatment of samples.

CONCLUSION

The mercuric ion sensor was assembled with a FIA system using immobilized mercuric reductase and catalase on porous glass carrier. With this sensor, Hg^{2+} ion concentrations higher than 0.025 µM (5ppb) of mercuric ion were detectable without complex pretreatment of the sample. The required sample volume is 0.2 ml and this volume is much smaller than the volume required by other Hg analysis methods.
Thus, this mercuric ion sensor is suitable for mercuric detection in low concentrations and small sample volumes without any pre-treatment of samples.

REFERENCES

[1] M.Uo, M.Numata, M.Suzuki, E.Tamiya, I.Karube and A.Makishima, "Preparation and Properties of Immobilized Mercuric Reductase on Porous Glass Carrier," J.Ceram.Soc.Japan, **100** [4] 430-433 (1992).

[2] B.Fox and C.T.Walsh, "Mercuric Reductase : Purification and Characterization of a Transposon-encorded Flavoprotein Containing an Oxidation-Reduction-Active Disulfide," J.Biol.Chem., **257** [5] 2498-2503 (1982).

[3] D.A.Lappi, F.E.Stolzenbach, N.O.Kaplan and M.D.Kamen, "Immobilization of Hydrogenase on Glass Beads," Biochem. Biophys. Res. Commun., **69** [4] 878-884 (1976).

[4] M.Uo, M.Numata, M.Suzuki, E.Tamiya, A.Makishima and I.Karube : under preparation.

[5] M.Ogata and H.Aikoh, "Mechanism of Metallic Mercury Oxidation *in vitro* by Catalase and Peroxidase," Biochem.Phermacol., **33** [3] 490-493 (1984).

[6] M.Uo, Y.Yamashika, K.Morita. I.Karube and A.Makishima, "Phase Separation of Halogen-Containing Sodium Borosilicate Glasses," J.Ceram.Soc.Japan, **100** [1] 17-21 (1992).

[7] D.E.Wylie, D.Lu, L.D.Carlson, R.Carlson, K.F.Babacan, S.M.Schuster and F.W.Wagner, "Monoclonal Antibodies Specific for Mercuric Ions," Proc.Natl.Acad.Sci. USA, **89**, 4104-4108 (1992)

[8] D.E.Wylie, L.D.Carlson, R.Carlson, F.W.Wagner and S.M.Schuster, "Detection of Mercuric Ions in Water by ELISA with a Mercury-Speciic Antibody," Anal. Biochem., **194**, 381-387 (1991).

BIMODAL POROUS CORDIERITE CERAMICS FOR YEAST CELL IMMOBILIZATION

Hisao ABE, Hideya SEKI, Akio FUKUNAGA, and Makoto EGASHIRA*

Ceramics Research Center of Nagasaki, 605-2 Hiekoba-go, Hasami-cho, Higashisonogi-gun, Nagasaki 859-37, Japan, and *Department of Materials Science and Engineering, Nagasaki University, 1-14 Bunkyo-machi, Nagasaki 852, Japan

Performance of the bimodal porous cordierite ceramics with large pores of 53-1000 μm and small pores of about 10 μm was evaluated as an immobilization support for yeast cells by measuring ethanol fermentation activity. The activity of the immobilized cells increased with an increase in the large pore volume percent, and the maximum activity was attained with the sample with large pores of 46 vol%; the activity was about 3.5 times larger than that of the sample without large pores; whereas, the effect of the size of large pores was slightly negative. These results were able to be understood in terms of substrate diffusion and surface area of the ceramic support.

INTRODUCTION

Application of inorganic porous materials for biocatalyst supports has advanced recently in many fields including the food industry [1,2], for fermentation [3,4], and for waste water treatment [5]. The advantages of the use of ceramic supports are: feasible treatment in immobilization, good mechanical durability during actual processing, and dimensional stability of the size-controlled pore. These advantages can be hardly realized by organic compounds often used as the supports for microbes. For immobilization of microbes, it is important to know the adsorption behavior of microbes on the material. From this viewpoint, the optimum pore size for immobilization of microbes was studied [6,7]. It is also well known that the zeta potential of both microbes and ceramics affects the initial stage of adsorption [8,9]. The honeycomb ceramic monolith was used as an immobilization support for a continuous fermentation process to improve the mass transfer in the system [1,10]. Though many studies have so far been reported on inorganic supports for immobilization, not much attention has been paid to the pore structure to enhance the fluid diffusion. Therefore, the authors examined the use of the bimodal porous cordierite ceramics, and evaluated the performance as an immobilization support for yeast cells.

EXPERIMENTAL

Immobilization Support Material

The bimodal porous cordierite ceramics, with large pores of 53-1000 μm and small pores of about 10 μm, were used for an immobilization support for yeast cells. The large pores were formed by adding coal powder to the raw materials. The details of preparation of the ceramics were reported in our previous paper [11]. Table 1 shows the size and volume of large pores in the specimens. The evaluation of the pore structure was made by Hg-porosimetry and observation with reflection microscope.

Table 1. Size and Volume of Large Pores in the Specimens

sample name	A	B	C	D	E	N
large pore size	1000-500	500-212	212-125	125-53	<53	-
large pore vol.	46	46	46	46	46	0

sample name	B1	B2	B3	B4		
large pore size	500-212	500-212	500-212	500-212	(μm)	
large pore vol.	10	21	31	41	(vol%)	

A cubic sample with a side length of 6 mm, cut from a large block of ceramics was used for the immobilization experiment.

Immobilization and Activity Measurement of Yeast Cells

Considering the feasibility in treatment and storage, dry yeast (Saccharomyces celevisiae) was used in this experiment. Nutrient broth (NB), composed of glucose(20 g), polypepton (5 g), meat extract(3 g), yeast extract(2 g), NaCl(2 g), and pure water(1000 ml), was used as a medium. The principal constituent of this medium is glucose, which causes ethanol fermentation in the presence of the yeast cells. The reaction is expressed by equation (1).

$$C_6H_{12}O_6 \rightarrow 2C_2H_5OH + 2CO_2 \qquad (1)$$

As equal amounts of ethanol and CO_2 are produced, the amount of ethanol produced can be determined by measuring CO_2 volume. The yeast cells were immobilized in the support by physical adsorption. In actual procedure, NB medium of 5 ml was pipetted into a 10 ml test tube, followed by sterilization in autoclave. A piece of the porous cordierite ceramic specimen and dry yeast cell of about 10 mg were added into the medium, and were slowly shaken under the conditions of 30 rpm and 35°C after impregnation of medium into the pores in vacuo. After incubation for 20 h, the medium was replaced 2 times by a new medium of the same composition and temperature, and the test tube was sealed with a rubber stopper. The gas phase of 0.1 ml in the test tube was sampled by a microsyringe every 10 min, and its composition was analyzed by gas chromatograph. The amount of the CO_2 produced was determined by the calculation explained later, assuming that gases in the test tube were CO_2, air, and water vapor. The measurement was performed every day from the 2nd to 6th day after immobilization. After each measurement, the immobilized cells in the sample were incubated again. The yeast cells after the 6th day measurement were observed by scanning electron microscope (SEM).

The volume of CO_2 produced in a test tube was calculated by the following equations (2) to (4).

$$V_{c(0)} = V_0 X_{c(1)} \qquad (2)$$

$$P_{i-1} = (V_0 - 0.1(i-1) + \sum V_{c(i-1)})/V_0 \qquad (3)$$

$$\sum_{i=2}^{N} V_{c(i-1)} = (0.1 \sum_{i=2}^{N} X_{c(i-1)} + (V_0 - 0.1i + 0.1) X_{c(i)} - V_{c(0)}) P_{i-1}/(1 - X_{c(i)}) \qquad (4)$$

Here, V_0 is a volume of gas phase in the test tube, and $X_{c(0)}$ and $V_{c(0)}$ the mean volume fraction and volume of CO_2 at a time 0 min, respectively. P_i is the pressure of gas phase in a test tube at each sampling time, and $X_{c(i)}$ and $V_{c(i)}$ the volume fraction and the volume of CO_2 produced in the period between i-th and (i-1)-th times' sampling.

Measurement of Diffusion Coefficient and Surface Area

The diffusion coefficient of the substrate glucose in the cordierite ceramics was measured. The bead sample prepared from the same sample as used in the immobilization experiment was used in this experiment. The measuring method adopted here is similar to that by Horowitz et al. [12,13]. That is, the sample was first dipped in a glucose solution of 2.5 wt% (C_0) for about 17 h to be equilibrated in a glucose concentration at 4°C. The sample filled with glucose solution was transferred subsequently into a test tube with 5 ml pure water and kept at each prescribed time from 1 to 25 min at 25°C. The diffusion coefficient D_e of the samples were calculated from equations (5) and (6), using the concentration of glucose measured by the total organic carbon analyzer.

$$C_{g(tj)}/C_0 = \sum_{i=j+1}^{N} C_{s(ti)} / \sum_{i=1}^{N} C_{s(ti)} \qquad (5)$$

$$C_{g(tj)}/C_0 = \frac{6}{\pi^2} \sum_{n=1}^{\infty} (1/n^2) \exp(-D_e n^2 \pi^2 t_j/r^2) \qquad (6)$$

Here, Cg(ti) and Cs(ti) are the concentrations of glucose in the porous ceramics and in the solution in a test tube at a time t_i, respectively, r is the radius of the bead sample, and N is the total number of test tubes.

The surface area of the porous ceramics is too small to be measured by nitrogen adsorption method. Therefore, Kr gas was used as an adsorbate. The sample was tested without crushing, using the large glass cell, to prevent any increase of the value. The value of area occupied by Kr atom $2.1*10^{-19}$ m^2 was used for the calculation of surface area (15,16).

RESULT AND DISCUSSION

Properties of the Support and Immobilized Yeast Cells

A typical cross sectional view of the ceramic support observed by reflection microscope is shown in Fig.1. The large pores were gradually brought into contact with each other with an increase in coal powder content. As the result, the median pore size measured by Hg-porosimetry was shifted to larger sizes with increasing coal powder content. However, the median pore size ranged from 20 to 80 μm, smaller by an order of magnitude than those of added coal particles, so that all large pores were found to be inkbottle type. Figure 2 shows an example of SEM photograph of the yeast cells on the 6th day after immobilization. The photograph suggests that yeast cells enter into the inner part and are proliferated there. Changes in the

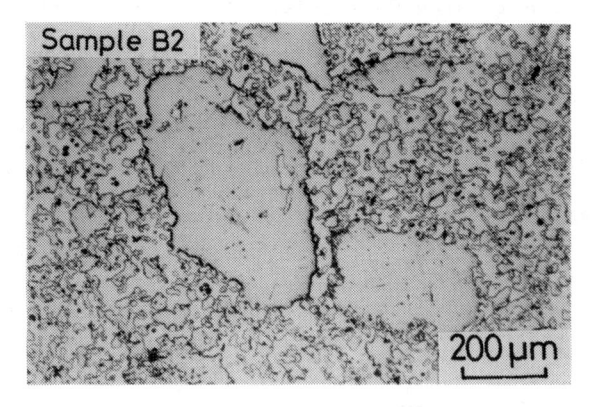

Figure 1 Bimodal pore structure of the tested cordierite ceramics.

Figure 2 Immobilized yeast cells in the support ceramics (Sample E).

volume of CO_2 produced from the cells with time are shown in Fig.3. Though the volume of CO_2 changes increasingly before 30 min, it becomes linear between 30 and 60 min. The slope of this linear part corresponds to the rate of production of CO_2, and also to that of ethanol as expressed in equation (1). Hence, the slope is dealt as the activity of the immobilized cells. The initial part of the curve till 30 min may be an induction period for saturation of CO_2 from the cells into the medium.

Effect of Large Pore in the Support

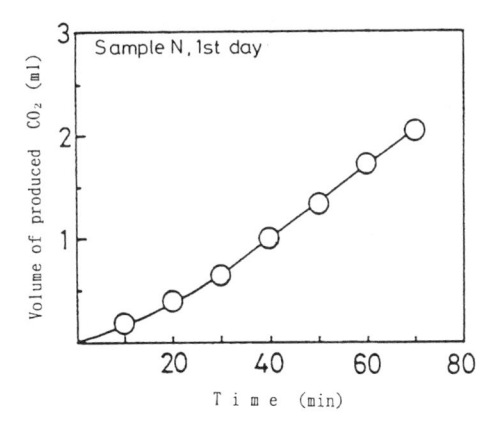

Figure 3 Production of CO_2 from the yeast cells immobilized in the ceramics.

Porous Materials

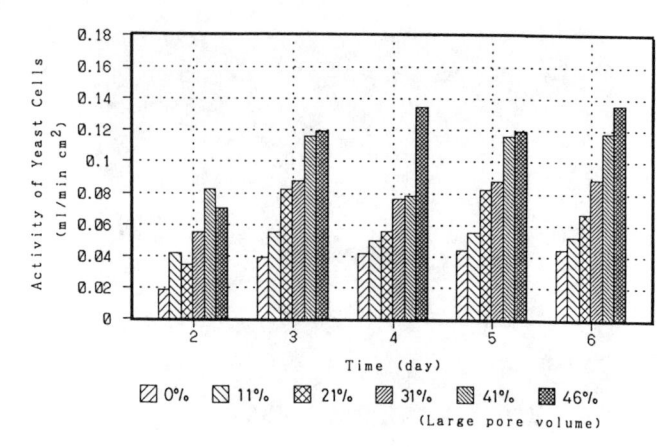

Figure 4 Changes of the activity with the large pore volume in the ceramics. Large pore size : 212-500 μm.

Figure 4 shows the activity change of the cells on the sample with variable large pore volume. The activity becomes almost constant between the 3rd and 6th day, and tends to increase with an increase in the large pore volume. The value on Sample B (large pore : 212-500 μm, 46 vol%) is about 3.5 times larger than that of Sample N (no large pore). Figure 5 shows the activity change with time on the samples with various pore sizes (Sample A, B, C, D, and E). The activity shows a maximum value between the 3rd and 6th day, but there is no dominant tendency in the activity change. Hence, the effect of the large pore size on the activity is not apparent from this

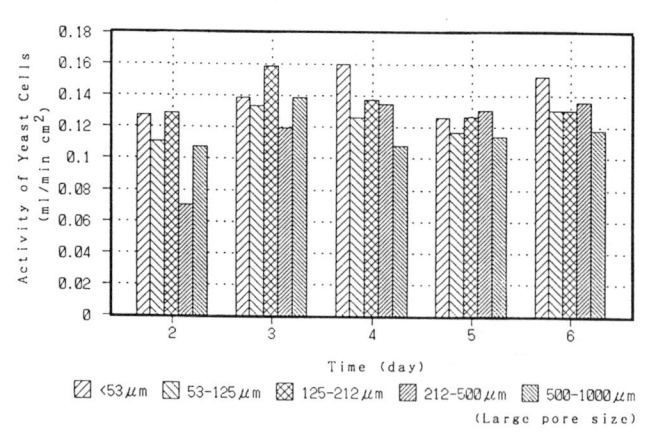

Figure 5 Changes of the activity with the large pore size in the ceramics. Large pore vplume : 46 %.

Porous Materials

result.

According to the SEM observation, yeast cells are proliferated in the surface layer of every sample. The cell population gradually decreases in the inner part of the ceramics. In Sample N, the colonies are never found at a depth of 1 mm. Here, the existence of colony means proliferation of the cells, while, in Sample B, the colonies are observed even at a depth of 2 mm.

Messing et al. insisted that the suitable pore size for immobilizing the yeast cell should be larger than the smallest dimension of the cells, but smaller than four times of the largest dimension [6]. As the cell size was about 5 μm, the suitable pore size is considered to be in the range of 5-20 μm, according to the above rule. However, this pore size range is not sufficient for the cells to enter into the inner part of the support. In addition, the results mentioned above show that the large pores are effective for proliferation of the cells at the inner part. As a reason for the activity improvement, the existence of large pores results in an increase in the diffusion rate. From this viewpoint, the diffusion coefficient of substrate glucose in the samples was measured and shown in the next section.

The Effect of Diffusion Coefficient on the Activity

Table 2. Diffusion Coefficient of the Samples

sample name	A	B	C	D	E	N
diffusion coefficient	37	24	15	16	11	2.1

sample name	B1	B2	B3	B4		
diffusion coefficient	2.1	4.3	10	12	($\times 10^6$ cm^2/s)	

Table 2 lists the diffusion coefficient of the tested samples. First, in Sample N (without large pore), D_e is 2.1×10^{-6} cm^2/s, a little smaller than those of the organic compounds dextran gels or polyacrylamide gels, 3.3×10^{-6} or 3.0×10^{-6} [13,14], respectively. But it steeply increases with an increase in the large pore volume. De in Sample B (large pore volume:46%) is 24×10^{-6} 12 times larger than that of Sample N. This may result from the shortening of the diffusion pathway by the existence of the large pore. In the samples with a constant large pore volume of 46%, D_e increases with increasing size of large pores, and shows the maximum value 37×10^{-6}. Even in sample E (large pore size:<53 μm), D_e is 11×10^{-6} which is almost the same as that of Sample B4 (large pore volume:41%, size:212-500 μm). The observed D_e values are larger than the molecular diffusion coefficient of glucose of 6.8×10^{-6} (at 25°C). This result suggests that the diffusion is controlled not only by molecular diffusion but by stirring outside.

Next, the effect of the diffusion rate on the activity is discussed. Figure 6 shows the relationship between the

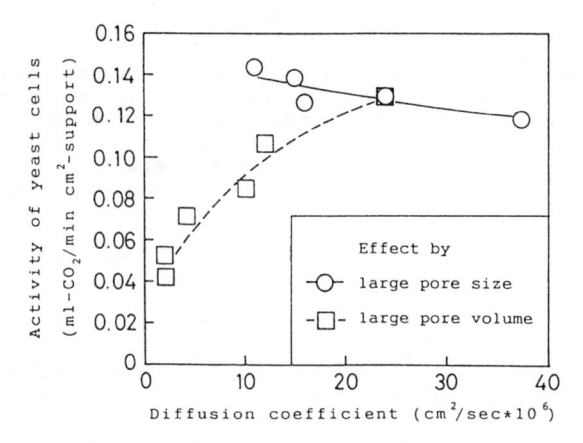

Figure 6 Relationship between activity and diffusion coefficient on the tested ceramics.

activity and the diffusion coefficient. An increase in diffusion coefficient with a change in large pore volume results in an increase in the activity. This result suggests that the diffusion of the solute is a rate-determining step in the tested range of pore size. On the other hand, the activity of the samples with constant large pore volume of 46%, decreases a little with an increase in the diffusion coefficient. Therefore, it is noted that the effect of the diffusion of substrate on the activity is not absolute, and that not only a substrate diffusion but other factors affect the activity.

Effect of Surface Area

In general, high catalytic activity is attained by a support with large surface area. In the case of the microbes, the material with large surface area should be a suitable support too, if the pore size is large enough for the microbes. Therefore, the surface area of the support was investigated as a factor influencing the activity. As the specific surface area of the tested samples was too small to be measured, even in use of Kr as an adsorbate, so it was difficult to compare a small difference in the actual surface area data. Hence, the surface area was calculated from the observed value of Sample N (contains small pores only). Assuming that shape of all large pores is spherical, the surface area can be calculated by the following equation (7).

$$S_V = S_0 D_a (1-\varepsilon)(100-P_L)/100 + 600 P_L/d \qquad (7)$$

Here, S_v is the surface area per unit volume. S_0, D_a, and ε are the specific surface area, apparent density, and porosity of Sample N, respectively. P_L is a large pore volume percent, $d(\mu m)$ the diameter of the large pore. Measured values of S_0, D_a, and ε were 0.143 m^2/g, 2.60 g/cm^3, and 0.431, respectively. The reason for use of the unit S_v (cm^2/cm^2) is of its correspondence to the activity unit (ml/min cm^2), that is, both units are based on the same bulk volume. Calculation results are shown in Table 3. The surface area Sv decreased with an

Table 3. Calculation Result of Surface Area of the Samples

sample name	A	B	C	D	E	N
Sv	1180	1220	1300	1450	2160	2110*

sample name	B1	B2	B3	B4	
Sv	1920	1700	1510	1310	(cm^2/cm^3)

*observed value

increase in the large pore volume, except Sample E (large pore size:<53 μm). This result is attributed to a decrease in the volume of small pore matrix due to the existence of large pores. Therefore, the existence of large pores is considered to decrease the surface area and hence the activity.

As described above, it is considered that two factors, diffusion coefficient and surface area, are highly dependent upon the existence of large pores, and each has an opposite effect on the activity. Hence, the highest activity may be attained in the pore structure satisfying both a better substrate diffusion and a larger surface area. It is also important to control the pore structure according to the conditions in the system, such as substrate viscosity and the mechanical strength required.

SUMMARY

The ethanol fermentation activity of immobilized yeast cells was evaluated on the bimodal porous cordierite ceramics with large- and small-sized pores, and the relationship between the activity and the pore structure was investigated. The activity of the immobilized cells increased with an increase in the large pore volume percent, and the maximum activity was attained with the sample with large pores of 46 vol%; the activity was about 3.5 times larger than that of the sample without large pores. Whereas, the effect of the size of large pores was relatively small. These results were able to be understood in terms of substrate diffusion and surface area of the ceramic support.

REFERENCES

[1] C. Ghommidh, J.M. Navarro, and G. Durand, "A Study of Acetic Acid Production by Immobilized Acetobacter Cell:Oxygen Transfer", Biotech. Bioeng.,24,605-17(1982)

[2] H. Horitsu, Y. Maseda, and K. Kawai,"A New Process for Soy Sauce Fermentation by Immobilized Yeast", Agric. Biol. Chem., 54(2)295-300(1990)
[3] K. Nakanishi, H. Murayama, K. Sato, S. Nagatsuka, T. Yasui, and S. Mitsui, "Continuous Beer Brewing with Yeast Immobilized on Granular Ceramics", Hakko Kougaku Kaishi, 67(6),509-14(1989)
[4] T. Tsoutsas, M. Kanellaki, C. Psarianos, A. Kalliafas, and A.A. Koutinas, "Kissiris:A Mineral Sapport for the Promotion of Ethanol Fermentation by Saccharomyces celevisiae", J. Ferment. Bioeng.,69(2)93-97(1990)
[5] K. Kida, S. Morimura, Y. Sonoda, M. Obe, and T. Kondo, "Support Media for Microbial Adhesion in an Aerobic Fluidized-bed Reactor", J. Ferment. Bioeng.,69(6)354-59(1990)
[6] R. A. Messing, R.A. Oppermann, Biotech. Bioeng.,"Pore Dimension for Accumulating Biomass. I. Microbes that Reproduce by Fission or by Budding", Biotech. Bioeng. 21,49-58(1979)
[7] K. Iwasaki, N. Ueno,"Porous Alumina Ceramics for Immobilization of Soy Sauce",J. Ceram. Soc. Japan, 98(11)1186-90(1990)
[8] H. Morisaki, R. Hattori,"Force Works to Microbes on Interface";pp. 19-21 in Interface and Microbes, 2nd edition, Edited by T. Yamada, Gakkai Shuppan Center,Tokyo,Japan,1988
[9] K. Kawase, M. Matsubara, and T. Majima,"A Fundamental Study of Immobilization of Microbes on the Ceramics", Zairyo Gijutsu, 7(1)19-24(1989)
[10] F. Shiraishi,"Maltose production from soluble starch by β-Amylase and Debranching Enzyme Immobilization on Ceramics Monolith",Kagaku Kougaku,54(6)398-400(1990)
[11] H. Abe, H. Seki, A. Fukunaga, M. Egashira,"Preparation and Water Permeation Property of Bimodal Cordierite Ceramics",J. Ceram. Soc. Japan,100(1)33-37(1992)
[12] S.B. Horowitz, I.R. Fenichel,"Solute Diffusional Specificity in Hydrogen Bonding Systems",J. Phys. Chem.,68(11)3378-85(1964)
[13] K. Nakanishi, A. Adachi, S. Yamamoto, R. Matsuno, A. Tanaka, T. Kamikubo,"Diffusion of Saccharides and Amino Acid in Cross-linked Polymers",J. Agric. Biolog. Chem.,41(12)2455-62(1977)
[14] E. Burito, J.D. Juan,"Diffusion Coefficients of Carbohydrates in Modified κ-Carrageenan Gels with and without Escherichia Coli Immobilized",J. Ferment. and Bioeng.,69(2)135-37(1990)
[15] H. Yanai,"Measuring Method of Pore Structure";pp.69-81 in Kyuchaku Kougaku Youron. Edited by M. Nanjyo, Kyouritsu Shuppan, Tokyo,1977
[16] K. Kinoshita,"Small Surface Area Measurement by Continuous Flow Gas Adsorption Method", Shimadzu Hyouron,48(1)29-33(1991)

A NEW APPROACH THAT USES BIOREACTORS WITH INORGANIC CARRIERS (CERAMIC) IN THE PRODUCTION OF FERMENTED FOODS AND BEVERAGES

Hiroyuki Horitsu
Department of Biotechnology, Faculty of Agriculture, Gifu University, Gifu 501-11, Japan

Soy sauce, beer and sake were produced by using bioreactors with yeasts immobilized on ceramic carriers. The new process has the considerable advantage of shortening the fermentation period 10 times, compared with conventional methods. Likewise, it was confirmed that yeasts were adsorbed on ceramics by Zeta potential interaction.

INTRODUCTION

In recent years, many investigations on the introduction of bioreactors with immobilized enzymes or microbes in food industries have been conducted. Apart from soy sauce, beer, sake, wine, and vinegar industries, the use of bioreactors have spread to various other areas.

Above all, immobilizing techniques (depending largely on the composition of carriers) play the most important role in the bioreactor system. As yet, an organic matrix such as calcium alginate gel (1-3) and potassium κ-carrageenan gel have been used.

Meanwhile inorganic materials (4-6) have many unique characteristics in comparison with organic matrices, but only a few are known for their utilization as carriers. The author investigated the applicability of ceramic carriers for the production of soy sauce, beer and sake.

METHODS

It must be emphasized here that the manufacturing of soy sauce, beer and sake using bioreactors. Consequently, their experimental installations are similar. In the experiment, ceramic matrices were used in the bioreactors in order to immobilize microorganisms. Physical properties of the ceramic carriers (filamentous cylindrical and amorphous beads) are shown in Table 1 and 2. Figs. 1, 2 and 3 show the experimental apparatuses for soy sauce and beer fermentation.

Table 1. Physical Characteristics of the Filamentous Cylindrical Ceramic

Characteristics	
Ingredients (%, w/w) Al_2O_3 : SiO_2	95 : 5
Average pore size (μm)	60 - 150
Porosity (%, v/v)	90

Table 2. Physical Characteristics of Amorphous Ceramic Beads

Characteristics	
Diameter (mm)	4 - 6
Ingredients (%,w/w) Al_2O_3 : SiO_2	90 : 10
Distribution of pore size (μm)	10 - 500
Average pore size (μm)	80 - 90
Surface area (cm^2/g)	300
Water absorption capacity (%,w/w)	30
Porosity (%,v/v)	52 - 55

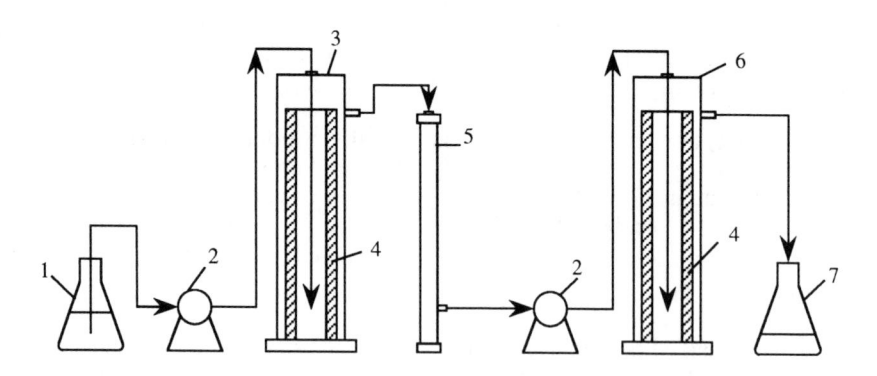

Figure 1 Schematic flow diagram for soy sauce production using reactors with immobilized yeasts on filamentous cylindrical ceramic carrier. (1) feed tank; (2) pump; (3) reactor I; (4) ceramic carrier; (5) ceramic filter; (6) reactor II; (7) reservoir.

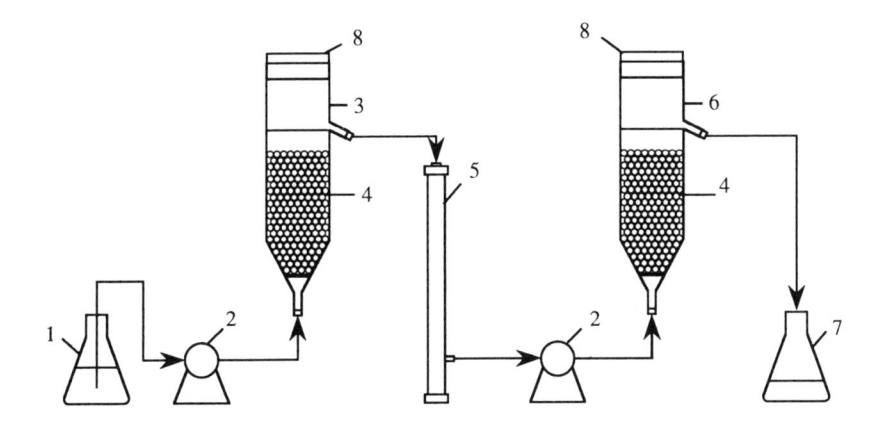

Figure 2 Schematic flow diagram for soy sauce production using reactors with immobilized yeasts on an amorphous ceramic beads carrier. (8) rubber plug. For other numbers, see Figure 1.

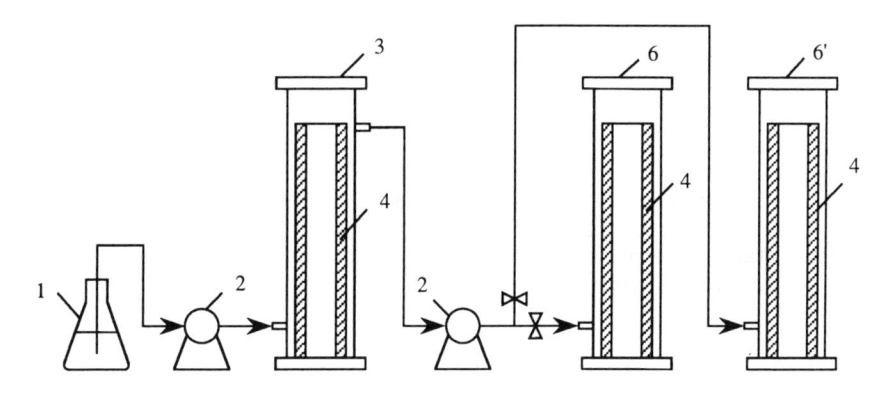

Figure 3 Schematic flow diagram for beer production using reactors with immobilized yeasts on a filamentous cylinder ceramic carrier. (6') reactor IIb. Other numbers are explained in Figure 1.

Two types of reactors were used: a stainless reactor (10 cm in diameter, 28 cm in height) shown in Fig. 1 and 3, and a glass cone cave style reactor shown in Fig. 2. The ceramic carriers were placed into each of them and then sterilized by

Porous Materials 383

autoclaving at 121°C for 15 min. As for regeneration, ceramic carriers were heated at 900°C for 1 h.

The feed solution for soy sauce fermentation was conventionally prepared and the inoculum was made by inoculating soy sauce koji into steam-cooked soybeans mixed with roasted wheat grains. Then, the obtained koji (inoculum mass with *Aspergillus oryzae*) was mixed with saline solution and kept at 55°C to produce high temperature digestion. After filtration, the filtrate was treated with *Pediococcus halophilus* for lactic acid fermentation. On the other hand, the feed solution for beer fermentation was prepared by adding hops to a saccharified-malt solution boiling. Finally, for the sake fermentation, the feed solution was made from a mixture of saccharified solution of both sake koji and saccharogenic enzyme added to saccharify steamed rice.

Soy sauce was analyzed according to the legal "Text for soy sauce" (7) in Japan whereas beer and sake analyses were performed according to the procedures described in "Notes on analysis methods from National Tax Bureau" (8). Aroma components were analyzed by head space gas chromatography. Cell fusion of soy sauce yeasts, *Zygosaccharomyces rouxii (Zygosacch. rouxii)*, and *Candida versatilis (C. versatilis)*, was done by a Shimadzu Somatic Hybridizer SSH-1 (an electric hybridization apparatus). Measurements of the Zeta-potential of yeasts and ceramics were carried out as follows: yeast cells *Zygosacch. rouxii* IFO 1877 and *C. versatilis* FRZ 451, cultivated at 30°C for 24 h, were harvested by centrifugation at 3000 rpm from a culture broth. The cells were washed twice with sterilized water and used for the measurement of the Zeta-potential. Ceramic beads were crushed in porcelain mortar and passed through a 70 mesh sieve. The yeast cells and ceramic powders were suspended in a bottle containing 200 ml of 0.01M KCl and shaken intensively over-night. The Zeta-potential was measured by using a System 7000 Acoustophoretic Titrator™ (Pen Kem) apparatus. The pH in the suspended cells and ceramic powder was varied using 1N-HCl or 1N-KOH.

RESULTS

Soy Sauce

The scanning electron micrographs of soy sauce yeasts, *Zygosacch. rouxii* and *C. versatilis*, which were both immobilized on filamentous cylindrical ceramic, is shown in Fig. 4 and the scanning electron micrographs of soy sauce yeasts, *Zygosacch. rouxii* and *C. versatilis* which were both immobilized on amorphous ceramic beads, is shown in Fig. 5.

Analyses of soy sauce produced by soy sauce yeasts and their fusant, using filamentous cylindrical ceramic and amorphous ceramic beads bioreactors, are shown in Table 3.

(a) (b)
Figure 4 SEM micrograph of yeasts adsorbed on filamentous cylindrical ceramic. (a) *Zygosacch. rouxii* (reactor I); (b) *C. versatilis* (reactor II).

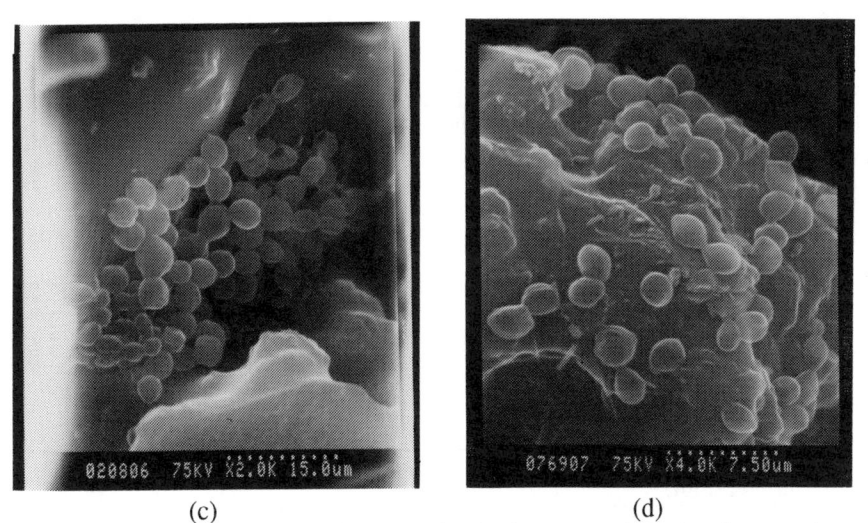

(c) (d)
Figure 5 SEM micrograph of yeasts adsorbed on amorphous beads ceramic (c) *Zygosacch. rouxii* (reactor I); (d) *C. versatilis* (reactor II).

Table 3. Analysis of Fermented Liquids in Various Reactors

	Cylinder (two stage)	Beads (two stage)	Fusant (one stage)	Commercial soy sauce
EtOH (%,v/v)	2.2	2.2	1.5	2.4
Formol-N (%)	1.2	1.1	1.2	1.0
Acidity (ml)	2.4	2.3	2.2	2.2
pH	4.9	5.0	5.0	4.8
Aroma components (mg/l)				
i-AmOH	14.7	10.0	12.1	15.3
i-BuOH	11	7	9	13
n-PrOH	4.9	4.5	4.2	5.0
Et-acetate	1.2	1.1	1.5	13
2-PE	16.0	10.0	9.6	8.4
4-EG	2.4	1.4	1.0	1.6
Fermentation period (days)	12	12	6	>120

The data of Table 3 indicate that the filamentous cylindrical ceramic is the most appropriate to use. Ethanol production by fusant strain was lower than that exhibited by wild soy sauce yeasts. Fermentation time of one stage-fermentation was half of that in two stage-fermentation, which we believe to be an advantage.

Beer

Analyses of beer produced by the bioreactor and that of beer produced conventionally are shown in Table 4. According to the data in Table 4, it is obvious that beer can also be manufactured in only 9 days by using a two stage-bioreactor. Also in this case, the filamentous cylinder ceramic was found to be the best.

Table 4. Comparison of Aroma Components of Beer Produced by Bioreactor and That of Beer Produced by Conventional Means

Components	bioreactor	conventional
EtOH (%,v/v)	5.7	4.8
Formol-N (%)	0.4	0.4
Total-N (%)	0.6	0.6
Acidity (ml)	1.3	1.4
pH	4.0	4.3
Acetaldehyde (mg/l)	8.0	4.0
Diacetyl (ml/l)	0.21	0.20
Aroma components (mg/l)		
i-AmOH	60	40
i-BuOH	14	7
n-PrOH	17	7
Et-acetate	12	13
Fermentation period (days)	9	>90

Sake

Using a bioreactor in sake processing, it was possible to obtain a fermented liquid with an ethanol concentration of 13% or more. The aroma component (Ginjyou) production could be maintained by controling the ratio height/ diameter of the liquid in the bioreactor. Compared with the conventional process (see Table 5), the fermentation time was only 4 days for the bioreactor versus 60 and 40 days for superrefined (grade) and regular (grade), respectively. This is about one-tenth and one-fifteenth of the fermentation time needed in the conventional process. Finally, the process presents an advantage of reusing carriers and bioreactors.

Table 5. Comparison of Aroma Components of Sake Produced by a Bioreactor and That of Conventionally Produced Sake

Components	Bioreactor	conventional super refined	regular
EtOH (%,v/v)	13.1	18.0	18.0
Formol-N (%)	1.1	0.9	1.6
Acidity (ml)	2.9	1.3	1.4
Acetaldehyde (mg/l)	14.5	N.D.	N.D.
Aroma components (mg/l)			
i-AmOH (A)	118	170	170
i-BuOH	56	70	64
n-PrOH	65	120	120
Et-acetate	137	120	30
i-Am-acetate (E)	7.7	1.5	0.5
Et-capronate	N.D.	10	3.0
E/A ratio (x100)	5.5	0.9	0.3
Fermentation period (day)	4	> 60	> 40

(N.D. ; not determined)

Adsorption Mechanism Between Yeast Cell Membrane and Ceramic Surface

The Zeta-potentials of the yeasts (*Zygosacch. rouxii* and *C. versatilis*) used in soy sauce fermentation are shown in Fig. 6. Both yeasts have a positive Zeta-potential. On the other hand, the Zeta-potential of the amorphous beads ceramic surface (shown in Fig. 7) is negative. These data suggest that yeast cell membranes are electrostatically bound on ceramic surfaces.

CONCLUSION

Using bioreactors with ceramics as carrier for immobilization of yeasts, we could produce soy sauce, beer, and sake in a short time (about one tenth of the conventional procedure time) in each case. It was also found that yeasts were adsorbed on ceramic carriers by zeta potential interaction.

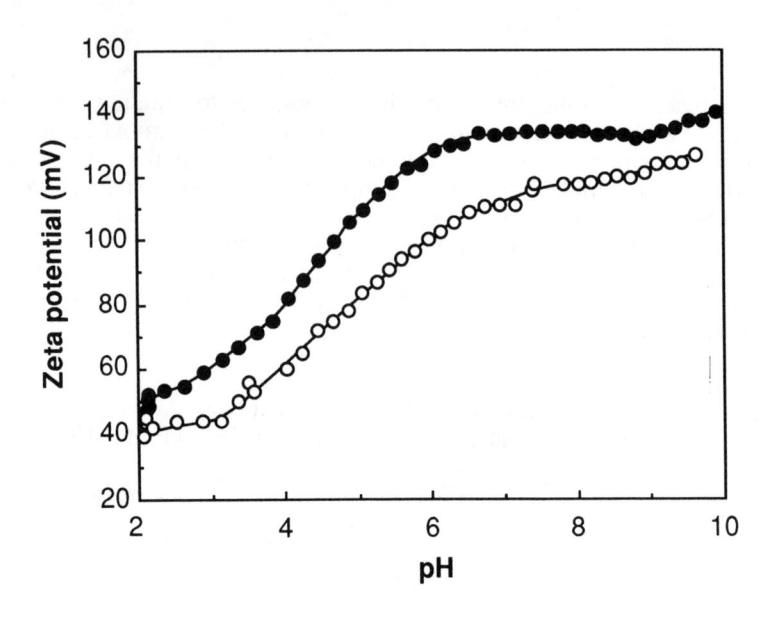

Figure 6 Zeta-potential of *Zygosacch. rouxii* and *C. versatilis*
● *Zygosacch. rouxii*; ○ *C. versatilis*.

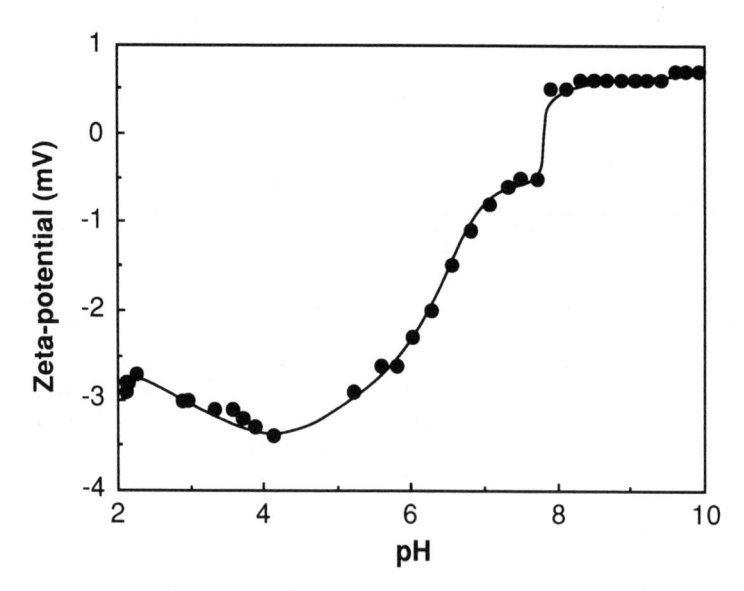

Figure 7 Zeta-potential of ceramics

REFERENCES

1) K. Osaki, Y. Okamoto, T. Akao, S. Nagata & H. Takamatsu. Fermentation of soy sauce with immobilized whole cells. J. Food Sci., 50, 1289 (1985).
2) H. Takamatsu, T. Iwase & T. Yokotsuka. Production manner of liquid seasonings. Japan Patent, 1120, 428 (Oct. 28, 1982).
3) T. Onaka, K. Nakanishi, T. Inoue & S. Kubo. Beer brewing with immobilized yeast. Bio/Technology, 3, 467 (1980).
4) H. Horitsu, Y. Maseda & K. Kawai. A new process for soy sauce fermentation by immobilized yeasts. Agric. Bio. Chem., 54, 295 (1990).
5) H. Horitsu, M. Y. Wang & K. Kawai. A modified process for soy sauce fermentation by immobilized yeasts. Agric. Bio. Chem., 55, 269 (1991).
6) H. Horitsu, E. Morishita, K. Takamizawa, K. Kawai, T. Suzuki & M. Maeda. A study of yeast adsorption on ceramic beads by zeta potential. Biosc. Biotech. Biochem., 56, 1501 (1992).
7) Text for soy sauce, Japan Soy Sauce Research, Tokyo (1985)
8) Notes on analysis methods from National Tax Bureau, Japan Brewing Association, Tokyo (1984)

POROUS CERAMIC CARRIER FOR BIOREACTOR

Mitsuo KAWASE, Yoshihiro KAMIYA and Masayuki KANENO
R&D Division of Engineering Business Group, NGK INSULATOR, LTD.
1 Maegata-cho, Handa 475, Japan

Physical adsorption of microbes on an inorganic carrier, using several types of ceramics as carriers and several types of yeast as microbe, was investigated to obtain useful knowledge for developing a ceramic carrier.

First, surface electric potential (zeta potential) of ceramics and microbes was measured, and performed microbe immobilization tests and evaluated the contribution of electrostatic force to microbe immobilization. Through the researches, we made clear that the electrostatic force played an imporant role in first step of microbe immobilization on ceramic surface and that most of the microbes and ceramics had negative zeta potential.

Secondly, we tried to shift the zeta potential of ceramics to positive by chemical modification, and we obtained a positive zeta potential carrier by introducing a positive functional group, such as amino group, to the ceramic surface with the help of silane coupling agent. For example, the zeta potential of cordierite was −20.4 mv and could be shifted to approximately +140 mv by 3-(2-Amino ethylaminopropyl) trimethoxysilane.

Finally, we tested the microbe immobilization ability of the modified ceramic carrier. As a result of the test, we found out that the micobe immobilizaiton ability of the ceramic carrier was improved by approximately ten times by chemical modification with 3-(2-Amino ethylaminopropy) trimethoxysilane.

INTRODUCTION

Recently, research on bioreactors, which efficiently produce medical, food, and industrial raw materials, have been actively conducted. In general, a bioreactor is a reactor for bio-chemical production, containing a so-called "biological catalyst", (such as microorganisms, cells, and enzymes). This catalyst efficiently converts raw materials to valuable medicine, food and chemicals. One of the typical types of bioreactors is the "fixed type bioreactor", in which the biological catalyst is immobilized on the carrier. Since efficiency of the reactor is dependent on the amount of immobilized microbe, adopting a high-affinity carrier is an important factor in determining the reactor performance. When developing such an excellent carrier, the mechanism of microbe immobilization must be understood. "Coulomb force" and "Van der Waals force" were considered to play a important role in the first step of microbe immobilization, especialy physical adsorption. However, the microbe immobilization mechanism of physical adsorption is still not well understood.

In this report, the effect of the surface electric charge of microorganisms and of the carrier on the immobilization was evaluated. The affinity (immobilization ability) of microorganisms to the ceramic carrier with an emphasis on the Coulomb force was analyzed. In addition, experiments were conducted to improve the affinity of the ceramic carrier to microoganisms, by shifting the ceramic surface potential using chemical modifications.

EXPERIMENTS

Samples

Microorganisms

The microbes used in this research were *saccharomyces cerevisae* (used for alcohol fermentation and baking bread), *bacillus subtilis* (mainly used as host for genetic recombination), and *escherichia coli* (which lives in the colon of mammals and is also used as host for genetic recombination).

Ceramics

The types of ceramics used in this research were cordierite (used as catalyst carrier for purifying car exhaust gas), alumina (frequently used as fire brick material for furnaces), and mullite (used as catalyst carrier for boiler denitration).

Zeta Potential Measurement

Microbes and ceramic particles in solution have a particle surface electric charge according to the characters. When a direct current electric field is applied to the charged particles, the particles start moving to the reverse pole. The water molecules around the surface of the particles move together with them, and form a sliding surface between themselves and the resting water molecules nearby. The potential on this sliding surface is the zeta potential. The zeta potential of microorganisms and that of ceramics were calculated as described below.

Zeta Potential of Microoraganisms

Microbes were collected from the medium by centrifugal separation, and washed with physiological saline. Next, the microbes were added

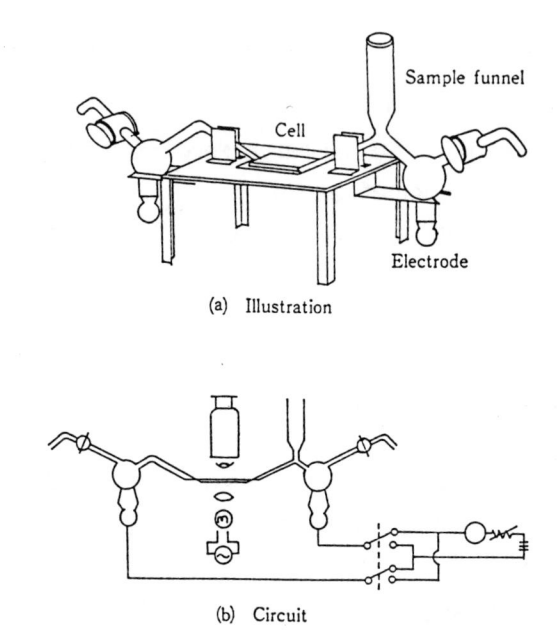

Figure 1 Electrophoretic apparatus (Brigg's Cell)

(a) Illustration

(b) Circuit

electrolyte solution (HCL-KCL:0.53 ms) of the specified pH to suspend it to a turbidity of $OD_{570}=0.3$. The microbial suspension was charged into the Brigg's cell illustrated in Fig. 1, and a direct current electric field was applied to the suspension. The zeta potential was calculated using Eg. (1) below.

The "OD_{570}" (Optical Density) is the absorbance of 570 nm visible rays and the index to suspending microbe concentration. $OD_{570}=1.0$ of yeast, for example, is equivalent to a micrbe concentration of approximately 8.0 × 10⁷ cells/ml.

$$\zeta = \frac{4\pi\mu}{D} \cdot \frac{A}{i} \cdot C \cdot u \cdot 9 \times 10 \qquad (1)$$

where ζ : zeta potential (mV)
μ : coefficient of water viscosity (Poise)
D : permittivity of solution (−)
A : sectional area of cell perpendicular to potential gradient (cm²)
i : current passing through cell (mA)
C : electric conductivity f suspension
u : migration velocity particle (cm/sec.)

Zeta Potential of Ceramics

Ceramics were pulverized and added to the electrolyte solution of a specified pH to make a suspension. After three minutes, relatively large precipitating ceramic particles were removed and the OD_{570} of the suspension was adjusted to 0.3 by dilution with distilled water. The migration velocity of the suspending particles (5 − 10 μ m) was measured using the Brigg's cell as the measurement of the zeta potential of microbes, thereby calculating the zeta potential from Eg. (1).

Shift of Surface Electric Charge of Ceramics

Silane coupling agents (products of Shin-Etsu Chemical Co., Ltd.) were used as shown in Table 1. A fixed amount of each ceramic was placed in a separable flask, and refluxed using toluene, in which a silane coupling agent had been dissolved for coupling. The surface electric charge of the modified ceramics was shifted according to the functional group peculiar to the coupling agent.

Table 1. Silane Coupling Agents for Chemical Modifications of Ceramics

Chemical name	M a r k
3-Phenylaminopropyltrimetoxysilane	LS-4500
N-(3-Diethoxymethylsilylpropyl)sccinimide	LS-4530
3-Aminopropyltriethoxysilane	LS-3150
3-Piperazinopropyltrimethoxysilane	LS-3600
3-Dimethylaminopropyldiethoxymethylsilane	LS-3675
Tris(2-methoxyethoxy)vinylsilane	LS-4080
3-(2-Aminoethylaminopropyl)trimethoxysilane	LS-2480
Dimethyloctadecyl-3-trimethoxylsilylpropylammonium	LS-6985
3-2-(2-Aminoethylaminoethylamino)propyltrimethoxysilane	LS-3750

Immobilization of Microbe on Ceramic Carrier

Yeast, which had been shake-cultured at 30 ℃ for two days, was collected by centrifugal separation, and washed with physiological saline. This yeast was added to 3% glucose-3% ethanol solution for immobilization, suspended it, and adjusted it to the specified OD_{570}. 10 mℓ of the yeast suspension was put in a test tube, 1 g of ceramics (with particle size ranging between 1.0mm and 1.4 mm) was added, and the test tube was shaken at 4 ℃ for two hours. Subsequently, the OD_{570} of the solution was measured, and decrease in the OD_{570} was considered as the number of yeast immobilized on the ceramic carrier.

RESULTS AND DISCUSSION

Surface Electric Charge of Microoranisms

Table 2 shows the zeta potentials of microorganisms at a pH of 3.0.

Most of the microorganisms tested, including *saccharomyces cerevisiae, bacillus,* and *escherichia coli,* had a negative zeta potential. Yeast was used as the microbe sample in the subsequent immobilization experiment, because all types of yeast (positive zeta potential yeast and negative potential yeast) are very much alike in shape (spherical) and size (several μm).

Table 2. Zeta Potential of Microorganisms at pH 3.0

Microorganisms	Zeta potential (mV)
Saccharomyces cerevisiae KSC-44	− 128.4
Saccharomyces cerevisiae IFO2091	− 102.8
Saccharomyces cerevisiae K-7	+ 9.1
Saccharomyces cerevisiae Z-73	+ 13.3
Bacillus subtilis	− 60.9
Escherichia coli	− 47.7
Brevibacterium ammoniagenes IAM1641	− 26.1

Surface Electric Charge of Ceramics

Non-treated Ceramics

Table 3 shows the zeta potential of each type of ceramic material. As the table shows, some types had a positive zeta potential and some had a negative. The absolute zeta potential was small, ranging from 15 mV to 32 mV. Therefore the large affinity between microbes and ceramics can not be expected.

Table 3. Zeta Potential of Ceramics at pH 3.0

Chemical component	Mark	Zeta potential (mV)
Zirconia : ZrO_2 68.5 (%)	SZN-T1	+ 15.7
Alumina : Al_2O_3 99.1 (%)	NEO-AMN	+ 27.6
Cordierite : $2MgO \cdot 2Al_2O_3 \cdot 5SiO_2$	DHC-141	− 32.1
Murite : $3Al_2O_3 \cdot 2SiO_2$	MC-1	− 20.4

Modified Ceramics

Using the silane coupling agents listed in Table 1, the amino group was introduced to the surface of the cordierite (DHC-141). The result is shown in Table 4. It is clear that introducing each type of amino group can shift the zeta potential of the ceramics up to +236 mV.

Table 4. Zeta Potential of Modified Ceramics (at pH 3.0)

Sliane coupling	Zeta potential (mV)
LS-4500	+ 45.1
LS-4530	+ 47.9
LS-3600	+ 56.2
LS-3675	+ 70.0
LS-4080	+ 89.7
LS-3150	+112.7
LS-2480	+141.7
LS-6985	+191.3
LS-3750	+235.9

Experiment in Microbe Immobilization

Immobilization of Microorganisms on Non-treated Ceramics

Ceramics and yeast with both positive and negative surface electric charges were used in immobiliation experiments, using various combinations of ceramics and yeast to clarify the effect of the Coulomb force on microbe immobilization. Alumina (positive electric charge, NEO-AMN) and cordierite (negative electric charge, NEO-141) were used, and *saccharomyces cerevisiae Z-73* (positive electric charge) was used for the microorganism. Table 5 shows the number of immobilized cells on each carrier estimated from the variation in OD_{570} before and after immobilization.

As Table 5 discloses, on the ceramics of positive zeta potential, more yeast of negative zeta potential was immobilized than yeast of positive zeta potential, and vice versa. Hence, the Coulomb force between the microorganisms and the carrier was expected to promote the immobilization. The DHC-141 has a much larger specific surface area (37 m^2/g) compared with the NEO-AMN (0.6 m^2/g), and consequently the site for microbe immobilization largely differs. For this reason, it is impossible to compare the number of immobilized cells between carriers.

Table 5. Immobilization of Yeasts on Ceramics with Different Zeta Potential

Yeasts \ Ceramics	NEO-AMN (+27.9 mV)	DHC-141 (−32.1 mV)
Z-73 (+13.2mV)	0.0	1.9
KSC-44 (−128.4mV)	0.44	0.54

Immobilization of Microorganisms on Modified Ceramics

Attempts were made to immobilize microorganisms on the modified ceramic carrier, of which the zeta potential had been shifted to positive with the help of the silane coupling agents. The results of experiments with the *saccharomyces cerevisiae* IFO2091 (-102.8 mV) and the KSC-44 (-128.4 mV) are shown in Fig.2. The cordierite (DHC-141) were used as the carrier. Although the data show a slight scattering, the more positive the surface electric charge of the carrier is, the larger the number of immobilized cells is when yeast with a negative surface charge, such as the IFO2091 or KSC-44, is used. When the LS-2480 was used for silane coupling agent, for example, the zeta potential of ceramics shifted to +141.7 mV, thereby immobilizing the negative yeast approximately ten times as much as the non-treated ceramics (-32.1 mV).

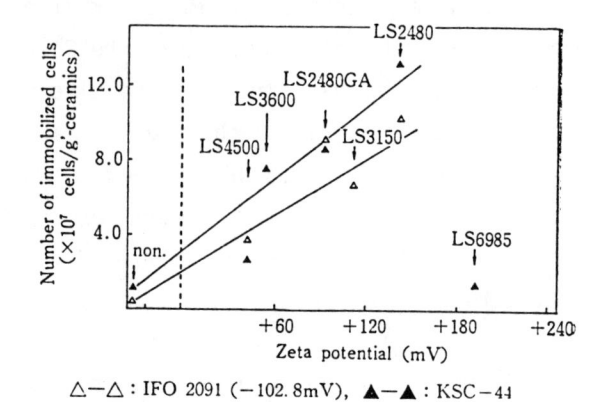

Figure 2 Effect of zeta potential on number of immobilized cells with negative yeasts(pH 3.0)

△—△ : IFO 2091 (−102.8mV), ▲—▲ : KSC−44 (−128.4mV)

CONCLUSIONS

This reseach showed the following :
(1) Most of microorganisms and ceramics have a negative zeta potential at a pH of 3.0.
(2) The zeta potential of surface-treated ceramics with silane coupling agents can be shifted up to +240 mV.
(3) The Coulomb force induced by the surface electric charge has significant effect in the early stages of microbe immobilization.
(4) Ceramics, with a surface electric charge shifted to positive, are capable of immobilizing microorganisms approximately ten times as much as non-treated ceramics.

REFERENCE

1) H.Morisaki and R.Hattori "Surfaces and Microorganisms", p.13, Gakkai Publishing Center, Tokyo (1986)

AMYLOSE SEPARATION FROM STARCH AND GLUCOSE SEPARATION FROM STARCH SACCHARIFIED SOLUTION BY FILTRATION THROUGH A CERAMIC MEMBRANE

Fujio TAKAHASHI and Yasuzo SAKAI
Department of Applied Chemistry, Faculty of Engineering, Utsunomiya University, 2753 Ishii-machi, Utsunomiya 321, Japan

A ceramic membrane has been developed for separation of amylose from starch. Yield of amylose was 3.0 g from 2 dm^3 of 1.1% corn starch after filtering for 50 h under 0.3 kg/cm^2 of gauge pressure and 3.0 m/s of feed velocity at 60°C. Glucose separation from starch saccharified solution was conducted with the ceramic membrane coated with hydrous zirconium oxide. Three dm^3 of 0.9% starch saccharified suspension hydrolyzed by 1 g of crude glucoamylase was filtered continuously through the ceramic membrane under 2 kg/cm^2 of gauge pressure and 0.8 m/s of feed velocity at 55°C. The glucose concentration in the permeated solution was kept constant at about 0.35 mol/dm^3 from 15 to 40 h. Various conditions were examined for coating hydrous zirconium oxide on the surface of the membrane. A maximum rejection of α-chymotrypsin was obtained in the region of pH 3.9.

INTRODUCTION

Ceramic membranes have found wide application for the separation of various substances.[1-3] The ceramic membrane has the following advantages: sterilization at high temperature because of heat resistant material; better durability than polymer membranes; difficult to thicken the cake on the ceramic membrane under external pressure; and a lower reduction in volume flux through the membrane.[4]

Starch is composed of 20~25% of amylose (MW 0.5~2 ×10^6) and 80~75%

of amylopectin (MW 15 ~400 ×10^6). Wide utilization of amylose, which is a type of linear polymer chains by $\alpha 1\to 4$ glycoside bond, might be expected, if it could be produced at low cost. The usual method requires organic solvents such as pentanol, 1-butanol and others to separate amylose from starch with many processes.[5] Consequently, it is significant to propose an easy and efficient separation method of amylose from starch.

A membrane reactor is useful for the preparation of glucose from starch or cellulose, because it provides easy separation of product from substrate and enzyme, and because immobilization of enzyme is not necessary.[6-8] The membrane reactor method has an advantage to conduct both enzymic reaction and separation of product simultaneously. The saccharification of starch is catalyzed by glucoamylase, which is an enzyme cleaving away successive glucose units beginning from the nonreducing end to yield glucose. In other words, the molecular weight of starch decreases slowly.

To apply the ceramic membrane as a molecular sieve, pores of the membrane were impregnated by hydrous zirconium oxide in order to reject high molecular compounds, of which molecular weight is above 10^5. It has been reported that zirconium colloid particles not only deposit on but also penetrate into the ceramic support.[4]

This paper presents two results, which were already reported; amylose separation from starch[9]; and glucose separation from starch saccharified solution by filtration through the ceramic membrane.[10] New results are presented on applying a coating of hydrous zirconium oxide onto the surface of the ceramic membrane in order to reduce the pore size.

EXPERIMENTAL

A Membrane Reactor Equipped with Ceramic Membrane

A tube of ceramic membrane with one end closed was made of porous alumina. This tube (5 mm in outside diameter, 50 cm long, 75 cm^2 of surface, 0.05 μm in average pore size) was loaded into a metal tube (7.5 mm in inside diameter), and was connected to a pump and a feed tank as shown in Fig. 1.

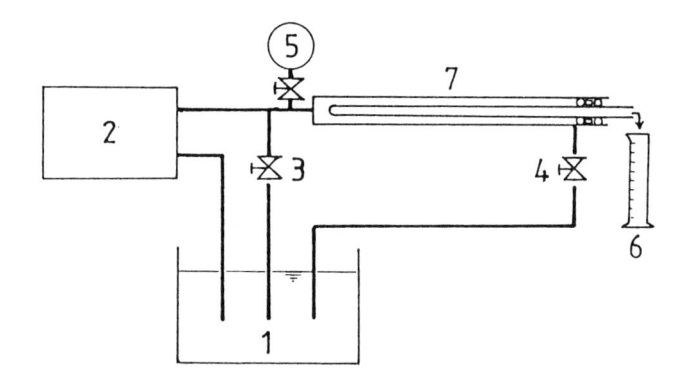

Figure 1 The schematic diagram of equipment. 1: Feed tank, 2: Circulating pump, 3: Valve (Flow control), 4: Valve (Pressure control), 5: Pressure gauge, 6: Measuring cylinder, 7: Membrane module.

A sample solution was added into the feed tank, and pressure was applied to the ceramic membrane from outside to inside of the tube. Only 0.02~1 % portion of the total liquid passed through the ceramic membrane, and the rest flowed along the outside of ceramic membrane and flowed back to the feed tank.

Coating Hydrous Zirconium Oxide on the Surface of Ceramic Membrane

Using the equipment in Fig. 1, a $ZrOCl_2$ solution was passed through the ceramic membrane in order to clog the holes of the membrane under external pressure. A suspension of $Zr(OH)_4$ was obtained by adding NaOH into $ZrOCl_2$ solution to increase the pH. The size of particles of hydrous zirconium oxide was controlled by changing the rate of pH increase. This is observed by viscosity, SV_{30}, and microscopic measurement. The rejection ratio was determined by using small sized-compound, α-chymotrypsin (MW. 25,000), because the rejection of starch should be higher than that of α-chymotrypsin.

Materials and Analysis

A requisite concentration of suspension was prepared by boiling a mixture of corn starch (nitrogen: 0.03%, water: 8.5%) and distilled water.

Glucose was analyzed by the Hanes' method, and amylose was determined by the following methods: (1) After hydrolysis of starch by β-amylase, maltose was determined by the Hanes' method. (2) Color developed by iodo-starch reaction was determined by spectrophotometry (λ_{max} 620 nm). (3) Weight measurement after freeze drying.

RESULTS AND DISCUSSION

Amylose Separation from Starch

The Effect of Gauge Pressure and Feed Velocity on Volume Flux and Amylose Entration

At a constant feed velocity, the velocity of filtering amylose decreased with increasing gauge pressure. At constant gauge pressure, the velocity of filtering amylose increased with increasing feed velocity. The results at 50 ℃ are summarized in Fig. 2. The volume flux and the amylose concentration in the filtrate at 60℃ increased as compared with 50℃.

The filtrates obtained during 0～15, 45～60 and 275～300 min were

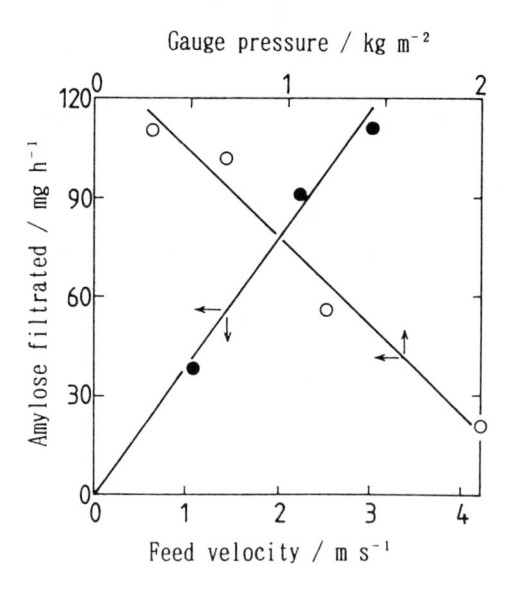

Figure 2 The effect of feed velocity and gauge pressure on the amount of amylose filtered at 50 ℃ during 1 h from 2 till 3 h.

Table 1. Analysis of Starch in the Filtrate

Filtrate (min)	Weight of solid material (mg/cm^3)	Maltose obtained by β-amylase (mg/cm^3)(A)	Optical density of (620 nm)(B)	A/B
0~15	1.06	0.93	0.308	3.0
45~60	0.80	0.79	0.264	3.0
275~300	0.64	0.64	0.214	3.0

analyzed. The results are shown in Table 1. The weight of maltose prepared was in accord with that of solid material in the filtrate except for 0~15 min. Since amylose should convert to 100% maltose by hydrolyzing with β-amylase, the solid material was considered to consist only of amylose for the filtered time of 0~15 min, which could be some amount of other compounds. Total amount of amylose in 1.04 dm^3 of filtrate was estimated to be 0.77 g from the spectrophotometric measurement.

Amylose Separation

Amylose separation from 2 dm^3 of 1.1% of corn starch was carried out under 0.3 kg/cm^2 of gauge pressure and 3.0 m/s of feed velocity at 60℃. The volume of suspension in the feed tank was kept constant by frequently adding the same volume of distilled water as that of the filtrate. The separation experiment was continued during 50 h. Total volume of the filtrate was 7.4 dm^3. The filtrate was condensed into one tenth the volume under reduced pressure and then dried by freeze drying. The yield of amylose was 3.0 g after drying at 110℃ during 2 h. A residual starch in the feed tank was 19.4 g, which was determined by freeze drying.

The volume flux and amylose concentration of filtrate with time are shown in Fig. 3. Both the volume flux and the amylose concentration of filtrate decreased somewhat rapidly during 0~7 h. This fact suggested that the deposition of starch might be progressing on the surface of ceramic membrane with completion occurring after 7 h. The amylose concentration in the filtrate decreased with time. The rate of decrease was on the first order with respect to the amylose concentration.

Figure 3 shows that the amylose concentration after 31 h decreased by one half as compared to the concentration at 7 h. The amount of amylose filtrated from 7 to 31 h was 1.44 g. This indicates that another 1.44 g of amylose remains in the feed tank. Up to 7 h from the beginning, 0.74 g of amylose was filtrated. Therefore the total amount of amylose was 1.44 ×2 + 0.74 = 3.62 g in the feed tank at the start of filtration. The amylose, which is possible to be filtered, corresponds to 15% of the total weight of starch, and is lower than the usual content (20~25%). In the connection, the values of amylose rejection at 0, 7 and 31 h were 0.50, 0.60 and 0.62, respectively, from the concentration of filtrate and that of filterable amylose in the feed tank.

Coating Hydrous Zirconium Oxide on the Surface of Ceramic Membrane

The ceramic membrane was coated with a hydrous zirconium oxide suspension. Various concentrations of NaOH solution were added into a mixture of 0.1 mol/dm^3 ZrOCl$_2$ and 0.5 mol/dm^3 NaCl to prepare suspensions of various pH. The solubility of zirconium hydroxide and the

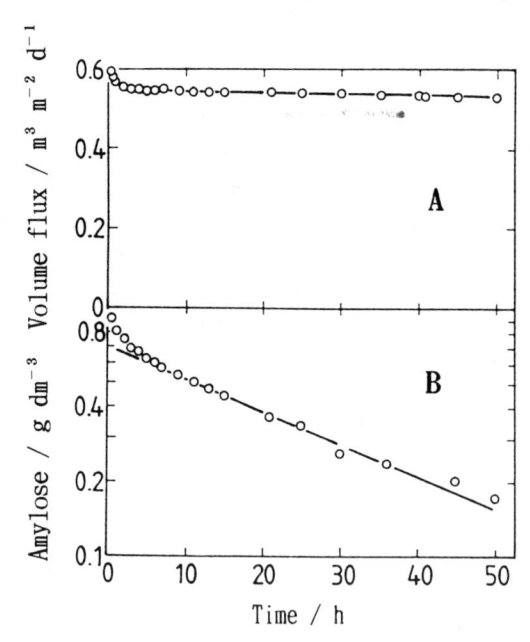

Figure 3 Continuous process for the filtration of amylose. A: Volume flux with time, B: Amylose concentration in the filtrate with time.

Table 2. The Molar Ratio of NaOH to ZrOCl$_2$, pH, SV$_{30}$ and Viscosity of Various Mixture of NaOH and ZrOCl$_2$.

Molar ratio (NaOH/ZrOCl$_2$)	pH (30 min after mixing)	SV$_{30}$ (%)	Viscosity (cP)
1.0	2.13	100	1.05
1.2	2.35	100	1.25
1.4	2.73	100	1.50
1.5	3.60	100	1.80
1.6	3.92	94.2	2.05
1.7	5.53	69.3	1.90
1.8	6.66	46.3	1.50
2.0	9.66	41.7	1.40

viscosity of solution changed with pH. The nucleation and the crystal growth were known to be influenced by the solubility of zirconium hydroxide and the viscosity of solution. SV$_{30}$ and the viscosity of solution were determined by the change in pH. SV$_{30}$ denotes the layer volume of zirconium hydroxide suspension settled after 30 min. The results are summarized in Table 2. No precipitation occurred below pH 3.6. A maximum viscosity was obtained at pH 3.9 due to precipitation of hydrous zirconium oxide. A maximum rejection of α-chymotrypsin was obtained in the region of pH 3.9 as shown in Fig. 4.

The particle size of precipitates at each pH was determined by using various membrane filters of different pore size. The average particle size was higher than 0.2 μm at pH 2.2 and higher than 0.5 μm at pH 2.4 ~ 6.7 as shown in Table 3. Since the average pore size of the ceramic membrane used was 0.05 μm, most particles of hydrous zirconium oxide are more likely to deposit onto the ceramic membrane rather than penetrate.

Glucose Separation from Starch Saccharified Suspension

After adding 1 g of crude glucoamylase into a starch suspension containing 270 g in 3 dm^3, the hydrolysis was carried out in the feed tank at 55℃ during 60 h. The starch saccharified suspension became viscous gradually as hydrolysis occurred. The glucose was filtered continuously

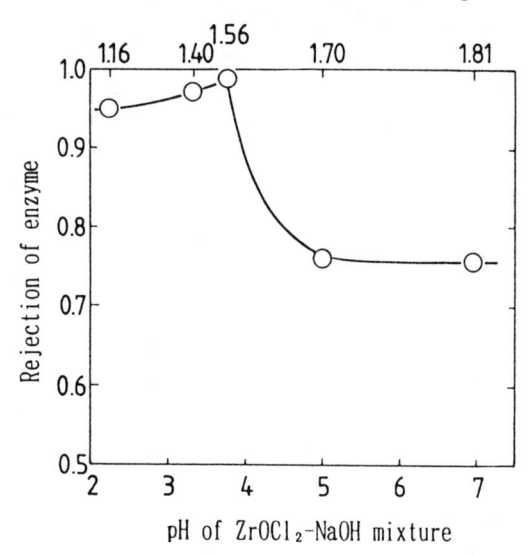

Figure 4 Rejection of enzyme by ceramic membrane coated with hydrous zirconium oxide. Coated zirconium oxide under 2 kg/cm² of gauge pressure and 1 m/s of feed velocity during 1 h at 30℃. Enzyme: 0.2 g/dm³ α-chymotrypsin.

Table 3. Particle Size of Precipitates of Hydrous Zirconium Oxide

Molar ratio (NaOH/ZrOCl₂)	pH	Pore size of Millipore filter (μm)	Permeation rate (%) *
1.0	2.20	5.0	61.9
1.0	2.20	3.0	52.5
1.0	2.20	0.8	0.5
1.0	2.20	0.2	0.0
1.2	2.41	5.0	0.2
1.4	2.80	5.0	0.1
1.6	4.11	5.0	0.02
1.7	5.50	5.0	0.02
1.8	6.71	5.0	0.0

*Into a mixture of 0.1 mol/dm³ ZrOCl₂ and 0.5 mol/dm³ NaOH, an equi-volume of NaOH solution was added at 30℃ to make a suspension. After 30 min, the suspension was filtered by membrane filter. The permeation rate was obtained from the optimum densities (660 nm) ratio of filtrate to suspension.

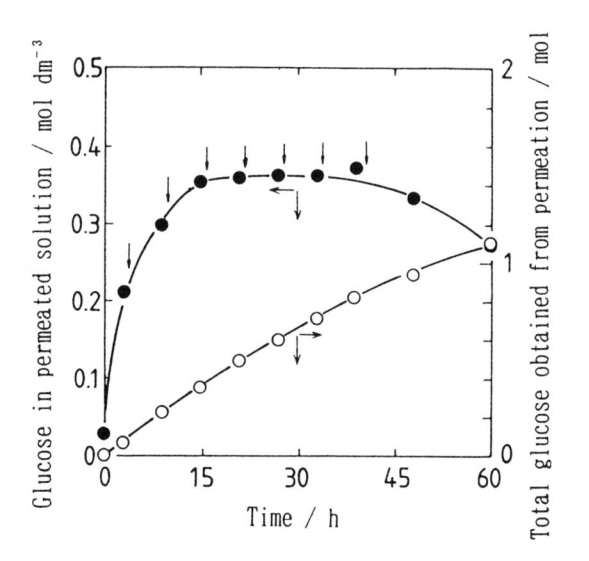

Figure 5 Starch saccharification and glucose separation by the continuous process. In order to keep the starch concentration constant, the starch suspension without the enzyme was added at the mark (↓) in figure.

through the coated ceramic membrane under 2 kg/cm² of gauge pressure and 0.8 m/s of feed velocity. The ceramic membrane was coated with different concentrations of $ZrOCl_2$. The flux decreased with an increase in $ZrOCl_2$ concentration. For the filtration, the flux should be as high as possible. On the other hand, the rejection of enzyme went lower, if the $ZrOCl_2$ concentration decreased. Therefore, the ceramic membrane was coated by 0.05 mol/dm³ $ZrOCl_2$ in 0.25 mol/dm³ NaCl to achieve both optimum flux and rejection. The result is shown in Fig. 5. Since the hydrolysis and the separation started at the same time, the glucose concentration in the filtrate was found to increase during first 15 h. The glucose concentration in the filtrate remained constant from 15 to 40 h. This result indicates that the rate of glucose formation is equal to that of glucose permeation. The rejection of enzyme was measured by analysis of nitrogen in the filtrate. The rejection of enzyme was initially 0.97 and then dropped to 0.87.

CONCLUSIONS

Amylose was separated from starch by filtration with the ceramic membrane. The method is very simple as compared with a conventional preparation of amylose. For the separation of glucose from starch saccharified solution, the ceramic membrane should be treated with a coating of hydrous zirconium oxide, which was coated at pH 3 ~4 under 2 kg/cm^2 of gauge pressure and 1 m/s of feed velocity during 1 h at 30℃.

REFERENCES

1) F. Takahashi and K. Otani, Ultrafiltration of return sludge through ceramic membrane,*Nippon Kagaku Kaishi* **1985**(5), 1000-1003.
2) Y. Sakai, H. Suzuki, S. Wakabayashi, and F. Takahashi, Preparation of inosinic acid from RNA using a membrane reactor equipped with hydrous zirconium(Ⅳ) oxide dynamic membrane layered on ceramic membrane, *Nippon Kagaku Kaishi,* **1988**(11), 1880-1884.
3) S. Noguchi, G. Shimura, K. Kimura, and H. Samejima, Production of 5'-mononucleotides using immobilized 5'-phosphodiesterase and 5'-AMP deaminase, *J.Solid-Phase Biochem.*, **1**(2), 105-118 (1976).
4) S. Nakao, T. Nomura, S. Kimura, and A. Watanabe, Formation and characteristics of inorganic dynamic membranes for ultrafiltration, *J. Chem. Eng. Jpn.*, **19**(3), 221-226 (1986).
5) Chief editors, K. Imabori and T. Yamakawa, "Seikagaku jiten"(Handbook of Biochemistry) pp69, Tokyo Kagaku Dojin (1984).
6) T. A. Butterworth, D. I. C. Wang, and A. J. Sinkey, Application of ultrafiltration for enzyme reaction during concentrations enzyme reaction, *Biotechnol. Bioeng.*, **12**, 615-631 (1970).
7) I. Ohlson, G. Tragardh, and B. Hahn-Hagerdal, Evalution of UF and RO in a cellulose saccharification process, *Desalination*, **51**, 93-101 (1984).
8) I. Ohlson and G. Tragardh, Enzymatic hydrolysis of sodium-hydroxide-pretreated shallow in an ultrafiltration membrane reactor, *Biotechnol. Bioeng.*, **26**, 647-653 (1984).
9) F. Takahashi and Y. Sakai, Facile method of amylose separation from corn starch by filtration through ceramic membrane,*Nippon Kagaku Kaishi,* **1990**(7), 809-811.
10) F. Takahashi, Y. Sakai, S. Kobayashi, and S. Wakabayashi, Separation of glucose from starch saccharified solution by applying hydrous zirconium(Ⅳ) oxide-dynamic membrane layered on ceramic membrane, *Nippon Kagaku Kaishi,* **1988**(1), 86-89.

PREPARATION OF THIN POROUS SILICA MEMBRANES FOR SEPARATION OF NON-AQUEOUS ORGANIC SOLVENT MIXTURES BY PERVAPORATION

Masashi Asaeda, Kazuto Okazaki and Atsushi Nakatani
Dept. of Chemical Engineering, Hiroshima University, Higashi-Hiroshima 724, JAPAN

Thin porous silica membranes were prepared on cylindrical porous alumina substrate by the sol-gel technique. The controlled particle size of silica sol gave a relatively sharp pore distribution of the membrane at less than a few nanometers, which was observed by the dynamic method of humid air permeation. By the sol-gel technique proposed here a porous silica membrane of pore size less than 1 nanometer was prepared quite easily. With the membrane prepared, some separation measurements were carried out for non-aqueous organic solvent mixtures (ethanol/toluene, ethanol/cyclohexane and ethanol/benzene) at 35-70°C by the pervaporation method. A high permeation flux of ethanol with a relatively large separation factor was obtained. The stability of silica membrane against organic solvents at relatively high temperature was determined.

INTRODUCTION

One of the potential application fields of inorganic porous materials is in the area of separation membrans for various mixtures. Because of their high stability at high temperature and against most chemicals, they have recently attracted intensive attention of many researchers. Especially, inorganic membranes with nano-scale pores are quite attractive for separation of gaseous mixtures and organic solvent mixtures of low molecular weight. The pore size is clearly one of the most important factors for selective permeation. It must be less than a few Angstroms for separation of inorganic gaseous mixtures by the mechanism of molecular sieving with activated permeation, and less than several Angstroms for effective separation of organic solvent mixtures.

Several approaches are known for preparation of such inorganic porous membranes of ultra-fine pores [1] : sol-gel method, pyrolysis of polymer precursor, CVD method, phase separation and leaching, etc.. Of these approaches the sol-gel method has been shown quite attractive by one of the authors for preparation of inorganic membranes with ultra-fine pores of less than 1 nanometer. Helium or hydrogen can be separated quite effectively from its nitrogen mixtures at high temperature by a porous silica membrane prepared by the sol-gel method [2]. Also, separation of aqueous solutions of organic solvents can be effectively performed with sol-gel prepared porous membranes [3,4,5,6], which is probably due to a relatively large difference in molecular diameter of water and organic solvent molecules.

For separation of non-aqueous organic solvent mixtures the controlled pore size, the difference in molecular structure and size, and selective affinity to the pore wall seem to be the key factors for effective separation. In this paper a simple but quite effective sol-gel technique is proposed for preparation of porous silica membranes with relatively sharp pore distributions. The membrane prepared by this technique is applied to separation of non-aqueous organic solvent mixtures (ethanol/toluene, ethanol/cyclohexane and ethanol/benzene) by the pervaporation method.

PREPARATION OF POROUS SILICA MEMBRANE

A sol-gel method was applied for preparation of porous silica membrane on a cylindrical porous , α-alumina substrate (average pore diameter: 1 micron, outside diameter: 1 cm, porosity: about 50%). There can be considered many factors which seem to control the characteristics of the final porous membrane: characteristics of porous substrate, characteristics of colloidal sols, coating, drying and firing conditions.

Colloidal Silica Sol

The size distribution of colloidal particles is the crucial factor for the control of pore size or pore size distribution in the final porous membrane. In this work three kinds of colloidal silica sol were prepared from their polymer solutions of different concentrations listed in Table 1. The polymer solution prepared by hydrolysis of tetraethoxysilane (TEOS) with small amount of nitric acid was diluted with 500 g of water and then boiled for more than 8 hours under stirring to obtain colloidal sol. In Fig. 1 are shown the observed particle size distributions of the sols measured a few weeks after the preparation by the light scattering method (ELS-800, Otsuka Electronics).

Membrane Preparation

Porous silica membranes were prerared by the sol-gel method using the colloidal silica sols diluted with water to 1-0.5 equivalent wt.% of TEOS before hydrolysis.

Table 1. Preparation of colloidal silica sols

	tetraethoxysilane [g]	water [g]	nitric acid [cc]
Sol-A	20	10	0.5
Sol-B	10	20	0.5
Sol-C	5	25	0.5

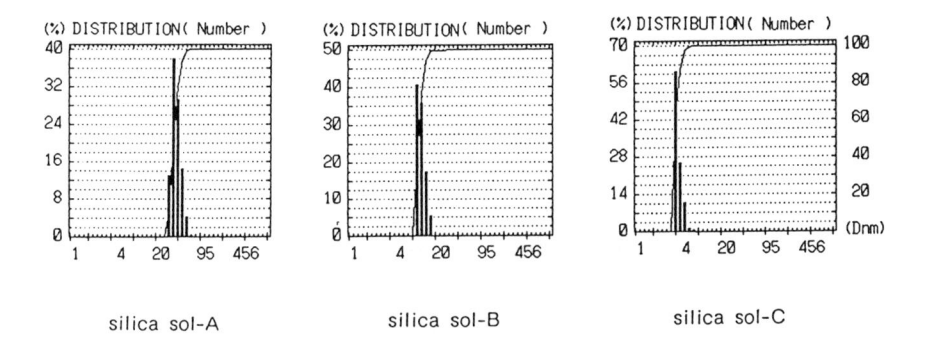

Figure 1 Particle size distribution of colloidal

Before coating with these sols the outer surface of a cylindrical porous α-alumina substrate was smoothed by polishing with sand paper (#1000) and then by depositing fine α-alumina powder with silica sol-A as its binder. The coating procedures are as follows: The coating was done while the substrate was hot at around 150° C by contacting with a cloth wet with the sol, and then it was fired at around 350° C for about 10 minutes before firing at around 570° C for several minutes. These procedures were repeated 3-4 times with sol-A, sol-B, sol-C in this order. In the final step of coating, the membrane was kept at 570° C for about 20 minutes. Some SEM micrographs of the membrane are shown in Fig. 2. Figure 2-a shows a cross section of the porous substrate without any treatment. In Fig. 2-b is shown a cross section after deposition of fine particles. The surface can be seen to be quite smooth after this treatment. A cross section after 3 times coating of sol-A is shown in Fig. 2-c. The thickness of the coated layer seems to be less than 2 μm at this stage of coating. Judging from the pore distribution results shown in Fig. 4 (membrane-A), no huge pores seem to be left at this stage. A cross section of the final membrane (membrane-C) is shown in Fig. 3, where a thin layer of about 1 μm in

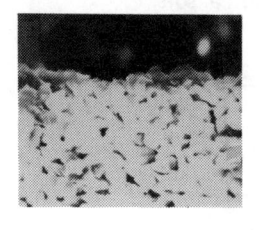

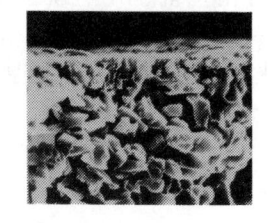

(a) substrate (b) after particle deposition (c) after sol-A coating

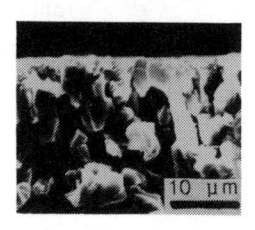

Figure 2　SEM micrographs of membrane cross section

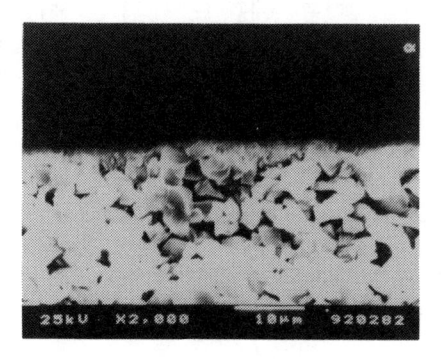

Figure 3　SEM micrograph of membrane cross section
(membrane-C)

thickness can be seen deposited at the top.

Pore Size Distribution

Determination of pore size distribution of the membrane is essential, not only for study of the separation mechanism, but also for preparation of membranes for effective separation of a given mixture. Because separation is dependent not on the average pore diameter but on the constriction or neck of a pore, the BET method to obtain pore size distribution or average pore size cannot always give adequate information about the pore size distribution effective for separation. For this reason

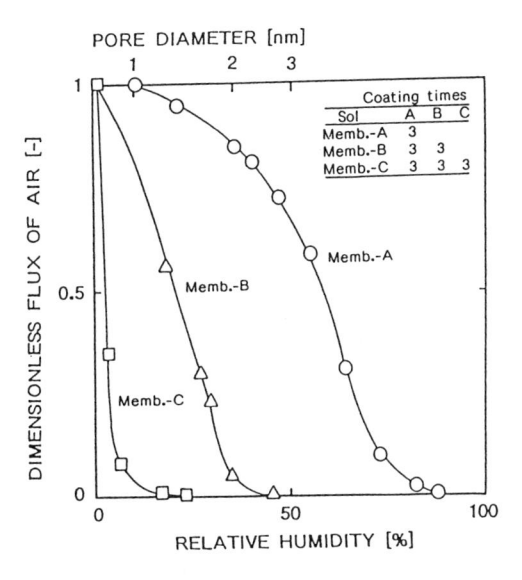

Figure 4 Pore size distribution by the dynamic method

we applied the dynamical method of permeability measurements of humid gas, where the permeation of inert gas can be blocked by the condensed water in the pores of diameter less than the one given by the Kelvin's equation of capillary condensation at a specified humidity. In Fig. 4 are shown some examples of the observed pore distribution after coating with the sols-A, -B and -C. Though these results can be considered not to give the exact pore size distribution, especially in the extremely small pore size region, because of the silanol formation on the pore wall and successive adsorption of water molecules (which possibly gives smaller apparent pore size), they can give some constructive information for preparation of porous membranes.

SEPARATION MEASUREMENTS

The pervaporation method was used for separation of organic solvent mixtures (ethanol/toluene, ethanol/cyclohexane and ethanol/benzene) at 35 – 70° C with the porous silica membrane-C prepared by the sol-gel procedures proposed here. A schematic diagram of the experimental apparatus is shown in Fig. 5. An organic solvent mixture was circulated at high speed around the membrane module

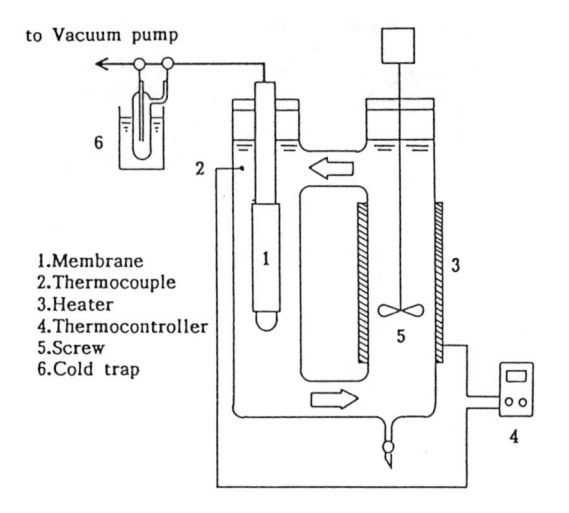

```
to Vacuum pump

                          6

1. Membrane
2. Thermocouple
3. Heater
4. Thermocontroller
5. Screw
6. Cold trap
```

Figure 5   Schematic diagram of pervaporation apparatus

(upstream) by a screw (5) at a specified constant temperature which was controlled by a thermocontroller (4). The inside of the cylindrical membrane (1) (downstream) was evacuated by a vacuum pump after removing the permeates by a cold trap (6) cooled by liquid nitrogen. The permeation flux was determined by weighing the condensed permeates, the concentration of which and that of the upstream mixture were determined by a gas chromatograph.

RESULTS AND DISCUSSION

In Figs. 6, 7 and 8 are shown some observed separation results for pervaporation of ethanol/toluene mixture at $70°C$, $35°C$ and $50°C$ with membrane-C, respectively. In these figures the observed flux of each component, the separation factor and the concentration of toluene in the permeates are shown against the toluene concentration in the upstream mixture. The observed ethanol flux is quite large compared with that of toluene, which shows that the membrane-C is ethanol selective for separation of ethanol/toluene mixture. The small leakage of toluene can be attributed to the relatively large pores left in the membrane, as shown by the tail in the observed pore size distribution curve in Fig. 4. The toluene leakage

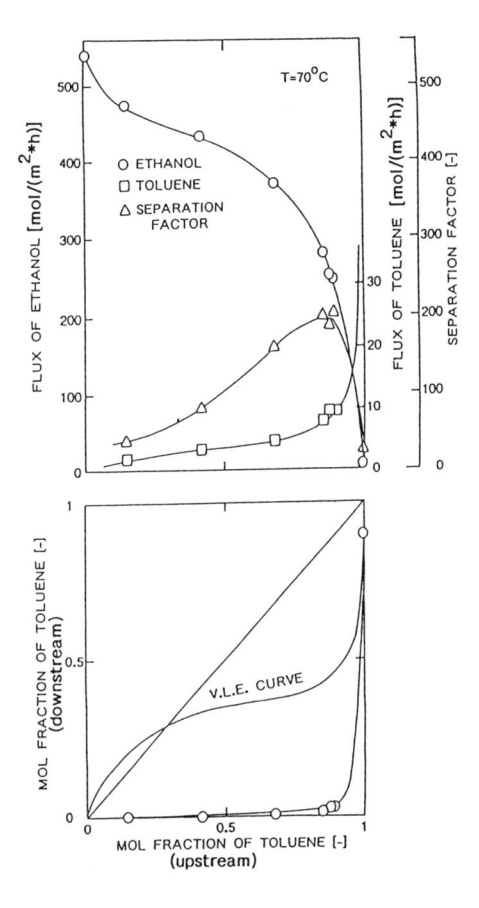

Figure 6   Separation results of ethanol/toluene mixture at 70°C

increases linearly with the increase of the upstream toluene concentration up to around 80 mol %, then it increases abruptly (Fig. 6). This suggests that at lower toluene concentration ethanol molecules adsorbed on the pore surface can reduce the effective pore diameter to make the permeation of toluene molecules difficult. At higher toluene concentration, however, some of the adsorbed ethanol molecules desorb because of the lower concentration of ethanol, leaving pores of larger effective diameter through which toluene molecules can manage to permeate. As the separation temperature becomes lower, this linearity of toluene leakage tends to persist to a higher toluene concentration in the upstream as shown in Figs. 7 and 8, since the desorption of ethanol becomes harder even at lower concentration of

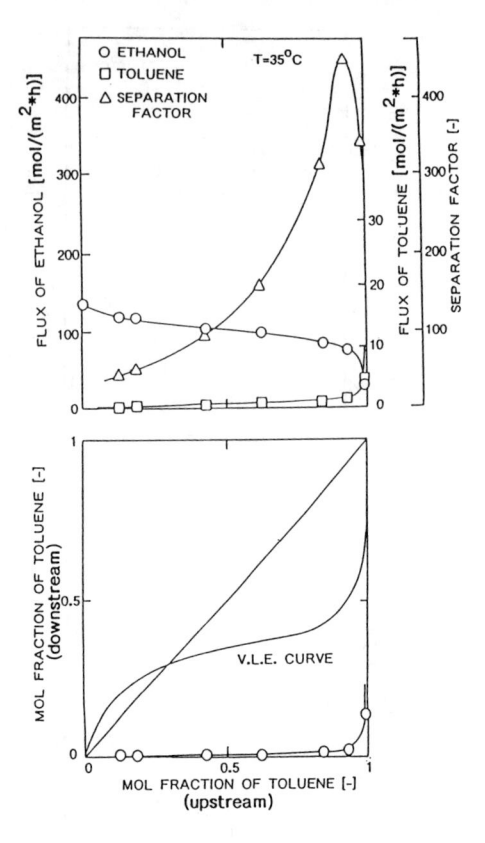

Figure 7   Separation results of ethanol/toluene mixture at 35°C

ethanol in the mixture. The flux of ethanol depends largely on the vapor–liquid equilibrium data or its partial vapor pressure at each concentration and temperature. The flux decreases as the separation temperature becomes lower because of smaller partial vapor pressure of ethanol.

In Fig. 9 are shown some pervaporation results for separation of ethanol/cyclohexane mixture at 50° C. In this case the leakage of cyclohexane is quite small, even in the region of high cyclohexane concentration, to give a high separation performance quite simillar to the previous results [7]. This is probably due to the non–flat molecular structure of cyclohexane and also due to its weak interaction or small affinity with the silica surface covered with silanol group.

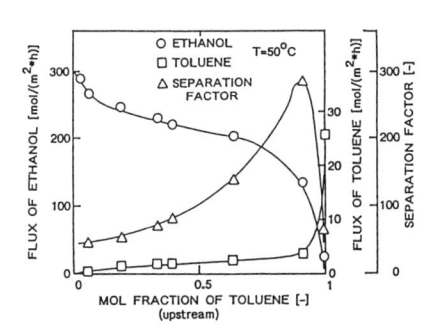

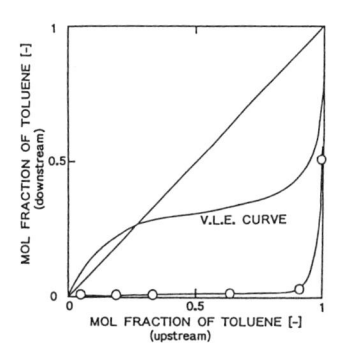

Figure 8   Separation results of ethanol/toluene mixture at 50°C

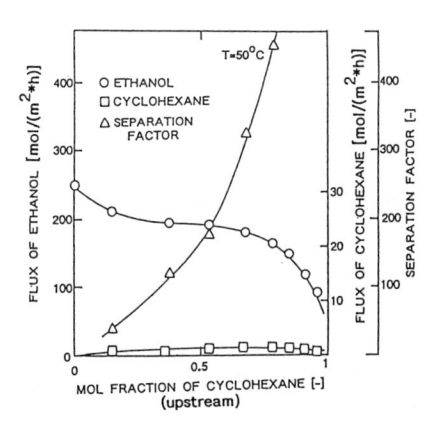

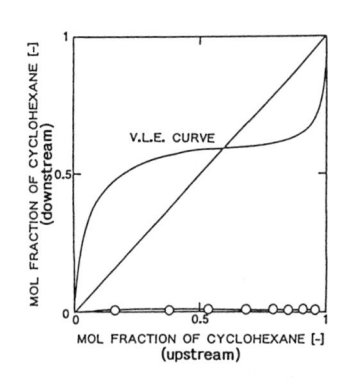

Figure 9   Separation results of ethanol/cyclohexane mixture at 50°C

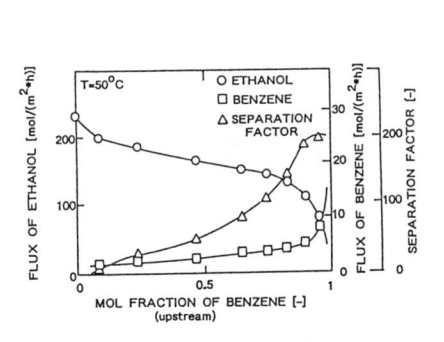

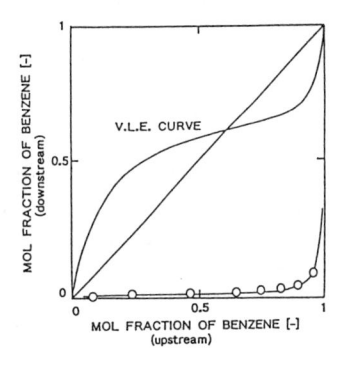

Figure 10   Separation results of ethanol/benzene mixture at 50°C

Porous Materials

Benzene can be more permeable than cyclohexane as shown in Fig. 10 to give a poorer separation performance for pervaporation of ethanol/benzene mixture. This is probably because of its flat molecular structure and also because of its larger affinity with silica surface compared with that of cyclohexane. From these considerations it can be concluded that the separation of non-aqueous organic solvent mixtures examined here with the porous silica membrane can be done mostly by the pore size and also by the selective adsorption of one of the components in the mixtures, and that the adsorption seems to be so loose that the adsorbed molecules can move in the adsorbed layer.

CONCLUSIONS

A simple sol-gel method was proposed for preparation of thin porous silica membranes of pore size less than 1 nanometer. With a membrane prepared by the method some pervaporation measurements were carried out for separation of non-aqueous organic solvent mixtures (ethanol/toluene, ethanol/cyclohexane and ethanol/benzene) at 35-70° C to obtain high permeation flux of ethanol with relatively large separation factors. The membrane was found to be quite stable after several days' measurements. The effective separation can be attributed to the appropriate pore size and the selective adsorption of ethanol on the silica surface.

REFERENCES

1) H. P. Hsieh; "Inorganic Membranes," in New Membrane Materials and Processes for Separation, AIChE Symposium Series No. 261, Vol. 84, 1-18 (1988)
2) S. Kitao, H. Kameda and M. Asaeda; "Gas Separation by Thin Porous Silica Membrane of Ultra Fine Pores at High Temperature", Maku (Membrane), No. 4, Vol. 15, 222-227 (1990)
3) M. Asaeda and L. D. Du; " Separation of Alcohol/Water Gaseous Mixtures by Thin Ceramic Membrane", J. Chem. Eng. Japan, No.1, Vol. 19, 72-77 (1986)
4) M. Asaeda, L. D. Du and M. Fuji; "Separation of Alcohol/Water Gaseous Mixtures by an Improved Ceramic Membrane", J. Chem. Eng. Japan, No. 1, Vol. 19, 84-85 (1986)
5) S. Kitao and M. Asaeda; " Separation of Organic Acid/Water Mixtures by Thin Porous Silica Membrane", J. Chem. Eng. Japan, No. 3, Vol. 23, 367-370 (1990)
6) S. Sakohara, S. Kitao, M. Ishizaki and M. Asaeda; " Separation of Solvent/Water mixtures by Porous Membrane" Inorganic Membranes, Proceedings of 1st. International Conference on Inorganic Membrane, Monplier in France, 231-236 (1989)
7) M. Asaeda and S. Kitao; " Separation of Molecular Mixtures by Inorganic Porous Membrane of Ultra Fine Pores ", Inorganic Membranes, Proceedings of 2nd International Conference on Inorganic Membranes, Monplier in France, 295-300 (1991), A.J.Burggraaff, J.Charpin and L.Cot, Trans Tech Publications

APPLICATIONS OF MICROPOROUS MATERIALS TO MEMBRANE REACTORS

Takao KOKUGAN and Gen KEITOH
Tokyo Univ. Agri. & Tech.
Nakamachi Koganei-shi, Tokyo 184, Japan
Tel.+0423-81-4221, Fax:+0423-81-4201

We studied a pure gas membrane reactor in order to use a closed system, such a chemical heat pump. In the pure gas system, the cut,that is, the ratio of permeate rate to total rates of retentate and permeate, significantly affects the reaction conversion rate. A dehydrogenation of cyclohexane was used as a reaction system. To widely vary the cut, we prepared the membranes with various permeabilities by coating the surface of porous ceramic tube, using the sol-gel method. A microporous Vycor glass has a larger separation factor and a smaller cut. A membrane coated 10x has a smaller separation factorand a larger cut. The reaction conversion rate is larger for this membrane at a suitable cut than for Vycor glass membrane. The cut is an important factor for a closed-system membrane reactor.

INTRODUCTION

Once reaction reaches an equilibrium state, no further reaction occurs. But if a product is removed from the reaction system by some means, the reaction shifts to the product side and a higher conversion rate can be obtained than the equilibrium one. This is known as the law of Le Chatelier-Braun.

In order to achieve this law, it is necessary for the reactor to be selectively permeable for the molecular products. A membrane is used as a means to remove the molecule selectively from the reaction system, and therefore it is called a membrane reactor. The concept of a membrane reactor is shown in Fig. 1, and it has the advantages of having a higher conversion rate, requiring no separator, and having energy efficiency.

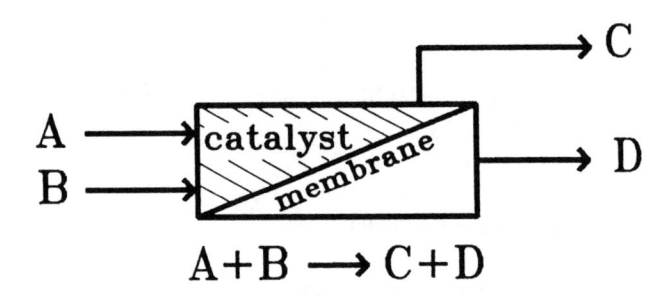

Figure 1 Concept of a membrane reactor

Generally reactions are carried out under the conditions of high temperature and high pressure. The membrane material for membrane reactors has been limited to microporous ceramics, glasses, or metals until now. Organic materials are not suitable for high temperatures beyond 150 ℃[1].

The applications of a membrane reactor to chemical processes were studied[2]. Membrane reactors for hydrogen generation have been investigated for a dilute system[3,4], because the reaction conversion rate is improved by adding an inert gas to the reactant. But a product concentration is low in the dilute system, and it is not suitable for a closed system; for example, a chemicalheat pump. Instead of a dilute system, the membrane reactor of a new system with a pure gas in the reaction side and reduction of pressure by vacuum pump in a permeate side is reported in this paper.

THEORETICAL CONSIDERATION

There are two methods to estimate the degree of conversion improvement by removing molecular products from the reaction system. One is the thermodynamics method and the other is a computer simulation. The latter is rigorous, but complicated and a case study approach. The detailed procedures were reported by Itoh[3], Mohan[5] and so on.

The former stands on the assumption that the system is in equilibrium, and it is approximate. But it is simple, and from it we can understand easily the effect of removing the molecular on reaction conversion rate. We consider the former approach here[6]. Let us consider the following reaction.

$$C_6H_{12} = C_6H_6 + 3H_2 \tag{1}$$

Equilibrium constant K_p is calculated from the following equation from thermodynamics.

$$\Delta G^0 = -RT \ln K_p \tag{2}$$

ΔG^0 is the Gibbs's free energy difference by reaction at standard state and K_p is defined by Eq.(3).

$$K_p = (PB)(3PH)^3/(PC) \tag{3}$$

$$= (c_0 a)(3c_0 a)^3/\{c_0(1-a)\}[\pi/\{c_0(1+3a)+n\}]^3 \tag{4}$$

c_0, a, n and P_i in Eq.(4) are feed concentration of cyclohexane, equilibrium conversion rate, molar number of inert gas and partial pressure of i-th component, respectively. π is the total pressure in the reaction system. If the reaction is carried out at 1 atm and 200 °C, ΔG^0 is 7.68 kcal/mol, then the equilibrium conversion rate is only 6.4 % for a pure system.

Membrane reactors can selectively permeate products. Microporous Vycor glass has a mean pore size of 40 Å in diameter. Hydrogen can permeate at a rate of 4-6 times faster than benzene and cyclohexane through the microporous Vycor glass.

The conversion rate was calculated for two cases. One is the ideal case, that is, only hydrogen can permeate through the membrane. The other is the case that hydrogen, benzene and cyclohexane are removed from the reaction system with the ratio of each permeability, that is, 0.707, 0.319 and 0.194, respectively. If one mole of cyclohexane is reacted under the conditions of constant pressure (1 atm) and constant temperature (473 K), the calculated conversion rates are shown in Fig. 2. In this figure solid and fine lines represent the value for the ideal case and the microporous Vycor glass, respectively. The conversion rates are increased by removing hydrogen, and maximum conversion rate is 100 % at 3 moles of removed hydrogen for ideal case, and 62.2% at 1.38 moles for microporous Vycor glass, theoretically. This approximate estimation is especially valid for a fast reaction.

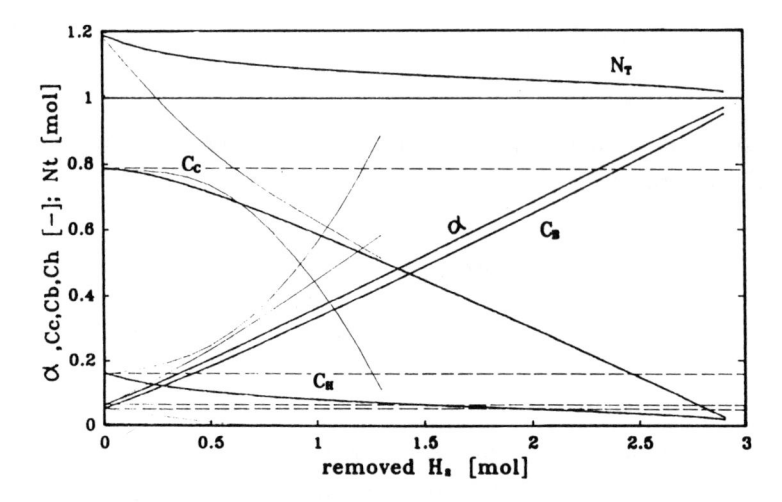

Figure 2 The effect of removing of H_2 on conversion

PERMEATION MECHANISM

Knudsen diffusion prevails as the permeation mechanism for a microporous membrane. The flux by Knudsen diffusion is given in Eq.(5).

$$J_i = [k(8/3)r/(2\pi M_i RT)^{1/2}] \{dp/dx\} \tag{5}$$

Permeability of a gas is in inverse proportion to the square root of the molecular weight, then the ratios of permeability of H_2, C_6H_6 and C_6H_{12} gases are $(1/2)^{1/2}$, $(1/78)^{1/2}$ and $(1/84)^{1/2}$, respectively. If the driving forces are equal to H_2, C_6H_6 and C_6H_{12} gases, the ratios of flux are 0.761, 0.122 and 0.117, respectively.

But surface diffusion prevails at low temperature, especially for condensable gases such as benzene and cyclohexane. The decrease in separation is attributed to surface diffusion. Itoh et al[3]. report the ratios of permeability of H_2, C_6H_6 and C_6H_{12} gases to be 1, 0.451 and 0.274 at 483 K, respectively. These are equivalent to 1, 2.82 and 1.78 times of Knudsen diffusion, respectively.

PREPARATION OF MEMBRANE

A thin layer of microporous glass was fabricated on the surface of a porous ceramic tubing, using the method described by Sakka[7]. The procedure is shown schematically in Fig. 3. A metal alkoxide solution was used in the coating process.

ALKOLATE METHOD

$Si(OC_2H_5)_4 : C_2H_5OH : H_2O : HCl =$
$25 : \quad 37.5 : 23.6 : 0.3$
$+ H_3BO_3$ or B_2O_3

↓ MIXING AND STIRRING

PARTIALLY HYDROLYZED SOLUTION

↓
COATING
↓
DRYING
↓
HEATING
↓
TREATMENT BY ACID
↓
POROUS GLASS MEMBRANE WITH CERAMIC TUBE
(THICKNESS 1um/coating)

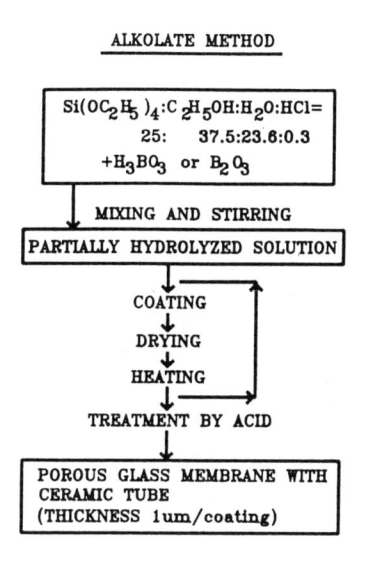

Figure 3 Procedure for preparation of membranes

A porous ceramic tubing with an outer diameter of 10 mm and an inner diameter of 7.8 mm was dipped in the metal alkoxide solution from the sealed end and pulled out at a rate of 2 mm/s. The solution coated on the ceramic tubing was dried 10 minutes or so at room temperature, then placed in an electric furnace and heated to 550 °C for 60 minutes, then removed and cooled. After the procedure of coating, drying and heating was repeated several times, the tubing was dipped in an aqueous 3M HCl acid solution at 80 °C for 6 hours to leach out the acid soluble phase. This produced a microporous membrane. The acid-leached membrane was washed with pure water at room temperature for more than 10 hours and then dried at 80-100 °C for over night.

APPARATUS AND EXPERIMENTAL PROCEDURE

A double-cylinder reactor of 180 mm height was used. The insidetube is made with a permeable membrane of 10 mm in outer diameter and outside tube is made with a Pyrex glass of 28 mm in inner diameter. An alumina-based catalyst containing of 0.5 wt % platinum was packed in the space between the tubes. Pressure was reduced in the permeate side to the maximum pressure of 15 kPa. Microporous Vycor glass and prepared membranes by coating on the surface of porous ceramic tube with sol-gel method were used. The schematic diagram of the experimental apparatus is shown in Fig. 4.

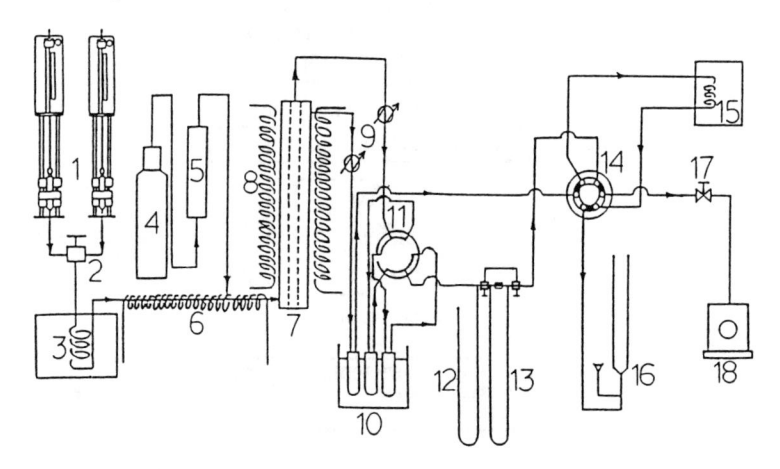

1.Micro-feeder, 2.Three-way stop valve, 3.Heater, 4.H_2 cylinder, 5.Flow-meter, 6.Heating cable, 7.Membrane reactor, 8.Mantle heater, 9.Condenser by city water, 10.Cooler by ice water, 11.Six-way valve, 12.Manometer, 13.Capillary flow-meter, 14.Rotary valve, 15.Gas chromatograph, 16.Soap-film flow-meter, 17.Needle valve, 18.Vacuum pump.

Figure 4 Schematic diagram of experimental apparatus.

RESULTS AND DISCUSSION

Permeability of Gases

The measured permeabilities for microporous Vycor glass and prepared membranes are given in Fig. 5. The permeability of a membrane coated 13 times is nearly equal to that of microporous Vycor glass. The permeability of a membrane coated with fewer than 13 layers is larger than the one of the microporous glass, the converse is also true. The more suitable membrane for this reaction system must have larger permeabilities for H_2 and C_6H_6 gases and a smaller one for C_6H_{12} gas. By coating the ceramic tubing via the sol-gel method, the membrane performs better when coated with fewer than 13 layers with respect to C_6H_6 gas, and performs better when having more than 13 layers with respect to C_6H_{12} gas.

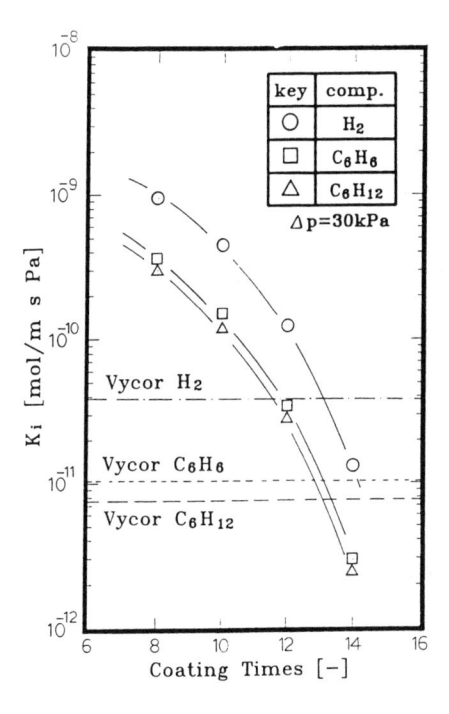

Figure 5 Permeabilities of gases for prepared membranes

Porous Materials

Relation of Conversion Rate with Reduced Pressure

Taking the feed flow rate F_0 as parameters, Fig. 6 shows the relationship of conversion rate with pressure difference for microporous Vycor glass. It is found that a greater reduction in pressure difference increases the the conversion rate. Increasing the feed flow rate decreases the conversion rate. All data of conversion rates obtained from the present membrane reactor were alone the equilibrium conversion rate of 6.4%. At a feed flow rate of 1.2 mmol/min the conversion rate was double that of equilibrium.

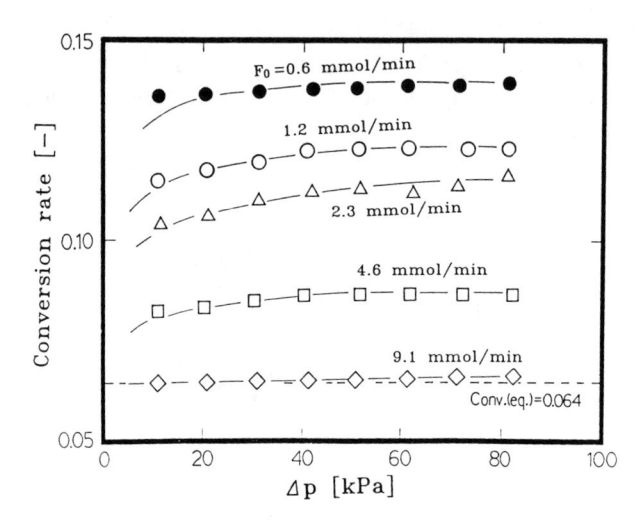

Figure 6 Relation of conversion rate with pressure difference

Effect of the Ratio of Permeate Flow Rate to Total Flow Rate on Conversion Rate

The cut(β) is defined by Eq.(6).

$$\beta = P / R = \text{sum} (P_i) / \text{sum} (R_i + P_i) \qquad (6)$$

where P_i and R_i are permeated and residual gas flow rates of i-th component, respectively. Fig. 7 shows the effects of the cut on conversion rate. Theoretically, the membrane reactor is equivalent to an equilibrium reactor when P = 0, or β = 0. The cut for Vycor glass can not increase to 0.25 at a feed flow rate of 1.2 mmol/min, due to its permeability and the flow rate.

The cut for the membrane reactor with membrane coated with 10 layers can be varied from 0 to 1 at a feed flow rate of 1.2 mmol/min, because its permeability is large. The conversion rate for this membrane reactor is better for Vycor glass at a cut β=0.6, though the selectivity of the membrane is not better than that of Vycor glass. We must be determined to obtain greater conversion rate by the optimum cut. However, this type of membrane has lower selectivity.

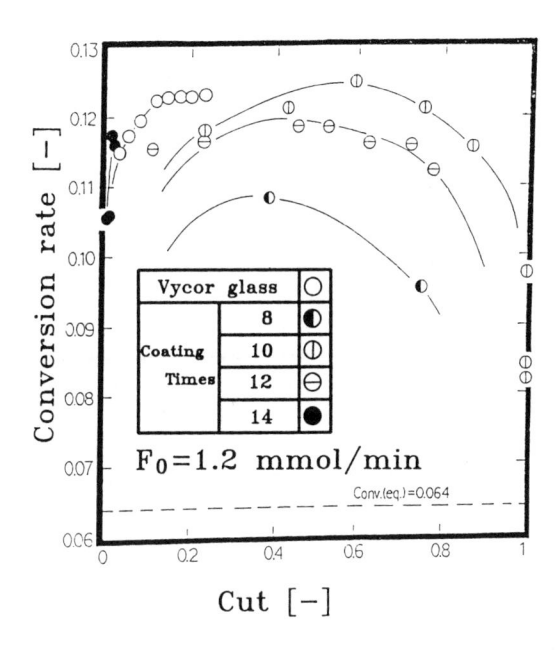

Figure 7 The effect of the cut on conversion rates

Effect of Feed Flow Rate on Conversion Rate

Fig. 8 shows the maximum conversion rates operating at optimum cut for various membranes. The maximum conversion rate depends on the feed flow rate and the cut. The maximum conversion rate is larger for the microporous Vycor glass than for a reactor with coated membrane at a lower flow rate. But at a higher flow rate, the maximum conversion rate for the reactor with a membrane coated with ten layers is larger than that of the microporous Vycor glass. The cut giving maximum conversion rate is smaller for the reactor with membrane of smaller permeability and is larger for the reactor with large permeability.

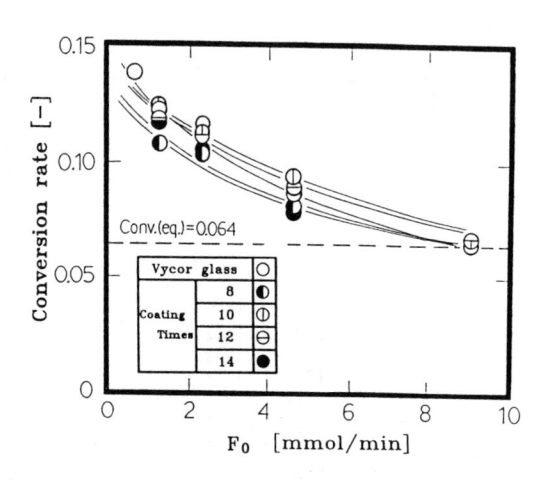

Figure 8 The effect of feed flow rate on conversion

CONCLUSION

Microporous Vycor glass has a larger separation factor and a smaller cut than the membrane coated with 10 layers. The reaction conversion rate for the membrane coated with 10 layers is larger at the cut $\beta=0.6$ than for the Vycor glass membrane. The cut is an important factor for the closed system membrane reactor. Membranes with not only a larger separation factor but also a larger permeability have been developed8). From the present work, it is necessary to design a membrane reactor under the optimum conditions of cut, feed flow rate, and so on for pure gas and reduced pressure systems.

REFERENCES

1) H.P.Hsieh:'Inorganic membrane', AIChE J. Symp. Ser., 261, 84, 1 (1988)
2) T.Kokugan, S.Ihida, Y.Oki and N.Tokunaga: 'Application of hydrogen generation membrane reactor by a pure gas system', Proceedings of the 56-th annual meeting of Soc. Chem. Eng. Japan. c-302 (1991)
3) N.Itoh, Y.Shindo, K.Haraya, K.Kobatake T.Hakuta and H.Yoshitome: 'Simulation of reaction process with membrane separation' Kagaku-Kougaku Ronbunshu, 9, 572 (1983)
4) K.C.Cannon and J.J.Hacskaylo: 'Evaluation of palladium- impregnation on the performance of a Vycor glass catalytic membrane reactor', J. Membrane Science, 65, 259 (1992)
5) K.Mohan and R.Govind:'Analysis of a co-current membrane reactor', Ind. Eng. Chem. Res., 27, 2064 (1988)
6) T.Kokugan: 'Application of inorganic membrane to reactor with separation', Chemical Engineering, 35, 292 (1990)
7) K.Kamiya and Y.Sakka: 'Preparation of porous materials by sol-gel method', J. Ceramic Japan, 86, 562 (1978)
8) E.Kikuchi and S.Uemiya:'Membrane reactor using microporous glass-supported thin film of palladium', Chemist Letters, 45, 489 (1989)

INDEX

Abe, H., 371
Abe, Y., 181
Adachi, T., 7
Adsorption, 155, 295
Adsorption-desorption isotherms, 203
Aeration, 343
Agricultural fields, 353
Al_2O_3 fibers, 147
Ambient pressure, 71
Amylose separation, 401
Apparent porosity, 263
Applications, 3, 81
Aqueous sol-gel, 41
Ar gas, 127
Argon separation, 335
Arai, H., 273
Asaeda, M., 411
Atkinson, A., 41

Beer, 381
Bending strength, 127, 147, 243
Beverages, 381
Bimodal porous cordierite ceramics, 371
Bimodally distributed alumina powders, 111
Bioreactors, 391
Borosilicate glass, 213
Brinker, C.J., 71
Bubbling test, 343
Building materials, 223

$CaO-Al_2O_3-SiO_2$ system, 127
Cardhouse structure, 169
Casting rate, 325
Catalysis, 3
Catalysts, 155
   supports, 27
$CaTi_4(PO_4)_6$, 181
Ceramic carriers, 381, 391
Ceramic foam, 285
Ceramic slurries, 89
Chang, K.L., 101
Chemical heat pump, 421

Chemical vapor deposition, 305
Chemisorbed waters on silica, 203
Chou, K.-S., 101
Chromatographic media, 41
Closed porosity, 127
CO_2, 213
Compressive strength, 263
Contact angle, 343
Conversion rate, 421
Copper, 233
Corrosion resistance, 263
Crack formation, 27
Crystal morphology, 273
Cut, 421
Cyclic voltammetry, 253
Cyclohexane dehydrogenation, 421

Dehumidification, 295
Densification, 127, 191
Deshpande, R., 71
Desiccants, 295
Diamagnetic materials, 335
Diffusion, 273, 371

Egashira, M., 371
Eguchi, K., 273
Electric conductivity, 253
Electric resistivity, 243
Electrochemical detector, 315
Electrochemical properties, 253
Electrostatic deposition, 305
Electrostatic force, 305
Electrostatic formation, 305
Enzymatic cycling reaction, 361
Enzyme immobilization, 181
Enzyme supports, 27

Fermentation activity, 371
Fiber reinforcement, 223
Filters, 27, 285
Filtration, 3
Flow injection analysis, 361
Food, 381
Fractal dimension, 223

Free Si content, 243
Frost durability, 223
Fukunaga, A., 371
Fukushima, T., 223

Gas pressure drops, 305
Gas sensors, 41
Gas separation, 3
Gel precipitation, 41
Glass powder, 127
Glass capillary tube, 343
Glucose separation, 401
Grease filters, 285
Gypsum mold, 325

Hakuman, M., 203
Harvest period, 353
Hashizuka, Y., 325
Hattori, K., 147
Hayashi, S., 335
Hexaaluminate, 273
High compressive strength, 61
High open porosity, 233
High pore volume, 61
High-pressure gas, 117
High strength, 233
Hokii, T., 263
Horitsu, H., 381
Hosono, H., 181
Hot isostatic pressing, 111, 117, 233, 315
    capsule-free, 111, 117, 315
    pressure, 315
HREM, 273
Humidity sensor, 181, 253
Hydrophobic gels, 71
Hydrothermal hot-pressing, 61
Hydrous zirconium oxide, 401

Immobilization, 371
Immobilized yeasts, 381
Inorganic membranes, 213
Inoue, H., 273
Intercalation, 41
Interconnected pores, 51
Interfacial energy, 343
Ioku, K., 61
Ion exchanger, 181

Ishizaki, K., 111, 117, 233, 315, 335
Isotherms, 295
Iwata, T., 285

Jain, R., 335
Joh, H., 203

Kaji, H., 51
Kamiya, Y., 391
Kanaoka, C., 263
Kaneno, M., 391
Kani, A., 243
Karube, I., 361
Katayama, S., 243
Katsuki, H., 137
Kawase, M., 391
Keitoh, G., 421
Koga, T., 243
Kojima, T., 353
Kokubu, T., 253
Kokugan, T., 421
Komarneni, S., 155, 295
Kondo, Y., 325
Kozuka, H., 7
Kubota, M., 353
Kurushima, T., 127, 147
Kuzjukēvičs, A., 111

Layer silicates, 155
Le Chatelier-Braun law, 421
$LiTi_2(PO_4)_3$, 181
Liu, H.C., 101
Low-density aerogels, 71
Low-melting-point materials, 263

Machida, M., 273
Macropore structure design, 51
Maebashi, N., 81
Makishima, A., 361
Malla, P.B., 295
Masuda, S., 305
Masuda, T., 285
Matsuda, O., 137
Mean pore size, 263
Mechanical properties, 27
Meissner effect, 335
Membrane reactors, 401, 421
Membrane separation, 213

Membranes, 3, 305, 401
  microfiltration, 81
  rotating disk type, 81
Mercuric ion sensors, 361
Mercuric reductase, 361
Mesopores, 169
Metal filtration, 89
Metal intercalated clays, 155
Microbe immobilization, 391
Microfiltration, 27
Micropores, 169
Microporous ceramics, 421
Microporous glass-ceramics, 181
Microporous Vycor glass, 421
Microstructure evolution, 101
Misra, S.N., 89
Monolayer capacity of physisorbed H_2O, 203
Montmorillonite, 155, 169

Nakai, K., 203
Nakamura, M., 223
Nakanishi, K., 51
Nakatani, A., 411
Nanko, M., 117
Nanocomposites, 155
Naono, H., 203
Needlelike mullite crystals, 137
Negative pressure difference system, 353
Nishioka, M., 61
Nitrogen gas, 203
Nitrogen sorption, 295
Numata, M., 361

Okada, S., 325
Okazaki, K., 411
Okumoto, Y., 315
Open cell structures, 89
Open porosity, 127, 315
Organic solvent mixtures separation, 411
Organic-inorganic compounds, 27
Organic molecules, 27
Osada, H., 243

Paramagnetic materials, 335
Pd-TPGC electrode, 253

Penetrating pores, 315
Penetration pressure, 325
Permeable refractories, 263
Pervaporation, 411
Photocatalysts, 27
Photodecomposition, 253
Photoelectrochemical properties, 253
Photoinduced current, 253
Photonic materials, 191
Pillared clay, 169
Plantations, 353
Polymeric preforms, 89
Polyol process, 155
Polyurethane foams, 89
Pore size, 233
Pore structure, 191, 223
Porous alumina, 111, 325
Porous ceramic filter, 101
Porous gels, 27
Porous glasses, 191, 213, 361
Porous media, 343
Porous mullite ceramics, 137
Porous platinum, 315
Porous Si-SiC, 243
Porous silica, 27
Porous silica gel, 203
Porous silica membrane, 411
Processing methods, 3

Reduction, 155

Sakai, Y., 401
Sake, 381
Sakka, S., 7
Sakurai, M., 203
Sawai, Y., 335
Scrap tile material, 127
Seki, H., 371
Sensors, 3
Separation, 27
Separation efficiency, 305
Shashi Mohan, A.L., 89
Sheppard, L.M., 3
Shibata, S., 191
Shibayama, M., 325
Shiomitsu, T., 273
Shirasu-type glass, 191
SiC whiskers, 147

Silica, 51
Silica gels, 27, 61, 295
Silicon nitride, 305
SIMS, 273
Sintering, 117, 191
Sintering behavior, 127, 147
Sintering gas pressures, 127
Slip casting, 325
Smith, D.M., 71
Soga, N., 51
Sol-gel method, 27, 411, 421
Sol-gel transition, 51
Sol particles, 169
Soy sauce, 81, 381
Spinodal decomposition, 51
Strength ratio, 147
Superconducting filter, 335
Supercritical drying, 169
Surface area, 273, 371
Surface corona discharge, 305
Surface diffusion, 117

Takagi, H., 137
Takahama, K., 169
Takahashi, F., 401
Takata, A., 117, 233
Tanaka, H., 213
Tantry, P.K., 89
Thermal conductivity, 137
Thermal insulation, 71
Thermal shock resistance, 137, 147
$Ti(HPO_4)_2 \cdot 2H_2O$, 181
TiO_2-SiO_2 porous glass-ceramics, 253
Titanium phosphate, 181
Tomita, K., 285
Tsuchinari, A., 263

Ueno, N., 343
Ultrafine particles, 305
Umino, M., 191
Uo, M., 361
Urano, S., 223

Water vapor, 203
Wettability, 325
Wetting behavior, 343

Yamada, A., 315

Yamamoto, H., 305
Yamanaka, S., 169
Yamane, M., 191, 253
Yamasaki, N., 61
Yanagisawa, K., 61
Yano, T., 191
Yasumori, A., 191
Yazawa, T., 213
Young's modulus, 233

Zeta potential, 391
ZrO_2 ceramics, 147
ZrO_2 fibers, 147